Smart Meter Data Analytics

Yi Wang · Qixin Chen · Chongqing Kang

Smart Meter Data Analytics

Electricity Consumer Behavior Modeling, Aggregation, and Forecasting

Science Press
Beijing

Springer

Yi Wang
Department of Electrical Engineering
Tsinghua University
Beijing, China

Qixin Chen
Department of Electrical Engineering
Tsinghua University
Beijing, China

Chongqing Kang
Department of Electrical Engineering
Tsinghua University
Beijing, China

ISBN 978-981-15-2623-7 ISBN 978-981-15-2624-4 (eBook)
https://doi.org/10.1007/978-981-15-2624-4

Jointly published with Science Press
The print edition is not for sale in China. Customers from China please order the print book from: Science Press.

This Springer imprint is published by the registered company Springer Nature Singapore Pte Ltd.
The registered company address is: 152 Beach Road, #21-01/04 Gateway East, Singapore 189721, Singapore

To Our Alma Mater
–Tsinghua University

Foreword

Smart grid is a cyber-physical-social system where the power flow, data flow, and business flow are deeply coupled. Enlightened consumers facilitated by smart meters form the foundation of a smart grid. Countries around the world are in the midst of massive smart meter installations for consumers on the pathway towards grid digitalization and modernization. It enables the collection of extensive fine-grained smart meter data, which could be processed by data analytical techniques, especially now widely available machine learning techniques. Big data and machine learning terms are widely used nowadays. People from different industries try to apply advanced machine learning techniques to solve their own practical issues. The power and energy industry is no exception. Smart meter data analytics can be conducted to fully explore the value behind these data to improve the understanding of consumer behavior and enhance electric services such as demand response and energy management.

This book explores and discusses the applications of data analytical techniques to smart meter data. The contents of the book are divided into three parts. The first part (Chaps. 1–2) provides a comprehensive review of recent developments of smart meter data analytics and proposes the concept of "electricity consumer behavior model". The second part (Chaps. 3–5) studies the data analytical techniques for smart meter data management, such as data compression, bad data detection, data generation, etc. The third part (Chaps. 6–12) conducts application-oriented research to depict the electricity consumer behavior model. This part includes electrical consumption pattern recognition, personalized tariff design for retailers, socio-demographic information identification, consumer aggregation, electrical load forecasting, etc. The prospects of future smart meter data analytics (Chap. 13) are also provided as the end of the book. The authors offer model formulations, novel algorithms, in-depth discussions, and detailed case studies in various chapters of this book.

One author of this book, Prof. Chongqing Kang, is a professional colleague. He is a distinguished scholar and pioneer in the power and energy area. He has done extensive work in the field of data analytics and load forecasting. This is a book worth reading; one will see how much insight can be gained from smart meter data alone. There are definitely broader qualitative understanding that can be gained from massive data collected in the realm of generation, transmission, distribution, and end use of the smart grid.

September 2019

Prof. Saifur Rahman
Joseph Loring Professor and Founding Director
Advanced Research Institute at Virginia Tech
Arlington, VA, USA

President of the IEEE Power and Energy Society
New York, NY, USA

Preface

Decarbonization, decentralization, and digitalization (3D) are three pathways to future power and energy systems modernization. Most of the developments of the power and energy industry mainly focus on the generation and transmission sectors, while there is still a long way to go for distribution and demand sectors. Distribution systems in the electric power system have recently seen an important influx of exciting smart grid technologies such as distributed energy resources (DERs), multiple energy systems integration, control infrastructure, and data-gathering equipment. Increasing renewable energy integration and improving energy efficiency are two effective approaches for decarbonization. However, increasing penetration of renewable energy integration challenges the reliability, economy, and flexibility (REF) of the power and energy systems. A large number of DERs such as distributed photovoltaic (PV) and electric vehicles make the distribution systems more decentralized and complex. Broad interaction between consumers and systems can help provide flexibility to the power system and realize personalized consumer service. Meanwhile, data acquisition devices such as smart meters are gaining popularity, which enables an immense amount of fine-grained electricity consumption data to be collected. The "cyber-physical-social" deep coupling characteristic of the power system becomes more prominent. Breakthroughs are needed to analyze the behavior of electricity consumers.

Data analytics and machine learning techniques such as deep learning, transfer learning, graphical models, sparse representation, etc., have been greatly and considerably developed in recent years. It seems natural to figure out how to apply these state-of-the-art techniques to consumer behavior analysis and distribution system operation. However, it is a predicament in the power industry that even though an increasing and huge number of smart meter data are collected and accessible to retailers and distribution system operators (DSOs), these data are not yet fully utilized for a better understanding of consumer behavior and an enhancement on the efficiency and sustainability of the power systems.

This book aims to make the best use of all of the data available to process and translate them into actual information and incorporate into consumer behavior modeling and distribution system operations. The research framework of the smart meter data analytics in this book can be summarized in the following figure.

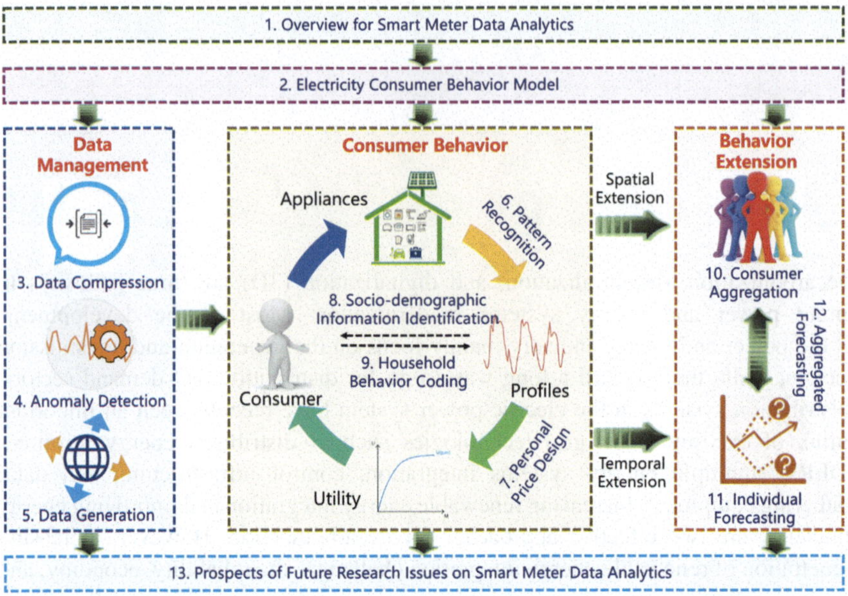

This book consists of 13 chapters. It begins with an overview of recent developments of smart meter data analytics and an introduction on the electricity consumer behavior model (ECBM). Since data management is the basis of further smart meter data analytics and its applications, three issues on data management, i.e., data compression, anomaly detection, and data generation, are subsequently studied. The main components of electricity consumer behavior model include the consumer himself, appliances, load profiles, and the corresponding utility function. The following works try to model the relationships among these components and discover the inherent law within the behavior. Specific works include pattern recognition, personalized price design, socio-demographic information identification, and household behavior coding. On this basis, this book extends consumer behavior in both spatial and temporal scales. Works such as consumer aggregation, individual load forecasting, and aggregated load forecasting are introduced. Finally, prospects of future research issues on smart meter data analytics are provided.

To help readers have a better understanding of what we have done, we would like to make a simple review of the 13 chapters in the following.

Chapter 1 conducts an application-oriented review of smart meter data analytics. Following the three stages of analytics, namely, descriptive, predictive, and prescriptive analytics, we identify the key application areas as load analysis, load forecasting, and load management. We also review the techniques and methodologies adopted or developed to address each application.

Chapter 2 proposes the concept of ECBM and decomposes consumer behavior into five basic aspects from the sociological perspective: behavior subject, behavior environment, behavior means, behavior result, and behavior utility. On this basis, the research framework for ECBM is established.

Chapter 3 provides a highly efficient data compression technique to reduce the great burden on data transmission, storage, processing, application, etc. It applies the generalized extreme value distribution characteristic for household load data and then utilizes it to identify load features including load states and load events. Finally, a highly efficient lossy data compression format is designed to store key information of load features.

Chapter 4 applies two novel data mining techniques, the maximum information coefficient (MIC) and the clustering technique by fast search and find of density peaks (CFSFDP), to detect electricity abnormal consumption or thefts. On this basis, a framework of combining the advantages of the two techniques is further proposed to boost the detection accuracy.

Chapter 5 proposes a residential load profiles generation model based on the generative adversarial network (GAN). To consider the different typical load patterns of consumers, an advanced GAN based on the auxiliary classifier GAN (ACGAN) is further to generate profiles under typical modes. The proposed model can generate realistic load profiles under different load patterns without loss of diversity.

Chapter 6 proposes a K-SVD-based sparse representation technique to decompose original load profiles into linear combinations of several partial usage patterns (PUPs), which allows the smart meter data to be compressed and hidden electricity consumption patterns to be extracted. Then, a linear support vector machine (SVM)-based method is used to classify the load profiles into two groups, residential customers and small- and medium-sized enterprises (SMEs), based on the extracted patterns.

Chapter 7 studies a data-driven approach for personalized time-of-use (ToU) price design based on massive historical smart meter data. It can be formulated as a large-scale mixed-integer nonlinear programming (MINLP) problem. Through load profiling and linear transformation or approximation, the MINLP model is simplified into a mixed-integer linear programming (MILP) problem. In this way, various tariffs can be designed.

Chapter 8 investigates how much socio-demographic information can be inferred or revealed from fine-grained smart meter data. A deep convolutional neural network (CNN) first automatically extracts features from massive load profiles. Then SVM is applied to identify the characteristics of the consumers. Different socio-demographic characteristics show different identification accuracies.

Chapter 9 uses smart meter data to identify energy behavior indicators through a cross-domain feature selection and coding approach. The idea is to extract and connect customers' features from the energy domain and demography domain. Smart meter data are characterized by typical energy spectral patterns, whereas household information is encoded as the energy behavior indicator. The proposed approach offers a simple, transparent, and effective alternative to a challenging cross-domain matching problem with massive smart meter data and energy behavior indicators.

Chapter 10 proposes an approach for clustering of electricity consumption behavior dynamics, where "dynamics" refer to transitions and relations between consumption behaviors, or rather consumption levels, in adjacent periods. To tackle the challenges of big data, the proposed clustering technique is integrated into a divide-and-conquer approach toward big data applications.

Chapter 11 offers a format of short-term probabilistic forecasting results in terms of quantiles, which can better describe the uncertainty of residual loads, and a deep-learning-based method, quantile long-short-term-memory (Q-LSTM), to implement probabilistic residual load forecasting. Experiments are conducted on an open dataset. Results show that the proposed method overrides traditional methods significantly in terms of pinball loss.

Chapter 12 proposes an ensemble method to forecast the aggregated load with sub-profiles where the multiple forecasts are produced by different groupings of sub-profiles. Different aggregated load forecasts can be obtained by varying the number of clusters. Finally, an optimal weighted ensemble approach is employed to combine these forecasts and provide the final forecasting result.

Chapter 13 discusses some research trends, such as big data issues, novel machine learning technologies, new business models, the transition of energy systems, and data privacy and security.

To summarize, this book provides various applications of smart meter data analytics for data management and electricity consumer behavior modeling. We hope this book can inspire readers to define new problems, apply novel methods, and obtain interesting results with massive smart meter data or even other monitoring data in the power systems.

Beijing, China
September 2019

Yi Wang
Qixin Chen
Chongqing Kang

Acknowledgements

This book made a summary of our research about smart meter data analytics achieved in recent years. These works were carried out in the Energy Intelligence Laboratory (EILAB), Department of Electrical Engineering, Tsinghua University, Beijing, China.

Many people contributed to this book in various ways. The authors are indebted to Prof. Daniel Kirschen from the University of Washington; Prof. Furong Li and Dr. Ran Li from the University of Bath; Dr. Tao Hong from the University of North Carolina at Charlotte; and Dr. Ning Zhang, Dr. Xing Tong, Mr. Kedi Zheng, Mr. Yuxuan Gu, Mr. Dahua Gan, and Mr. Cheng Feng from Tsinghua University, who have contributed materials to this book.

We also thank Mr. Yuxiao Liu, Mr. Qingchun Hou, Mr. Haiyang Jiang, Mr. Yinxiao Li, Mr. Pei Yong, Mr. Jiawei Zhang, Mr. Xichen Fang, and Mr. Tian Xia at Tsinghua University for their assistance in pointing out typos and checking the whole book.

In addition, we acknowledge the innovative works contributed by others in this increasingly important area especially through IEEE Power & Energy Society Working Group on Load Aggregator and Distribution Market, and appreciate the staff at Springer for their assistance and help in the preparation of this book.

This book is supported in part by the National Key R&D Program of China (2016YFB0900100), in part by the Major Smart Grid Joint Project of National Natural Science Foundation of China and State Grid (U1766212), and in part by the Key R&D Program of Guangdong Province (2019B111109002). The authors really appreciate their supports.

Yi Wang
Qixin Chen
Chongqing Kang

Contents

1 Overview of Smart Meter Data Analytics 1
1.1 Introduction 1
1.2 Load Analysis 4
1.2.1 Bad Data Detection 5
1.2.2 Energy Theft Detection 6
1.2.3 Load Profiling 8
1.2.4 Remarks 9
1.3 Load Forecasting 11
1.3.1 Forecasting Without Smart Meter Data 11
1.3.2 Forecasting with Smart Meter Data 14
1.3.3 Probabilistic Forecasting 16
1.3.4 Remarks 18
1.4 Load Management 19
1.4.1 Consumer Characterization 19
1.4.2 Demand Response Program Marketing 21
1.4.3 Demand Response Implementation 22
1.4.4 Remarks 23
1.5 Miscellanies 25
1.5.1 Connection Verification 25
1.5.2 Outage Management 26
1.5.3 Data Compression 26
1.5.4 Data Privacy 27
1.6 Conclusions 28
References 28

2 Electricity Consumer Behavior Model 37
2.1 Introduction 37
2.2 Basic Concept of ECBM 39
2.2.1 Definition 39
2.2.2 Connotation 41

2.2.3 Denotation 42
2.2.4 Relationship with Other Models 43
2.3 Basic Characteristics of Electricity Consumer Behavior 45
2.4 Mathematical Expression of ECBM 47
2.5 Research Paradigm of ECBM 50
2.6 Research Framework of ECBM 51
2.7 Conclusions 57
References 57

3 Smart Meter Data Compression 59
3.1 Introduction 59
3.2 Household Load Profile Characteristics 61
3.2.1 Small Consecutive Value Difference 61
3.2.2 Generalized Extreme Value Distribution 62
3.2.3 Effects on Load Data Compression 64
3.3 Feature-Based Load Data Compression 66
3.3.1 Distribution Fit 66
3.3.2 Load State Identification 67
3.3.3 Base State Discretization 67
3.3.4 Event Detection 68
3.3.5 Event Clustering 69
3.3.6 Load Data Compression and Reconstruction 69
3.4 Data Compression Performance Evaluation 71
3.4.1 Related Data Formats 71
3.4.2 Evaluation Index 72
3.4.3 Dataset 72
3.4.4 Compression Efficiency Evaluation Results 73
3.4.5 Reconstruction Precision Evaluation Results 74
3.4.6 Performance Map 74
3.5 Conclusions 77
References 77

4 Electricity Theft Detection 79
4.1 Introduction 79
4.2 Problem Statement 81
4.2.1 Observer Meters 81
4.2.2 False Data Injection 81
4.2.3 A State-Based Method of Correlation 83
4.3 Methodology and Detection Framework 83
4.3.1 Maximum Information Coefficient 84
4.3.2 CFSFDP-Based Unsupervised Detection 85
4.3.3 Combined Detecting Framework 86

4.4 Numerical Experiments . . . 88
4.4.1 Dataset . . . 88
4.4.2 Comparisons and Evaluation Criteria . . . 89
4.4.3 Numerical Results . . . 90
4.4.4 Sensitivity Analysis . . . 93
4.5 Conclusions . . . 97
References . . . 97

5 Residential Load Data Generation . . . 99
5.1 Introduction . . . 99
5.2 Model . . . 101
5.2.1 Basic Framework . . . 101
5.2.2 General Network Architecture . . . 102
5.2.3 Unclassified Generative Models . . . 106
5.2.4 Classified Generative Models . . . 110
5.3 Methodology . . . 113
5.3.1 Data Preprocessing . . . 114
5.3.2 Model Training . . . 115
5.3.3 Metrics . . . 118
5.4 Case Studies . . . 122
5.4.1 Data Description . . . 122
5.4.2 Unclassified Generation . . . 123
5.4.3 Classified Generation . . . 125
5.5 Conclusion . . . 134
References . . . 134

6 Partial Usage Pattern Extraction . . . 137
6.1 Introduction . . . 137
6.2 Non-negative K-SVD-Based Sparse Coding . . . 139
6.2.1 The Idea of Sparse Representation . . . 139
6.2.2 The Non-negative K-SVD Algorithm . . . 140
6.3 Load Profile Classification . . . 141
6.3.1 The Linear SVM . . . 141
6.3.2 Parameter Selection . . . 142
6.4 Evaluation Criteria and Comparisons . . . 143
6.4.1 Data Compression-Based Criteria . . . 143
6.4.2 Classification-Based Criteria . . . 144
6.4.3 Comparisons . . . 145
6.5 Numerical Experiments . . . 146
6.5.1 Description of the Dataset . . . 146
6.5.2 Experimental Results . . . 147
6.5.3 Comparative Analysis . . . 152

6.6 Further Multi-dimensional Analysis 154
6.6.1 Characteristics of Residential & SME Users 154
6.6.2 Seasonal and Weekly Behaviors Analysis 156
6.6.3 Working Day and Off Day Patterns Analysis 158
6.6.4 Entropy Analysis 159
6.6.5 Distribution Analysis 160
6.7 Conclusions 161
References 161

7 Personalized Retail Price Design 163
7.1 Introduction 163
7.2 Problem Formulation 165
7.2.1 Problem Statement 165
7.2.2 Consumer Problem 166
7.2.3 Compatible Incentive Design 166
7.2.4 Retailer Problem 167
7.2.5 Data-Driven Clustering and Preference Discovering 168
7.2.6 Integrated Model 171
7.3 Solution Methods 172
7.3.1 Framework 172
7.3.2 Piece-Wise Linear Approximation 172
7.3.3 Eliminating Binary Variable Product 173
7.3.4 CVaR 173
7.3.5 Eliminating Absolute Values 174
7.4 Case Study 174
7.4.1 Data Description and Experiment Setup 174
7.4.2 Basic Results 175
7.4.3 Sensitivity Analysis 178
7.5 Conclusions and Future Works 183
Appendix I 183
Appendix II 184
References 185

8 Socio-demographic Information Identification 187
8.1 Introduction 187
8.2 Problem Definition 189
8.3 Method 190
8.3.1 Why Use a CNN? 190
8.3.2 Proposed Network Structure 191
8.3.3 Description of the Layers 192
8.3.4 Reducing Overfitting 195
8.3.5 Training Method 196

8.4 Performance Evaluation and Comparisons 196
8.4.1 Performance Evaluation 196
8.4.2 Competing Methods 197
8.5 Case Study 199
8.5.1 Data Description 199
8.5.2 Basic Results 199
8.5.3 Comparative Analysis 201
8.6 Conclusions 203
References 203

9 Coding for Household Energy Behavior 205
9.1 Introduction 205
9.2 Basic Idea and Framework 206
9.3 Load Profile Clustering 207
9.3.1 GMM-Based Typical Load Profile Extraction 208
9.3.2 X-Means-Based Load Profile Clustering 210
9.4 Socioeconomic Genes Identification Method 210
9.4.1 Socioeconomic Information Classification 210
9.4.2 The Concept of Socioeconomic Genes 213
9.4.3 Socioeconomic Genes Evaluation Indicators 213
9.4.4 Socioeconomic Gene Search Method 216
9.5 Load Profile Prediction 216
9.6 Case Studies 217
9.6.1 Consumer Load Profile Classification 218
9.6.2 Socioeconomic Gene Search Result 218
9.6.3 Consumer Load Profile Prediction 220
9.7 Conclusions 222
References 222

10 Clustering of Consumption Behavior Dynamics 225
10.1 Introduction 225
10.2 Basic Methodology 227
10.2.1 Data Normalization 228
10.2.2 SAX for Load Curves 229
10.2.3 Time-Based Markov Model 230
10.2.4 Distance Calculation 231
10.2.5 CFSFDP Algorithm 232
10.3 Distributed Algorithm for Large Data Sets 233
10.3.1 Framework 234
10.3.2 Local Modeling-Adaptive *k*-Means 235
10.3.3 Global Modeling-Modified CFSFDP 237
10.4 Case Studies 237
10.4.1 Description of the Data Set 237
10.4.2 Modeling Consumption Dynamics for Each Customer 238

10.4.3 Clustering for Full Periods 239
10.4.4 Clustering for Each Adjacent Periods 240
10.4.5 Distributed Clustering 242
10.5 Potential Applications 243
10.6 Conclusions 245
References 245

11 Probabilistic Residential Load Forecasting 247
11.1 Introduction 247
11.2 Pinball Loss Guided LSTM 249
11.2.1 LSTM 250
11.2.2 Pinball Loss 251
11.2.3 Overall Networks 252
11.3 Implementations 254
11.3.1 Framework 254
11.3.2 Data Preparation 254
11.3.3 Model Training 255
11.3.4 Probabilistic Forecasting 256
11.4 Benchmarks 256
11.4.1 QRNN 256
11.4.2 QGBRT 257
11.4.3 LSTM+E 257
11.5 Case Studies 257
11.5.1 Data Description 258
11.5.2 Residential Load Forecasting Results 258
11.5.3 SME Load Forecasting Results 261
11.6 Conclusions 264
References 268

12 Aggregated Load Forecasting with Sub-profiles 271
12.1 Introduction 271
12.2 Load Forecasting with Different Aggregation Levels 272
12.2.1 Variance of Aggregated Load Profiles 272
12.2.2 Scaling Law 274
12.3 Clustering-Based Aggregated Load Forecasting 276
12.3.1 Framework 276
12.3.2 Numerical Experiments 277
12.4 Ensemble Forecasting for the Aggregated Load 279
12.4.1 Proposed Methodology 279
12.4.2 Case Study 281
12.5 Conclusions 285
References 285

13 Prospects of Future Research Issues 287
13.1 Big Data Issues 287
13.2 New Machine Learning Technologies 289
13.3 New Business Models in Retail Market 289
13.4 Transition of Energy Systems 290
13.5 Data Privacy and Security 291
References 292

Chapter 1
Overview of Smart Meter Data Analytics

Abstract The widespread popularity of smart meters enables an immense amount of fine-grained electricity consumption data to be collected. Meanwhile, the deregulation of the power industry, particularly on the delivery side, has continuously been moving forward worldwide. How to employ massive smart meter data to promote and enhance the efficiency and sustainability of the power grid is a pressing issue. To date, substantial works have been conducted on smart meter data analytics. To provide a comprehensive overview of the current research and to identify challenges for future research, this chapter conducts an application-oriented review of smart meter data analytics. Following the three stages of analytics, namely, descriptive, predictive, and prescriptive analytics, we identify the critical application areas as load analysis, load forecasting, and load management. We also review the techniques and methodologies adopted or developed to address each application.

1.1 Introduction

Smart meters have been deployed around the globe during the past decade. Smart meters, together with the communication network and data management system, constitute the advanced metering infrastructure (AMI), which plays a vital role in power delivery systems by recording the load profiles and facilitating bi-directional information flow [1]. The widespread popularity of smart meters enables an immense amount of fine-grained electricity consumption data to be collected. Billing is no longer the only function of smart meters. High-resolution data from smart meters provide rich information on the electricity consumption behaviors and lifestyles of the consumers. Meanwhile, the deregulation of the power industry, particularly on the delivery side, is continuously moving forward in many countries worldwide. These countries are now sparing no effort on electricity retail market reform. Increasingly more participators, including retailers, consumers, and aggregators, are involved in making the retail market more prosperous, active, and competitive [2]. How to employ massive smart meter data to promote and enhance the efficiency and sustainability of the demand side has become an important topic worldwide.

Y. Wang et al., *Smart Meter Data Analytics*,
https://doi.org/10.1007/978-981-15-2624-4_1

In recent years, the power industry has witnessed considerable developments in data analytics in the processes of generation, transmission, equipment, and consumption. Increasingly more projects on smart meter data analytics have also been established. The National Science Foundation (NSF) of the United States provides a standard grant for cross-disciplinary research on smart grid big data analytics [3]. Several projects for smart meter data analytics are supported by the CITIES Innovation Center in Denmark. These projects investigate machine learning techniques for smart meter data to improve forecasting and money-saving opportunities for customers [4]. The Bits to Energy Lab which is a joint research initiative of ETH Zurich, the University of Bamberg, and the University of St. Gallen, has launched several projects for smart meter data analytics for customer segmentation and scalable efficiency services [5]. The Siebel Energy Institute, a global consortium of innovative and collaborative energy research, funds cooperative and innovative research grants for data analytics in smart girds [6]. Meanwhile, the National Science Foundation of China (NSFC) and the National Key R&D Program of China are approving increasingly more data-analytics-related projects in the smart grid field, such as the National High Technology Research and Development Program of China (863 Program) titled Key Technologies of Big Data Analytics for Intelligent Distribution and Utilization. ESSnet Big Data, a project within the European statistical system (ESS), aims to explore big data applications, including smart meters [7]. The work package in the ESSnet Big Data project concentrates on smart meter data access, handling, and deployments of methodologies and techniques for smart meter data analytics. National statistical institutes from Austria, Denmark, Estonia, Sweden, Italy, and Portugal jointly conduct this project.

Apart from academic research, data analytics has already been used in the industry. In June 2017, SAS published the results from its industrial analytics survey [8]. This survey aims to provide the issues and trends shaping how utilities deploy data and analytics to achieve business goals. There are 136 utilities from 24 countries that responded to the survey. The results indicate that data analytics application areas include energy forecasting, smart meter analytics, asset management/analytics, grid operation, customer segmentation, energy trading, credit and collection, call center analytics, and energy efficiency and demand response program engagement and marketing. More and more energy data scientists will be jointly trained by universities and industry to bridge the talent gap in energy data analytics [9]. Meanwhile, the privilege of smart meters and deregulation of the demand side are accelerating the birth of many start-ups. These start-ups attempt to collect and analyze smart meter data and provide insights and value-added services for consumers and retailers to make profits. More details regarding industrial applications can be found from the businesses of the data-analytics-based start-ups.

Analytics is known as the scientific process of transforming data into insights for making better decisions. It is commonly dissected into three stages: descriptive analytics (what do the data look like), predictive analytics (what is going to happen with the data), and prescriptive analytics (what decisions can be made from the data). This review of smart meter data analytics is conducted from these three aspects.

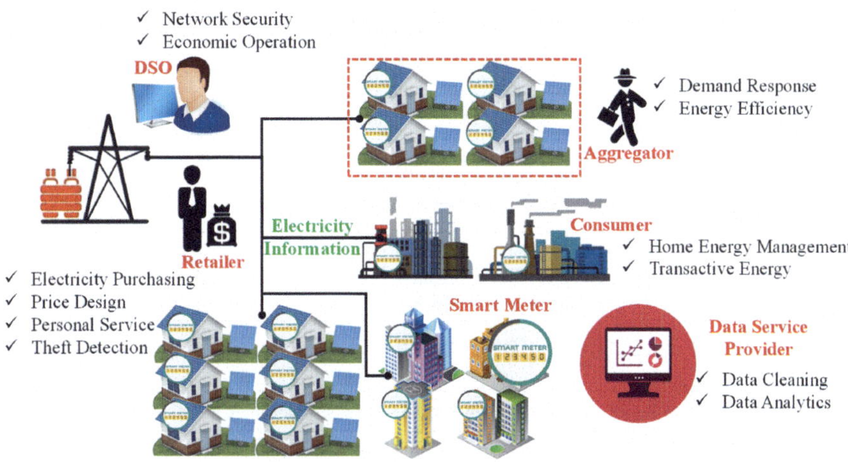

Fig. 1.1 Participators and their businesses on the demand side

Figure 1.1 depicts the five major players on the demand side of the power system: consumers, retailers, aggregators, distribution system operators (DSO), and data service providers. For retailers, at least four businesses related to smart meter data analytics need to be conducted to increase the competitiveness in the retail market. (1) Load forecasting, which is the basis of decision making for the optimization of electricity purchasing in different markets to maximize profits. (2) Price design to attract more consumers. (3) Providing good service to consumers, which can be implemented by consumer segmentation and characterization. (4) Abnormal detection to have a cleaner dataset for further analysis and decrease potential loss from electricity theft. For consumers, individual load forecasting, which is the input of future home energy management systems (HEMS) [10], can be conducted to reduce their electricity bill. In the future peer-to-peer (P2P) market, individual load forecasting can also contribute to the implementation of transactive energy between consumers [11, 12]. For aggregators, they deputize a group of consumers for demand response or energy efficiency in the ancillary market. Aggregation level load forecasting and demand response potential evaluation techniques should be developed. For DSO, smart meter data can be applied to distribution network topology identification, optimal distribution system energy management, outage management, and so forth. For data service providers, they need to collect smart meter data and then analyze these massive data and provide valuable information for retailers and consumers to maximize profits or minimize cost. Providing data services, including data management and data analytics, is an important business model when increasingly more smart meter data are collected and to be processed.

To support the businesses of retailers, consumers, aggregators, DSO, and data service providers, following the three stages of analytics, namely, descriptive, predictive and prescriptive analytics, the main applications of smart meter data analytics are classified into load analysis, load forecasting, load management, and so forth.

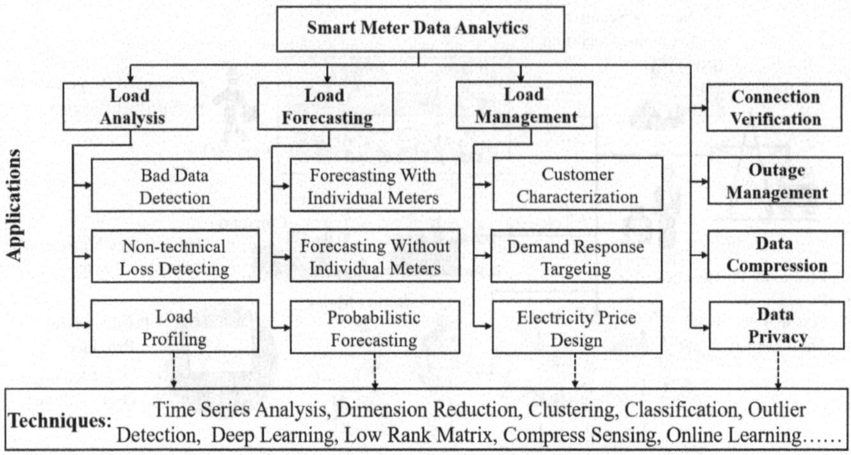

Fig. 1.2 Taxonomy of smart meter data analytics

The detailed taxonomy is illustrated in Fig. 1.2. The machine learning techniques used for smart meter data analytics include time series analysis, dimensionality reduction, clustering, classification, outlier detection, deep learning, low-rank matrix, compressed sensing, online learning, and so on. Studies on how smart meter data analytics works for each application and what methodologies have been applied will be summarized in the following sections.

This chapter attempts to provide a comprehensive review of the current research in recent years and identify future challenges for smart meter data analytics. Note that every second or higher frequency data used for nonintrusive load monitoring (NILM) are very limited at present due to the high cost of communicating and storing the data. The majority of smart meters collect electricity consumption data at a frequency of every 15 min to each hour. In addition, several comprehensive reviews have been conducted on NILM. Thus, in this chapter, works about NILM are not included.

1.2 Load Analysis

Figure 1.3 shows eight typical normalized daily residential load profiles obtained using the simple *k*-means algorithm in the Irish resident load dataset. The load profiles of different consumers on different days are diverse. Having a better understanding of the volatility and uncertainty of the massive load profiles is very important for further load analysis. In this section, the works on load analysis are reviewed from the perspectives of anomaly detection and load profiling. Anomaly detection is very important because training a model such as a forecasting model or clustering model on a smart meter dataset with anomalous data may result in bias or failure for parameter estimation and model establishment. Moreover, reliable smart meter data are

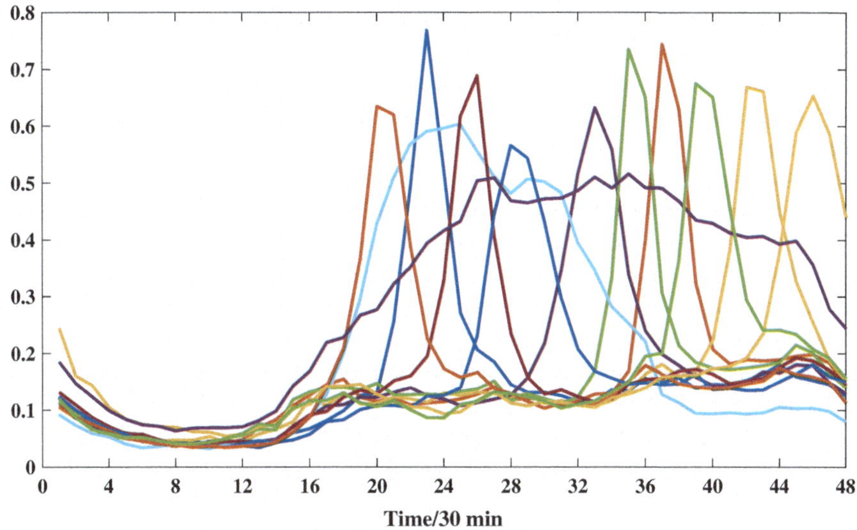

Fig. 1.3 Typical normalized daily residential load profiles

important for accurate billing. The works on anomaly detection in smart meter data are summarized from the perspective of bad data detection and NTL detection (or energy theft detection). Load profiling is used to find the basic electricity consumption patterns of each consumer or a group of consumers. The load profiling results can be further used for load forecasting and demand response programs.

1.2.1 Bad Data Detection

Bad data, as discussed here, can be missing data or unusual patterns caused by unplanned events or the failure of data collection, communication, or entry. Bad data detection can be divided into probabilistic, statistical, and machine learning methods [13]. The methods for bad data detection in other research areas could be applied to smart meter data. Only the works closely related to smart meter bad data detection are surveyed in this subsection. According to the modeling methods, these works are summarized as time-series-based methods, low-rank matrix technique-based methods, and time-window-based methods.

Smart meter data are essentially time series. An optimally weighted average (OWA) method was proposed for data cleaning and imputation in [14], which can be applied to offline or online situations. It was assumed that the load data could be explained by a linear combination of the nearest neighbor data, which is quite similar to the autoregressive moving average (ARIMA) model for time series. The optimal weight was obtained by training an optimization model. While in [15], the nonlinear relationship between the data at different time periods and exogenous inputs was modeled by combining autoregressive with exogenous inputs (ARX) and artificial neural network (ANN) models where the bad data detection was modeled as a hypothesis testing on the extreme of the residuals. A case study on gas flow data was performed and showed an improvement in load forecasting accuracy after ARX-based bad data detection. Similarly, based on the auto-regression (AR) model, the generalized extreme Studentized deviate (GESD) and the Q-test were proposed to detect the outliers when the number of samples is more and less than ten, respectively, in [16]. Then, canonical variate analysis (CVA) was conducted to cluster the recovered load profiles, and a linear discriminate analysis (LDA) classifier was further used to search for abnormal electricity consumption. Instead of detecting bad data, which forecasting method is robust to the cyber attack or bad data without bad data detection was investigated in [17].

The electricity consumptions are spatially and temporally correlated. Exploring the spatiotemporal correlation can help identify the outliers and recover them. A low-rank matrix fitting-based method was proposed in [18] to conduct data cleaning and imputation. An alternating direction method of multipliers (ADMM)-based distributed low-rank matrix technique was also proposed to enable communication and data exchange between different consumers and to protect the privacy of consumers. Similarly, to produce a reliable state estimation, the measurements were first processed by low-rank matrix techniques in [19]. Both off-line and on-line algorithms have been proposed. However, the improvement in state estimation after low-rank denoising has not been investigated. Low-rank matrix factorization works well when the bad data are randomly distributed. However, when the data are unchanged for a certain period, the low-rank matrix cannot handle it well.

Rather than detecting all the bad data directly, strategies that continuously detect and recover a part within a certain time window have also been studied. A clustering approach was proposed on the load profiles with missing data in [20, 21]. The clustering was conducted on segmented profiles rather than the entire load profiles in a rolling manner. In this way, the missing data can be recovered or estimated by other data in the same cluster. Collective contextual anomaly detection using a sliding window framework was proposed in [22] by combining various anomaly classifiers. The anomalous data were detected using overlapping sliding windows. Since smart meter data are collected in a real-time or near real-time fashion, an online anomaly detection method using the Lambda architecture was proposed in [23]. The proposed online detection method can be parallel processed, having high efficiency when working with large datasets.

1.2.2 Energy Theft Detection

Strictly speaking, smart meter data with energy theft also belong to bad data. The bad data discussed above are unintentional and appear temporarily, whereas energy theft may change the smart meter data under certain strategies and last for a relatively long time. Energy theft detection can be implemented using smart meter data and power system state data, such as node voltages. The energy theft detection methods with only smart meter data are summarized in this part from two aspects: supervised learning and unsupervised learning.

Supervised classification methods are effective approaches for energy theft detection, which generally consists of two stages: feature extraction and classification. To train a theft detection classifier, the non-technical loss was first estimated in [24]. k-means clustering was used to group the load profiles, where the number of clusters was determined by the silhouette value [25]. To address the challenge of imbalanced data, various possible malicious samples were generated to train the classifier. An energy theft alarm was raised after a certain number of abnormal detections. Different numbers of abnormal detections resulted in different false-positive rates (FPR) and Bayesian detection rates (BDR). The proposed method can also identify energy theft types. Apart from clustering-based feature extraction, an encoding technique was first performed on the load data in [26], which served as the inputs of classifiers including SVM and a rule-engine-based algorithm to detect the energy theft. The proposed method can run in parallel for real-time detection. By introducing external variables, a top-down scheme based on decision tree and SVM method was proposed in [27]. The decision tree estimated the expected electricity consumption based on the number of appliances, persons, and outdoor temperature. Then, the output of the decision tree was fed to the SVM to determine whether the consumer is normal or malicious. The proposed framework can also be applied for real-time detection.

Obtaining the labeled dataset for energy theft detection is difficult and expensive. Compared with supervised learning, unsupervised energy theft detection does not need the labels of all or partial consumers. An optimum-path forest (OPF) clustering algorithm was proposed in [28], where each cluster is modeled as a Gaussian distribution. The load profile can be identified as an anomaly if the distance is greater than a threshold. Comparisons with frequently used methods, including k-means, Birch, affinity propagation (AP), and Gaussian mixture model (GMM), verified the superiority of the proposed method. Rather than clustering all load profiles, clustering was only conducted within an individual consumer to obtain the typical and atypical load profiles in [29]. A classifier was then trained based on the typical and atypical load profiles for energy theft detection. A case study in this paper showed that extreme learning machine (ELM) and online sequential-ELM (OS-ELM)-based classifiers have better accuracy compared with SVM. Transforming the time series smart meter data into the frequency domain is another approach for feature extraction. Based on the discrete Fourier transform (DFT) results, the features extracted in the reference interval and examined interval were compared based on the so-called *Structure & Detect* method in [30]. Then, the load profile can be determined to be normal or malicious. The proposed method can be implemented in a parallel and distributed manner, which can be used for the on-line analysis of large datasets. Another unsupervised energy theft detection method is to formulate the problem as a load forecasting problem. If the metered consumption is considerably lower than the forecasted consumption, then the consumer can be marked as a malicious consumer.

An anomaly score was given to each consumption data and shown with different colors to realize visualization in [31].

1.2.3 Load Profiling

Load profiling refers to the classification of load curves or consumers according to electricity consumption behaviors. In this subsection, load profiling is divided into direct-clustering-based and indirect-clustering-based approaches. Various clustering techniques, such as k-means, hierarchical clustering, and self-organizing map (SOM), have been directly implemented on smart meter data [32–34]. Two basic issues about direct clustering are first discussed. Then, the works on indirect clustering are classified into dimensionality reduction, load characteristics, and variability and uncertainty-based methods according to the features that are extracted before clustering.

There are some basic issues associated with direct clustering. The first issue is the resolution of smart meter data. In [35], three frequently used clustering techniques, namely, k-means, hierarchical algorithms, and the Dirichlet process mixture model (DPMM) algorithm, were performed on the smart meter data with different frequencies varying from every 1 min to 2 h to investigate how the resolution of smart meter data influences the clustering results. The results showed that the smart meter data with a frequency of at least every 30 min is sufficiently reliable for most purposes. The second issue is that smart meter data are essentially time-series data. In contrast to traditional clustering methods for static data, k-means modified for dynamic clustering was proposed in [36] to address time-dependent data. The dynamic clustering allows capturing the trend of clusters of consumers. A two-stage clustering strategy was proposed in [37] to reduce the computational complexity. In the first stage, k-means was performed to generate the local representative load profiles; in the second stage, clustering was further performed on the clustering centers obtained in the first stage at the central processor. In this way, the clustering method can be performed in a distributed fashion and largely reduce the overall complexity.

Apart from direct clustering, increasingly more literature is focusing on indirect clustering, i.e., feature extraction is conducted before clustering. Dimensionality reduction is an effective way to address the high dimensionality of smart meter data. Principal component analysis (PCA) was performed on yearly load profiles to reduce the dimensionality of original data and then k-means was used to classify consumers in [38]. The components learned by PCA can reveal the consumption behaviors of different connection point types. Similarly, PCA was also used to find the temporal patterns of each consumer and spatial patterns of several consumers in [39]. Then, a modified k-medoids algorithm based on the Hausdorff distance and Voronoi decomposition method was proposed to obtain typical load profiles and detect outliers. The method was tested on a large real dataset to prove the effectiveness and efficiency. Deep-learning-based stacked sparse auto-encoders were applied for load profile compression and feature extraction in [40]. Based on the reduced and encoded load profile, a locality sensitive hashing (LSH) method was further proposed to classify the load profiles and obtain the representative load profiles.

Insights into the local and global characteristics of smart meter data are important for finding meaningful typical load profiles. Three new types of features generated by applying conditional filters to meter-resolution-based features integrated with shape signatures, calibration and normalization, and profile errors were proposed in [41] to cluster daily load curves. The proposed feature extraction method was of low computational complexity, and the features were informative and understandable for describing the electricity usage patterns. To capture local and global shape variations, 10 subspace clustering and projected clustering methods were applied to identify the contact type of consumers in [42]. By focusing on the subspace of load profiles, the clustering process was proven to be more robust to noise. To capture the peak load and major variability in residential consumption behavior, four key time periods (overnight, breakfast, daytime, and evening) were identified in [43]. On this basis, seven attributes were calculated for clustering. The robustness of the proposed clustering was verified using the bootstrap technique.

The variability and uncertainty of smart meter data have also been considered for load profiling. Four key time periods, which described different peak demand behaviors, coinciding with common intervals of the day were identified in [43], and then a finite mixture-model-based clustering was used to discover ten distinct behavior groups describing customers based on their demand and variability. The load variation was modeled by a lognormal distribution, and a Gaussian mixture model (GMM)-based load profiling method was proposed in [44] to capture the dynamic behavior of consumers. A mixture model was also used in [45] by integrating the C-vine copula method for the clustering of residential load profiles. The high-dimensional nonlinear correlations among consumptions of different time periods were modeled using the C-vine copula. This method has an effective performance in large datasets. While in [46], a Markov model was established based on the separated time periods to describe the electricity consumption behavior dynamics. A clustering technique consisting of fast search and find of density peaks (CFSFDP) integrated into a divide-and-conquer distributed approach was proposed to find typical consumption behaviors. The proposed distributed clustering algorithm had higher computational efficiency. The expectation-maximization (EM)-based mixture model clustering method was applied in [47] to obtain typical load profiles, and then the variabilities in residential load profiles were modeled by a transition matrix based on a second-order Markov chain and Markov decision processes. The proposed method can be used to generate pseudo smart meter data for retailers and protect the privacy of consumers.

1.2.4 Remarks

Table 1.1 provides the correspondence between the key techniques and the surveyed references in smart meter data analytics for load analysis.

Table 1.1 Brief summary of the literature on load analysis

Load analysis	Key words	References
Bad data detection	Time series analysis	[14–16]
	Low rank matrix	[18, 19]
	Time window	[20–23]
Energy theft detection	Supervised learning	[24, 26, 27]
	Unsupervised learning	[28–31]
Load profiling	Direct clustering	[32–37]
	Dimension reduction	[32, 38–40]
	Local characteristics	[41–43]
	Variability and uncertainty	[43–47]

For bad data detection, most of the bad data detection methods are suitable for business/industrial consumers or higher aggregation level load data, which are more regular and have certain patterns. The research on bad data detection on the individual consumer is still limited and not a trivial task because the load profiles of an individual consumer show more variation. In addition, since bad data detection and repairing are the basis of other data analytics application, how much improvement can be made for load forecasting or other applications after bad data detection is also an issue that deserves further investigation. In addition, smart meter data are essentially streaming data. Real-time bad data detection for some real-time applications, such as very-short-term load forecasting, is another concern. Finally, as stated above, bad data may be brought from data collection failure. Short period anomaly usage patterns may also be identified as bad data even though it is "real" data. More related factors, such as sudden events, need to be considered in this situation. Redundant data are also good sources for "real" but anomaly data identification.

For energy theft detection, with a longer time period of smart meter data, the detection accuracy is probably higher because more data can be used. However, using longer historical smart meter data may also lead to a detection delay, which means that we need to achieve a balance between the detection accuracy and detection delay. Moreover, different private data and simulated data have been tested on different energy theft detection methods in the existed literature. Without the same dataset, the superiority of a certain method cannot be guaranteed. The research in this area will be promoted if some open datasets are provided. Besides, in most cases, one paper proposes one energy theft detection method. Just like ensemble learning for load forecasting, can we propose an ensemble detection framework to combine different individual methods?

For load profiling, the majority of the clustering methods are used for stored smart meter data. However, the fact is that smart meter data are streaming data. Sometimes, we need to deal with the massive streaming data in a real-time fashion for specific applications. Thus, distributed clustering and incremental clustering methods can be further studied in the field of load profiling. Indirect load profile methods extract features first and then conduct clustering on the extracted features. Some clustering methods such as deep embedding clustering [48] that can implement feature extraction and clustering at the same time, have been proposed outside the area of electrical engineering. It is worth trying to apply these state-of-the-art methods to load profiling. Most load profiling methods are evaluated by clustering-based indices, such as similarity matrix indicator (SMI), Davies–Bouldin indicator (DBI) and Silhouette Index (SI) [49]. More application-oriented matrices such as forecasting accuracy are encouraged to be used to guide the selection of suitable clustering methods. Finally, how to effectively extract meaningful features before clustering to improve the performance and efficiency of load profiling is another issue that needs to be further addressed.

1.3 Load Forecasting

Load forecasts have been widely used by the electric power industry. Power distribution companies rely on short- and long-term forecasts at the feeder level to support operations and planning processes, while retail electricity providers make pricing, procurement and hedging decisions largely based on the forecasted load of their customers. Figure 1.4 presents the normalized hourly profiles of a week for four different types of loads, including a house, a factory, a feeder, and a city. Loads of a house, a factory, and a feeder are more volatile than the city-level load. In reality, the higher level the load is measured at, the smoother the load profile typically is. Developing a highly accurate forecast is nontrivial at lower levels.

Although the majority of the load forecasting literature has been devoted to forecasting at the top (high voltage) level, the information from medium/low voltage levels, such as distribution feeders and even down to the smart meters, offer some opportunities to improve the forecasts. A recent review of load forecasting was conducted in [50], focusing on the transition from point load forecasting to probabilistic load forecasting. In this section, we will review the recent literature for both point and probabilistic load forecasting with the emphasis on the medium/low voltage levels. Within the point load forecasting literature, we divide the review based on whether the smart meter data is used or not.

1.3.1 Forecasting Without Smart Meter Data

Compared with the load profiles at the high voltage levels, the load profiles aggregated to a customer group or medium/low voltage level are often more volatile and sensitive to the behaviors of the customers being served. Some of them, such as the load of a residential community, can be very responsive to the weather conditions. Some others, such as the load of a large factor, can be driven by specific work schedules. Although these load profiles differ by the customer composition, these

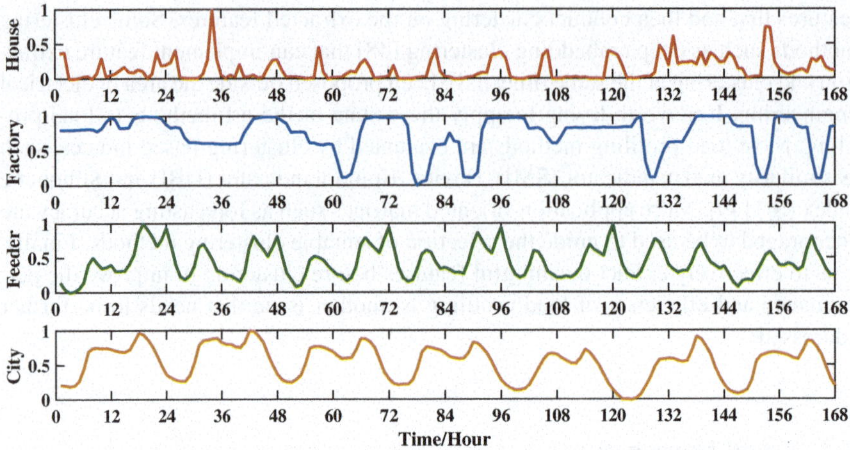

Fig. 1.4 Normalized hourly profiles of a week for four types of loads

load forecasting problems share some common challenges, such as accounting the influence from the competitive markets, modeling the effects of weather variables, and leveraging the hierarchy.

In competitive retail markets, electricity consumption is largely driven by the number of customers. The volatile customer count contributes to the uncertainties in the future load profile. A two-stage long-term retail load forecasting method was proposed in [51] to take customer attrition into consideration. The first stage was to forecast each customer's load using multiple linear regression with a variable selection method. The second stage was to forecast customer attrition using survival analysis. Thus, the product of the two forecasts provided the final retail load forecast. Another issue in the retail market is the consumers' reactions to the various demand response programs. While some consumers may respond to the price signals, others may not. A nonparametric test was applied to detect the demand-responsive consumers so that they can be forecasted separately [52]. Because the authors did not find publicly available demand data for individual consumers, the experiment was conducted using aggregate load in the Ontario power gird.

Since the large scale adoption of electrical air conditioning systems in the 1940s, capturing the effects of weather on load has been a major issue in load forecasting. Most load forecasting models in the literature include temperature variables and their variants, such as lags and averages. How many lagged hourly temperatures and moving average temperatures can be included in a regression model? An investigation was conducted in [53]. The case study was based on the data from the load forecasting track of GEFCom2012. An important finding is that a regression-based load forecasting model estimated using two to three years of hourly data may include more than a thousand parameters to maximize the forecast accuracy. In addition, each zone may need a different set of lags and moving averages.

Not many load forecasting papers are devoted to other weather variables. How to include humidity information in load forecasting models was discussed in [54], where the authors discovered that the temperature-humidity index (THI) might not be optimal for load forecasting models. Instead, separating relative humidity, temperature and their higher-order terms and interactions in the model, with the corresponding parameters being estimated by the training data, were producing more accurate load forecasts than the THI-based models. A similar investigation was performed for wind speed variables in [55]. Comparing with the models that include wind chill index (WCI), the ones with wind speed, temperature, and their variants separated were more accurate.

The territory of a power company may cover several micro-climate zones. Capturing the local weather information may help improve the load forecast accuracy for each zone. Therefore, proper selection of weather stations would contribute to the final load forecast accuracy. Weather station selection was one of the challenges designed into the load forecasting track of GEFCom2012 [56]. All four winning team adopted the same strategy: first deciding how many stations should be selected, and then figuring out which stations to be selected [57–60]. A different and more accurate method was proposed in [61], which follows a different strategy, determining how many and which stations to be selected at the same time instead of sequentially. The method includes three steps: rating and ranking the individual weather stations, combining weather stations based on a greedy algorithm, and rating and ranking the combined stations. The method is currently being used by many power companies, such as the North Carolina Electric Membership Corporation, which was used as one of the case studies in [61].

The pursuit of operational excellence and large-scale renewable integration is pushing load forecasting toward the grid edge. Distribution substation load forecasting becomes another emerging topic. One approach is to adopt the forecasting techniques and models with good performance at higher levels. For instance, a three-stage methodology, which consists of preprocessing, forecasting, and postprocessing, was taken to forecast loads of three datasets ranging from distribution level to transmission level [62]. A semi-parametric additive model was proposed in [63] to forecast the load of the Australian National Electricity Market. The same technique was also applied to forecast more than 2200 substation loads of the French distribution network in [64]. Another load forecasting study on seven substations from the French network was reported in [65], where a conventional time series forecasting methodology was used. The same research group then proposed a neural network model to forecast the load of two French distribution substations, which outperformed a time series model [66].

Another approach to distribution load forecasting is to leverage the connection hierarchy of the power grid. In [67], The load of a root node of any subtree was forecasted first. The child nodes were then treated separately based on their similarities. The forecast of a "regular" node was proportional to the parent node forecast, while the "irregular" nodes were forecasted individually using neural networks. Another attempt to make use of the hierarchical information for load forecasting was made in [68]. Two case studies were conducted, one based on New York City and its substations, and the other one based on PJM and its substations. The authors demonstrated the effectiveness of aggregation in improving the higher-level load forecast accuracy.

1.3.2 Forecasting with Smart Meter Data

The value that smart meters bring to load forecasting is two-fold. First, smart meters make it possible for the local distribution companies and electricity retailers to better understand and forecast the load of an individual house or building. Second, the high granularity load data provided by smart meters offer great potential for improving the forecast accuracy at aggregate levels.

Because the electricity consumption behaviors at the household and building levels can be much more random and volatile than those at aggregate levels, the traditional techniques and methods developed for load forecasting at an aggregate level may or may not be well suited. To tackle the problem of smart meter load forecasting, the research community has taken several different approaches, such as evaluating and modifying the existing load forecasting techniques and methodologies, adopting and inventing new ones, and a mixture of them.

A highly cited study compared seven existing techniques, including linear regression, ANN, SVM, and their variants [69]. The case study was performed based on two datasets: one containing two commercial buildings and the other containing three residential homes. The study demonstrated that these techniques could produce fine forecasts for the two commercial buildings but not the three residential homes. A self-recurrent wavelet neural network (SRWNN) was proposed to forecast an education building in a microgrid setting [70]. The proposed SRWNN was shown to be more accurate than its ancestor wavelet neural network (WNN) for both building-level load forecasting (e.g., a 694 kW peak education building in British Columbia, Canada) and state- or province-level load forecasting (e.g., British Columbia and California).

Some researchers tried deep learning techniques for the household- and building-level load forecasting. Conditional Restricted Boltzmann Machine (CRBM) and Factored Conditional Restricted Boltzmann Machine (FCRBM) were assessed in [71] to estimate energy consumption for a household and three submetering measurements. FCRBM achieves the highest load forecast accuracy compared with ANN, RNN, SVM, and CRBM. Different resolutions ranging from one minute to one week have been tested. A pooling-based deep recurrent neural network (RNN) was proposed in [72] to learn spatial information shared between interconnected customers and to address the over-fitting challenges. It outperformed ARIMA, SVR, and classical deep RNN on the Irish CER residential dataset.

Sparsity is a key character in household-level load forecasting. A spatiotemporal forecasting approach was proposed in [73], which incorporated a large dataset of many driving factors of the load for all surrounding houses of a target house. The proposed method combined ideas from Compressive Sensing and data decomposition to exploit the low-dimensional structures governing the interactions among the nearby houses. The Pecan Street data was used to evaluate the proposed method. Sparse coding was used to model the usage patterns in [74]. The case study was based on a dataset collected from 5000 households in Chattanooga, TN, where Including the sparse coding features led to 10% improvements in forecast accuracy. A least absolute shrinkage and selection (LASSO)-based sparse linear method was proposed to forecast individual consumption in [75]. The consumer's usage patterns can be extracted from the non-zero coefficients, and it was proven that data from other consumers contribute to the fitted residual. Experiments on real data from Pacific Gas and Electric Company showed that the LASSO-based method has low computational complexity and comparable accuracy.

A commonly used method to reduce noise in smart meter data is to aggregate the individual meters. To keep the salient features from being buried during aggregation, clustering techniques are often used to group similar meters. In [76], next-day load forecasting was formulated as a functional time series problem. Clustering was first performed to classify the historical load curves into different groups. The last observed load curve was then assigned to the most similar cluster. Finally, based on the load curves in this cluster, a functional wavelet-kernel (FWK) approach was used to forecast the next-day load curve. The results showed that FWK with clustering outperforms simple FWK. Clustering was also conducted in [77] to obtain the load patterns. Classification from contextual information, including time, temperature, date, and economic indicator to clusters, was then performed. Based on the trained classifier, the daily load can be forecasted with known contextual information. A shape-based clustering method was performed in [78] to capture the time drift characteristic of the individual load, where the cluster number was smaller than those obtained by traditional Euclidean-distance-based clustering methods. The clustering method is quite similar to k-means, while the distance is quantified by dynamic time warping (DTW). Markov models were then constructed to forecast the shape of the next-day load curve. Similar to the clustering method proposed in [78], a k-shape clustering was proposed in [79] to forecast building time-series data, where the time series shape similarity was used to update the cluster memberships to address the time-drift issue.

The fine-grained smart meter data also introduce new perspectives to the aggregation level load forecasting. A clustering algorithm can be used to group customers. Each customer group can then be forecasted with different forecasting models. Finally, the aggregated load forecast can be obtained by summing the load forecast of each group. Two datasets including the Irish CER residential dataset and another dataset from New York were used to build the case study in [80]. Both showed that forecast errors can be reduced by effectively grouping different customers based on their energy consumption behaviors. A similar finding was presented in [81] where the Irish CER residential dataset was used in the case study. The results showed that cluster-based forecasting can improve the forecasting accuracy and that the performance depends on the number of clusters and the size of the consumer.

The relationship between group size and forecast accuracy based on Seasonal-Naïve and Holt-Winters algorithms was investigated in [82]. The results showed that forecasting accuracy increases as group size increases, even for small groups.

A simple empirical scaling law is proposed in [83] to describe how the accuracy changes as different aggregation levels. The derivation of the scaling law is based on the Mean Absolute Percentage Error (MAPE). Case studies on the data from Pacific Gas and Electric Company show that MAPE decreases quickly with the increase of the number of consumers when the number of consumers is less than 100,000. When the number of consumers is more than 100,000, the MAPE has a little decrease.

Forecast combination is a well-known approach to accuracy improvement. A residential load forecasting case study showed that the ensembles outperformed all the individual forecasts from traditional load forecasting models [84]. By varying the number of clusters, different forecasts can be obtained. A novel ensemble forecasting framework was proposed in [85] to optimally combine these forecasts to further improve the forecasting accuracy.

Traditional error measures such as MAPE cannot reasonably quantify the performance of individual load forecasting due to the violation and time-shifting characteristics. For example, MAPE can easily be influenced by outliers. A resistant MAPE (r-MAPE) based on the calculation of the Huber M-estimator was proposed in [86] to overcome this situation. The mean arctangent absolute percentage error (MAAPE) was proposed in [87] to consider the intermittent nature of individual load profiles. MAAPE, a variation of MAPE, is a slope as an angle, the tangent of which is equal to the ratio between the absolute error and real value, i.e., the absolute percentage error (APE). An error measure designed for household-level load forecasts was proposed in [88] to address the time-shifting characteristic of household-level loads. In addition to these error measures, some modifications of MAPE and mean absolute error (MAE) have been used in other case studies [74, 75].

1.3.3 Probabilistic Forecasting

A probabilistic forecast provides more information about future uncertainties that what a point forecast does. As shown in Fig. 1.5, a typical point forecasting process contains three parts: data inputs, modeling, and data outputs (forecasts). As summarized in [50], there are three ways to modify the workflow to generate probabilistic forecasts: (1) generating multiple input scenarios to feed to a point forecasting model; (2) applying probabilistic forecasting models, such as quantile regression; and (3) augmenting point outputs to probabilistic outputs by imposing simulated or modeled residuals or making ensembles of point forecasts.

On the input side, scenario generation is an effective way to capture the uncertainties from the driving factors of electricity demand. Various temperature scenario generation methods have been proposed in the literature, such as direct usage of the previous years of hourly temperatures with the dates fixed [89], shifting the historical temperatures by a few days to create additional scenarios [90], and bootstrapping the historical temperatures [91]. A comparison of these three methods based on pinball loss function was presented in [92]. The results showed that the shifted-date method dominated the other two when the number of dates being shifted is within

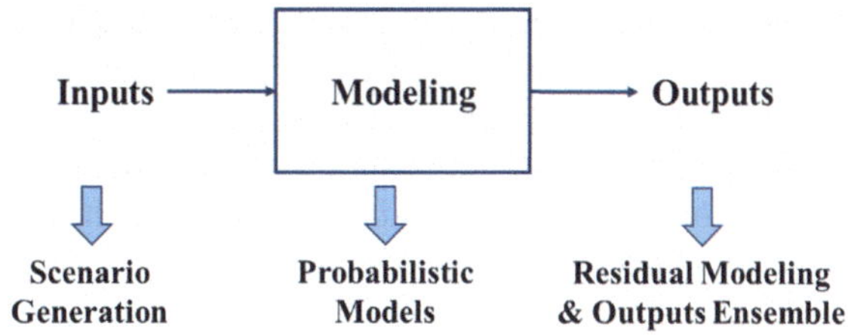

Fig. 1.5 From point forecasting to probabilistic forecasting

a range. An empirical formula was also proposed to select parameters for the temperature scenario generation methods. The idea of generating temperature scenarios was also applied in [93]. An embedding based quantile regression neural network was used as the regression model instead of the MLR model, where the embedding layer can model the effect of calendar variables. In this way, the uncertainties of both future temperature and the relationship between temperature and load can be comprehensively considered. The scenario generation method was also used to develop a probabilistic view of power distribution system reliability indices [94].

On the output side, one can convert point forecasts to probabilistic ones via residual simulation or forecast combination. Several residual simulation methods were evaluated in [95]. The results showed that the residuals do not always follow a normal distribution, though group analysis increases the passing rate of normality tests. Adding simulated residuals under the normality assumption improves probabilistic forecasts from deficient models, while the improvement is diminishing as the underlying model improves. The idea of combining point load forecasts to generate probabilistic load forecasts was first proposed in [96]. The quantile regression averaging (QRA) method was applied to eight sister load forecasts, a set of point forecasts generated from homogeneous models developed in [53]. A constrained QRA (CQRA) was proposed in [97] to combine a series of quantiles obtained from individual quantile regression models.

Both approaches mentioned above rely on point forecasting models. It is still an unsolved question whether a more accurate point forecasting model can lead to a more skilled probabilistic forecast within this framework. An attempt was made in [98] to answer this question. The finding is that when the two underlying models are significantly different w.r.t. the point forecast accuracy, a more accurate point forecasting model would lead to a more skilled probabilistic forecast.

Various probabilistic forecasting models have been proposed by statisticians and computer scientists, such as quantile regression, Gaussian process regression, and density estimation. These off-the-shelf models can be directly applied to generate probabilistic load forecasts [50]. In GEFCom2014, a winning team developed a quantile generalized additive model (quantGAM), which is a hybrid of quantile regression and generalized additive models [99]. Probabilistic load forecasting has also been conducted on individual load profiles. Combining the gradient boosting method and quantile regression, a boosting additive quantile regression method was proposed in [100] to quantify the uncertainty and generate probabilistic forecasts. Apart from the quantile regression model, kernel density estimation methods were tested in [101]. The density of electricity data was modeled using different implementations of conditional kernel density (CKD) estimators to accommodate the seasonality in consumption. A decay parameter was used in the density estimation model for recent effects. The selection of kernel bandwidths and the presence of boundary effects are two main challenges with the implementation of CKD that were also investigated.

1.3.4 Remarks

Table 1.2 provides the correspondence between the key techniques and the surveyed references in smart meter data analytics for load forecasting.

Forecasting the loads at aggregate levels is a relatively mature area. Nevertheless, there are some nuances in the smart grid era due to the increasing need of highly accurate load forecasts. One is on the evaluation methods. Many forecasts are being evaluated using widely used error measures such as MAPE, which does not consider the consequences of over- or under-forecasts. In reality, the cost to the sign and magnitude of errors may differ significantly. Therefore, the following research question rises: how can the costs of forecast errors be integrated into the forecasting processes? Some research in this area would be helpful to bridge the gap between forecasting and decision making. The second one is load transfer detection, which is a rarely touched area in the literature. Distribution operators may transfer the load from one circuit to another permanently, seasonally, or on an ad hoc basis, in response to maintenance

Table 1.2 Brief summary of the literature on load forecasting

Load forecasting	Key words	References
Without individual meters	Consumer attrition/demand response	[51, 52]
	Weather modeling & selection	[53–61]
	Traditional high accurate model	[62–66]
	Hierarchical forecasting	[67, 68]
With individual meters	Traditional methods	[69, 70]
	Sparse coding/deep learning	[71–75]
	Clustering	[76–79]
	Aggregation load	[80–82, 84, 85, 102, 103]
	Evaluation criteria	[74, 75, 86–88, 100]
Probabilistic forecasting	Scenario generation	[89–94]
	Residual modeling & output ensemble	[53, 95–97]
	Probabilistic forecasting models	[50, 99–101]

needs or reliability reasons. These load transfers are often poorly documented. Without smart meter information, it is difficult to physically trace the load blocks being transferred. Therefore, a data-driven approach is necessary in these situations. The third one is hierarchical forecasting, specifically, how to fully utilize zonal, regional, or meter load and local weather data to improve the load forecast accuracy. In addition, it is worth studying how to reconcile the forecasts from different levels for the applications of aggregators, system operators, and planners. The fourth one is on the emerging factors that affect electricity demand. The consumer behaviors are being changed by many modern technologies, such as rooftop solar panels, large batteries, and smart home devices. It is important to leverage the emerging data sources, such as technology adoption, social media, and various marketing surveys.

To comprehensively capture the uncertainties in the future, researchers and practitioners recently started to investigate in probabilistic load forecasting. Several areas within probabilistic load forecasting would need some further attention. First, distributed energy resources and energy storage options often disrupt the traditional load profiles. Some research is needed to generate probabilistic net load forecasts for the system with high penetration of renewable energy and large scale storage. Secondly, forecast combination is widely regarded in the point forecasting literature as an effective way to enhance the forecast accuracy. There is a primary attempt in [97] to combine quantile forecasts. Further investigations can be conducted on combining other forms probabilistic forecasts, such as density forecasts and interval forecasts. Finally, the literature of probabilistic load forecasting for smart meters is still quite limited. Since the meter-level loads are more volatile than the aggregate loads, probabilistic forecasting has a natural application in this area.

1.4 Load Management

How smart meter data contribute to the implementation of load management is summarized from three aspects in this section: the first one is to have a better understanding of sociodemographic information of consumers to provide better and personalized service. The second one is to target potential consumers for demand response program marketing. The third one is the issue related to demand response program implementation including price design for price-based demand response and baseline estimation for incentive-based demand response.

1.4.1 Consumer Characterization

The electricity consumption behaviors of the consumers are closely related to their socio-demographic status. Bridging the load profiles to socio-demographic status is an important approach to classify the consumers and realize personalized services. A naive problem is to detect consumer types according to the load profiles. The other two issues are identifying socio-demographic information from load profiles and predicting the load shapes using the socio-demographic information.

Identifying the type of consumers can be realized by simple classification. The temporal load profiles were first transformed into the frequency domain in [104] using fast Fourier transformation (FFT). Then the coefficients of different frequencies were used as the inputs of classification and regression tree (CART) to place consumers in different categories. FFT decomposes smart meter data based on a certain sine function and cosine function. Another transformation technique, sparse coding, has no assumption on the base signal but learns them automatically. Non-negative sparse coding was applied to extract the partial usage patterns from original load profiles in [105]. Based on the partial usage patterns, linear SVM was implemented to classify the consumers into residents and small and medium-sized enterprises (SME). The classification accuracy is considerably higher than the discrete wavelet transform (DWT) and PCA.

There are still consumers without smart meter installations. External data, such as the socio-demographic status of consumers, are applied to estimate their load profiles. Clustering was first implemented to classify consumers into different energy behavior groups, and then energy behavior correlation rate (EBCR) and indicator dominance index (IGD) were defined and calculated to identify the indicators higher than a threshold [106]. Finally, the relationship between different energy behavior groups and their socio-demographic status was mapped. Spectral clustering was applied to generate typical load profiles, which were then used as the inputs of predictors such as random forests (RF) and stochastic boosting (SB) in [107]. The results showed that with commercial and cartographic data, the load profiles of consumers can be accurately predicted. The stepwise selection was applied to investigate the factors that have a great influence on residential electricity consumption in [108]. The location, floor area, the age of consumers, and the number of appliances are main factors, while the income level and homeownership have little relationship with consumption. A multiple linear regression model was used to bridge the total electricity consumption, maximum demand, load factor, and ToU to dwelling and occupant socioeconomic variables in [109]. The factors that have a great impact on total consumption, maximum load, load factor, and ToU were identified. The influence of socioeconomic status of consumers' electricity consumption patterns was evaluated in [110]. RF regression was applied to combine socioeconomic status and environmental factors to predict the consumption patterns.

More works focus on how to mine the socio-demographic information of consumers from the massive smart meter data. One approach is based on a clustering algorithm. DPMM was applied in [111] for household and business premise load profiling where the number of clusters was not required to predetermined. The clustering results obtained by the DPMM algorithm have a clear corresponding relation with the metadata of dwellings, such as the nationality, household size, and type of dwelling. Based on the clustering results, multinomial logistic regression was applied to the clusters and dwelling and appliance characteristics in [112]. Each cluster was analyzed according to the coefficients of the regression model. Feature extraction and selection have also been applied as the attributes of the classifier. A feature set including the average consumption over a certain period, the ratios of two consumptions in different periods, and the temporal properties was established in [113]. Then, classification or regression was implemented to predict the socio-demographic status according to these features. Results showed that the proposed feature extraction method outperforms the biased random guess. More than 88 features from consumption, ratios, statistics, and temporal characteristics were extracted, and then correlation, KS-test, and η^2-based feature selection methods were conducted in [114]. The so-called extend CLASS classification framework was used to forecast the deduced properties of private dwellings. A supervised classification algorithm called dependent-independent data classification (DID-Class) was proposed to address the challenges of dependencies among multiple classification-relevant variables in [115]. The characteristics of dwellings were recognized based on this method, and comparisons with SVM and traditional CLASS proposed in [113] were conducted. The accuracy of DID-Class with SVM and CLASS is slightly higher than those of SVM and CLASS. To capture the intra-day and inter-day electricity consumption behavior of the consumers, a two-dimensional convolutional neural network (CNN) was used in [116] to make a bridge between the smart meter data and socio-demographic information of the consumers. The deep learning method can extract the features automatically and outperforms traditional methods.

1.4.2 Demand Response Program Marketing

Demand response program marketing is to target consumers who have a large potential to be involved in demand response programs. On one hand, 15 min or half-hour smart meter data cannot provide detail information on the operation status of the appliance; on the other hand, the altitude of consumers towards demand response is hard to model. Thus, the demand response potential cannot be evaluated directly. In this subsection, the potential of demand response can be indirectly evaluated by analyzing the variability, sensitivity to temperature, and so forth.

Variability is a key index for evaluating the potential of demand response. A hidden Markov model (HMM)-based spectral clustering was proposed in [117] to describe the magnitude, duration, and variability of the electricity consumption and further estimate the occupancy states of consumers. The information on the variability, occupancy states, and inter-temporal consumption dynamics can help retailers or aggregators target suitable consumers at different time scales. Both adaptive k-means and hierarchical clustering were used to obtain the typical load shapes of all the consumers within a certain error threshold in [118]. The entropy of each consumer was then calculated according to the distribution of daily load profiles over a year, and the typical shapes of load profiles were analyzed. The consumers with lower entropy have relatively similar consumption patterns on different days and can be viewed as a greater potential for demand response because their load profiles are more predictable. Similarly, the entropy was calculated in [46] based on the state transition matrix. It was stated that the consumers with high entropy are suitable for price-based demand response for their flexibility to adjust their load profile according to the change in price, whereas the consumers with low entropy are suitable for incentive-based demand response for their predictability to follow the control commands.

Estimation of electricity reduction is another approach for demand response potential. A mixture model clustering was conducted on a survey dataset and smart meter data in [47] to evaluate the potential for active demand reduction with wet appliances. The results showed that both the electricity demand of wet appliances and the attitudes toward demand response have a great influence on the potential for load shifting. Based on the GMM model of the electricity consumption of consumers and the estimated baseline, two indices, i.e., the possibility of electricity reduction greater than or equal to a certain value and the least amount of electricity reduction with a certain possibility were calculated in [119]. These two indices can help demand response implementers have a probabilistic understanding of how much electricity can be reduced. A two-stage demand response management strategy was proposed in [120], where SVM was first used to detect the devices and users with excess load consumption and then a load balancing algorithm was performed to balance the overall load.

Since appliances such as heating, ventilation and air conditioning (HVAC) have great potential for demand response, the sensitivity of electricity consumption to outdoor air temperature is an effective evaluation criterion. Linear regression was applied to smart meter data and temperature data to calculate this sensitivity, and the maximum likelihood approach was used to estimate the changing point in [121]. Based on that, the demand response potentials at different hours were estimated. Apart from the simple regression, an HMM-based thermal regime was proposed to separate the original load profile into the thermal profile (temperature-sensitive) and base profile (non-temperature-sensitive) in [122]. The demand response potential can be calculated for different situations, and the proposed method can achieve much more savings than random selection. A thermal demand response ranking method was proposed in [123] for demand response targeting, where the demand response potential was evaluated from two aspects: temperature sensitivity and occupancy. Both linear regression and breakpoint detection were used to model the thermal regimes; the true linear response rate was used to detect the occupancy.

1.4.3 Demand Response Implementation

Demand response can be roughly divided into price-based demand response and incentive-based demand response. Price design is an important business model to attract consumers and maximize profit in price-based demand response programs; baseline estimation is the basis of quantifying the performance of consumers in incentive-based demand response programs. The applications of smart meter data analytics in price design and baseline estimation are summarized in this subsection.

For tariff design, an improved weighted fuzzy average (WFA) k-means was first proposed to obtain typical load profiles in [124]. An optimization model was then formulated with a designed profit function, where the acceptance of consumers over price was modeled by a piecewise function. The similar price determination strategy was also presented in [125]. Conditional value at risk (CVaR) for the risk model was further considered in [126] such that the original optimization model becomes a stochastic one. Different types of clustering algorithms were applied to extract load profiles with a performance index granularity guided in [127]. The results showed that different clusterings with different numbers of clusters and algorithms lead to different costs. GMM clustering was implemented on both energy prices and load profiles in [128]. Then, ToU tariff was developed using different combinations of the classifications of time periods. The impact of the designed price on demand response was finally quantified.

For baseline estimation, five naive baseline methods, HighXofY, MidXofY, LowXofY, exponential moving average, and regression baselines, were introduced in [129]. Different demand response scenarios were modeled and considered. The results showed that bias rather than accuracy is the main factor for deciding which baseline provides the largest profits. To describe the uncertainty within the consumption behaviors of consumers, Gaussian-process-based probabilistic baseline estimation was proposed in [130]. In addition, how the aggregation level influences the relative estimation error was also investigated. k-means clustering of the load profiles in non-event days was first applied in [131], and a decision tree was used to predict the electricity consumption level according to demographics data, including household characteristics and electrical appliances. Thus, a new consumer can be directly classified into a certain group before joining the demand response program and then simple averaging and piecewise linear regression were used to estimate to baseline load in different weather conditions. Selecting a control group for baseline estimation was formulated as an optimization problem in [132]. The objective was to minimize the difference between the load profiles of the control group and demand response group when there is no demand response event. The problem was transformed into a constrained regression problem.

1.4.4 Remarks

Table 1.3 provides the correspondence between the key techniques and the surveyed references in smart meter data analytics for load management.

For consumer characterization, it is essentially a high dimensional and nonlinear classification problem. There are at least two ways to improve the performance of consumer characterization: (1) conducting feature extraction or selection; (2) developing classification models. In the majority of existing literature, the features for consumer characterization are manually extracted. A data-driven feature extraction method might be an effective way to further improve performance. The classification is mainly implemented by the shallow learning models such as ANN and SVM. We

Table 1.3 Brief summary of the literature on load management

Load management	Key words	References
Consumer characterization	Consumer type	[104, 105]
	Load profile prediction	[106–110]
	Socio-demographic status prediction	[111–116]
Demand response program marketing	Variability	[46, 117, 118]
	Electricity reduction	[47, 119, 120]
	Temperature sensitivity	[121–123]
Demand response implementation	Tariff design	[124–128]
	Baseline estimation	[129–132]

can try different deep learning networks to tackle high nonlinearity. We also find that the current works are mainly based on the Irish dataset [133]. Low Carbon London dataset may be another good choice. More open datasets are needed to enrich the research in this area.

For demand response program marketing, evaluating the potential for load shifting or reduction is an effective way to target suitable consumers for different demand response programs. Smart meter data with a frequency of 30 min or lower cannot reveal the operation states of the individual appliance; thus, several indirect indices, including entropy, sensitivity to temperature and price, are used. More indices can be further proposed to provide a comprehensive understanding of the electricity consumption behavior of consumers. Since most papers target potential consumers for demand response according to the indirect indices, a critical question is why and how these indices can reflect the demand response potential without experimental evidence? More real-world experimental results are welcomed for the research.

For demand response implementation, all the price designs surveyed above are implemented with a known acceptance function against price. However, the acceptance function or utility function is hard to estimate. How to obtain the function has not been introduced in the existing literature. If the used acceptance function or utility function is different from the real one, the obtained results will deviate from the optimal results. Sensitivity analysis of the acceptance function or utility function assumption can be further conducted. Except for traditional tariff design, some innovative prices can be studied, such as different tariff packages based on fine-grained smart meter data. For baseline estimation, in addition to deterministic estimation, probabilistic estimation methods can present more future uncertainties. Another issue is how to effectively incorporate the deterministic or probabilistic baseline estimation results into demand response scheduling problem.

1.5 Miscellanies

In addition to the three main applications summarized above, the works on smart meter data analytics also cover some other applications, including power network connection verification, outage management, data compression, data privacy, and so forth. Since only several trials have been conducted in these areas and the works in the literature are not so rich, the works are summarized in this miscellanies section.

1.5.1 Connection Verification

The distribution connection information can help utilities and DSO make the optimal decision regarding the operation of the distribution system. Unfortunately, the entire topology of the system may not be available especially at low voltage levels. Several works have been conducted to identify the connections of different demand nodes using smart meter data.

Correlation analysis of the hourly voltage and power consumption data from smart meters were used to correct connectivity errors in [134]. The analysis assumed that the voltage magnitude decreases downstream along the feeder. However, the assumption might be incorrect when there is a large amount of distributed renewable energy integration. In addition to consumption data, both the voltage and current data were used in [135] to estimate the topology of the distribution system secondary circuit and the impedance of each branch. This estimation was conducted in a greedy fashion rather than an exhaustive search to enhance computational efficiency. The topology identification problem was formulated as an optimization problem minimizing the mutual-information-based Kullback–Leibler (KL) divergence between each two voltage time series in [136]. The effectiveness of mutual information was discussed from the perspective of conditional probability. Similarly, based on the assumption that the correlation between interconnected neighboring buses is higher than that between non-neighbor buses, the topology identification problem was formulated as a probabilistic graph model and a Lasso-based sparse estimation problem in [137]. How to choose the regularization parameter for Lasso regression was also discussed.

The electricity consumption data at different levels were analyzed by PCA in [138] for both phase and topology identification where the errors caused by technical loss, smart metering, and clock synchronization were formulated as Gaussian distributions. Rather than using all smart meter data, a phase identification problem with incomplete data was proposed in [139] to address the challenge of bad data or null data. The high-frequency load was first obtained by a Fourier transform, and then the variations in high-frequency load between two adjacent time intervals were extracted as the inputs of saliency analysis for phase identification. A sensitivity analysis of smart meter penetration ratios was performed and showed that over 95% accuracy can be achieved with only 10% smart meters.

1.5.2 Outage Management

A power outage is defined as an electricity supply failure, which may be caused by short circuits, station failure, and distribution line damage [140]. Outage management is viewed as one of the highest priorities of smart meter data analytics behind billing. It includes outage notification (or last gasp), outage location and restoration verification.

How the outage management applications work, the data requirements and the system integration considerations were introduced in [141]. The outage area was identified using a two-stage strategy in [142]. In the first stage, the physical distribution network was simplified using topology analysis; in the second stage, the outage area was identified using smart meter information, where the impacts of communication were also considered. A smart meter data-based outage location prediction method was proposed in [143] to rapidly detect and recover the power outages. The challenges of smart meter data utilization and required functions were analyzed. Additionally, as a way to identify the faulted section on a feeder or lateral, a new multiple-hypothesis method was proposed in [144], where the outage reports from smart meters were used as the input of the proposed multiple-hypothesis method. The problem was formulated as an optimization model to maximize the number of smart meter notifications. A novel hierarchical framework was established in [145] for outage detection using smart meter event data rather than consumption data. It can address the challenges of missing data, multivariate count data, and variable selection. How to use data analytics method to model the outages and reliability indices from weather data was discussed in [94]. Apart from the data analytics method for outage management, more works on smart meter data-based outage managements have been adopted to the corresponding communication architectures [146, 147].

1.5.3 Data Compression

Massive smart meter data present more challenges with respect to data communication and storage. Compressing smart meter data to a very small size and without (large) loss can ease the communication and storage burden. Data compression can be divided into lossy compression and lossless compression. Different compression methods for electric signal waveforms in smart grids are summarized in [148].

Some papers exist that specifically discuss the smart meter data compression problem. Note that the changes in electricity consumption in adjunct time periods are much smaller than the actual consumption, particularly for very high-frequency data. Thus, combining normalization, variable-length coding, and entropy coding, and the differential coding method was proposed in [149] for the lossless compression of smart meter data. While different lossless compression methods, including IEC 62056-21, A-XDR, differential exponential Golomb and arithmetic (DEGA) coding, and Lempel Ziv Markov chain algorithm (LZMA) coding, were compared on REDD and SAG datasets in [150]. The performances on the data with different granularities were investigated. The results showed that these lossless compression methods have better performance on higher granularity data.

For low granularity (such as 15 min) smart meter data, symbolic aggregate approximation (SAX), a classic time series data compression method, was used in [46, 151] to reduce the dimensionality of load profiles before clustering. The distribution of load profiles was first fitted by generalized extreme value in [152]. A feature-based load data compression method (FLDC) was proposed by defining the base state and stimulus state of the load profile and detecting the change in load status. Comparisons with the piecewise aggregate approximation (PAA), SAX, and DWT were conducted. Non-negative sparse coding was applied to transform original load profiles into a higher dimensional space in [105] to identify the partial usage patterns and compress the load in a sparse way.

1.5.4 Data Privacy

One of the main oppositions and concerns for the installation of smart meters is the privacy issue. The socio-demographic information can be inferred from the fine-grained smart meter data, as introduced in Sect. 1.4. Several works in the literature discuss how to preserve the privacy of consumers.

A study on the distributed aggregation architecture for additive smart meter data was conducted in [153]. A secure communication protocol was designed for the gateways placed at the consumers' premises to prevent revealing individual data information. The proposed communication protocol can be implemented in both centralized and distributed manners. A framework for the trade-off between privacy and utility requirement of consumers was presented in [154] based on a hidden Markov model. The utility requirement was evaluated by the distortion between the original and the perturbed data, while the privacy was evaluated by the mutual information between the two data sequences. Then, a utility-privacy trade-off region was defined from the perspective of information theory. This trade-off was also investigated in [155], where the attack success probability was defined as an objective function to be minimized and ε-privacy was formulated. The aggregation of individual smart meter data and the introduction of colored noise were used to reduce the success probability.

Edge detection is one main approach for NILM to identify the status of appliances. How the data granularity of smart meter data influences the edge detection performance was studied in [156]. The results showed that when the data collection frequency is lower than half the on-time of the appliance, the detection rate dramatically decreases. The privacy was evaluated by the F-score of NILM. The privacy preservation problem was formulated as an optimization problem in [157], where the objective was to minimize the sum of the expected cost, disutility of consumers caused by the late use of appliances, and information leakage. Eight privacy-enhanced scheduling strategies considering on-site battery, renewable energy resources, and appliance load moderation were comprehensively compared.

1.6 Conclusions

In this chapter, we have provided a comprehensive review of smart meter data analytics in retail markets, including the applications in load forecasting, abnormal detection, consumer segmentation, and demand response. The latest developments in this area have been summarized and discussed. In addition, we have proposed future research directions from the prospective big data issue, developments of machine learning, novel business model, energy system transition, and data privacy and security. Smart meter data analytics is still an emerging and promising research area. We hope that this review can provide readers a complete picture and deep insights into this area.

References

1. Mohassel, R. R., Fung, A., Mohammadi, F., & Raahemifar, K. (2014). A survey on advanced metering infrastructure. *International Journal of Electrical Power & Energy Systems, 63*, 473–484.
2. Yang, J., Zhao, J., Luo, F., Wen, F., & Dong, Z. Y. (2017). Decision-making for electricity retailers: A brief survey. *IEEE Transactions on Smart Grid, 9*(5), 4140–4153.
3. National Science Foundation. (2016). Smart grids big data. https://www.nsf.gov/awardsearch/showAward?AWD_ID=1636772&HistoricalAwards=false.
4. Liu, X., Heller, A., & Nielsen P. S. (2017). CITIESData: A smart city data management framework. *Knowledge and Information Systems, 53*(3), 699–722.
5. Bits to energy lab projects. Retrieved July 31, 2017, from http://www.bitstoenergy.ch/home/projects/.
6. Siebel Energy Institute. (2016). Advancing the science of smart energy. http://www.siebelenergyinstitute.org/.
7. Wp3 overview. Retrieved July 31, 2017, from https://webgate.ec.europa.eu/fpfis/mwikis/essnetbigdata/index.php/WP3_overview.
8. SAS. (2017). Utility analytics in 2017: Aligning data and analytics with business strategy. Technical report.
9. Hong, T., Gao, D. W., Laing, T., Kruchten, D., & Calzada, J. (2018). Training energy data scientists: Universities and industry need to work together to bridge the talent gap. *IEEE Power and Energy Magazine, 16*(3), 66–73.
10. Keerthisinghe, C., Verbič, G., & Chapman, A. C. (2016). A fast technique for smart home management: ADP with temporal difference learning. *IEEE Transactions on Smart Grid, 9*(4), 3291–3303.
11. Pratt, A., Krishnamurthy, D., Ruth, M., Hongyu, W., Lunacek, M., & Vaynshenk, P. (2016). Transactive home energy management systems: The impact of their proliferation on the electric grid. *IEEE Electrification Magazine, 4*(4), 8–14.
12. Morstyn, T., Farrell, N., Darby, S. J., & McCulloch, M. D. (2018). Using peer-to-peer energy-trading platforms to incentivize prosumers to form federated power plants. *Nature Energy, 3*(2), 94.
13. Hodge, V., & Austin, J. (2004). A survey of outlier detection methodologies. *Artificial Intelligence Review, 22*(2), 85–126.
14. Peppanen, J., Zhang, X., Grijalva, S., & Reno, M. J. (2016). Handling bad or missing smart meter data through advanced data imputation. In *IEEE Power & Energy Society Innovative Smart Grid Technologies Conference (ISGT)*, pp. 1–5.

15. Akouemo, H. N., & Povinelli, R. J. (2017). Data improving in time series using ARX and ANN models. *IEEE Transactions on Power Systems*, *32*(5), 3352–3359.
16. Li, X., Bowers, C. P., & Schnier, T. (2010). Classification of energy consumption in buildings with outlier detection. *IEEE Transactions on Industrial Electronics*, *57*(11), 3639–3644.
17. Jian, L., Tao, H., & Meng, Y. (2018). Real-time anomaly detection for very short-term load forecasting. *Journal of Modern Power Systems and Clean Energy*, *6*(2), 235–243.
18. Mateos, G., & Giannakis, G. B. (2013). Load curve data cleansing and imputation via sparsity and low rank. *IEEE Transactions on Smart Grid*, *4*(4), 2347–2355.
19. Huang, H., Yan, Q., Zhao, Y., Wei, L., Liu, Z., & Li, Z. (2017). False data separation for data security in smart grids. *Knowledge and Information Systems*, *52*(3), 815–834.
20. Al-Wakeel, A., Jianzhong, W., & Jenkins, N. (2017). k-means based load estimation of domestic smart meter measurements. *Applied Energy*, *194*, 333–342.
21. Al-Wakeel, A., Jianzhong, W., & Jenkins, N. (2016). State estimation of medium voltage distribution networks using smart meter measurements. *Applied Energy*, *184*, 207–218.
22. Araya, D. B., Grolinger, K., ElYamany, H. F., Capretz, M. A., & Bitsuamlak, G. (2017). An ensemble learning framework for anomaly detection in building energy consumption. *Energy and Buildings*, *144*, 191–206.
23. Liu, X., Iftikhar, N., Nielsen, P. S., & Heller, A. (2016). Online anomaly energy consumption detection using lambda architecture. In *International Conference on Big Data Analytics and Knowledge Discovery*, pp. 193–209.
24. Jokar, P., Arianpoo, N., & Leung, V. C. (2016). Electricity theft detection in AMI using customers' consumption patterns. *IEEE Transactions on Smart Grid*, *7*(1), 216–226.
25. Wang, K., Wang, B., & Peng, L. (2009). Cvap: validation for cluster analyses. *Data Science Journal*, *8*, 88–93.
26. Depuru, S. S. S. R., Wang, L., Devabhaktuni, V., & Green, R. C. (2013). High performance computing for detection of electricity theft. *International Journal of Electrical Power & Energy Systems*, *47*, 21–30.
27. Jindal, A., Dua, A., Kaur, K., Singh, M., Kumar, N., & Mishra, S. (2016). Decision tree and SVM-based data analytics for theft detection in smart grid. *IEEE Transactions on Industrial Informatics*, *12*(3), 1005–1016.
28. Júnior, L. A. P., Ramos, Caio C. O., Rodrigues, D., Pereira, D. R., de Souza, A. N., da Costa, K. A. P. & Papa, J. P. (2016). Unsupervised non-technical losses identification through optimum-path forest. *Electric Power Systems Research*, *140*, 413–423.
29. Nizar, A. H., Dong, Z. Y., & Wang, Y. (2008). Power utility nontechnical loss analysis with extreme learning machine method. *IEEE Transactions on Power Systems*, *23*(3), 946–955.
30. Botev, V., Almgren, M., Gulisano, V., Landsiedel, O., Papatriantafilou, M., & van Rooij, J. (2016). Detecting non-technical energy losses through structural periodic patterns in AMI data. In *IEEE International Conference on Big Data*, pp. 3121–3130.
31. Janetzko, H., Stoffel, F., Mittelstädt, S., & Keim, D. A. (2014). Anomaly detection for visual analytics of power consumption data. *Computers & Graphics*, *38*, 27–37
32. Chicco, G. (2012). Overview and performance assessment of the clustering methods for electrical load pattern grouping. *Energy*, *42*(1), 68–80.
33. Zhou, K., Yang, S., & Shen, C. (2013). A review of electric load classification in smart grid environment. *Renewable and Sustainable Energy Reviews*, *24*, 103–110.
34. Wang, Y., Chen, Q., Kang, C., Zhang, M., Wang, K., & Zhao, Y. (2015). Load profiling and its application to demand response: A review. *Tsinghua Science and Technology*, *20*(2), 117–129.
35. Granell, R., Axon, C. J., & Wallom, D. C. (2015). Impacts of raw data temporal resolution using selected clustering methods on residential electricity load profiles. *IEEE Transactions on Power Systems*, *30*(6), 3217–3224.
36. Benítez, I., Quijano, A., Díez, J.-L., & Delgado, I. (2014). Dynamic clustering segmentation applied to load profiles of energy consumption from spanish customers. *International Journal of Electrical Power & Energy Systems*, *55*, 437–448.
37. Al-Jarrah, O. Y., Al-Hammadi, Y., Yoo, P. D., & Muhaidat, S. (2017). Multi-layered clustering for power consumption profiling in smart grids. *IEEE Access*, *5*, 18459–18468.

38. Koivisto, M., Heine, P., Mellin, I., & Lehtonen, M. (2013). Clustering of connection points and load modeling in distribution systems. *IEEE Transactions on Power Systems*, *28*(2), 1255–1265.
39. Chelmis, C., Kolte, J., & Prasanna, V. K. (2015). Big data analytics for demand response: Clustering over space and time. In *IEEE International Conference on Big Data*, pp. 2223–2232.
40. Varga, E. D., Beretka, S. F., Noce, C., & Sapienza, G. (2015). Robust real-time load profile encoding and classification framework for efficient power systems operation. *IEEE Transactions on Power Systems*, *30*(4), 1897–1904.
41. Al-Otaibi, R., Jin, N., Wilcox, T., & Flach, P. (2016). Feature construction and calibration for clustering daily load curves from smart-meter data. *IEEE Transactions on Industrial Informatics*, *12*(2), 645–654.
42. Piao, M., Shon, H. S., Lee, J. Y., & Ryu, K. H. (2014). Subspace projection method based clustering analysis in load profiling. *IEEE Transactions on Power Systems*, *29*(6), 2628–2635.
43. Haben, S., Singleton, C., & Grindrod, P. (2016). Analysis and clustering of residential customers energy behavioral demand using smart meter data. *IEEE Transactions on Smart Grid*, *7*(1), 136–144.
44. Stephen, B., Mutanen, A. J., Galloway, S., Burt, G., & Järventausta, P. (2014). Enhanced load profiling for residential network customers. *IEEE Transactions on Power Delivery*, *29*(1), 88–96.
45. Sun, M., Konstantelos, I., & Strbac, G. (2016). C-vine copula mixture model for clustering of residential electrical load pattern data. *IEEE Transactions on Power Systems*, *32*(3), 2382–2393.
46. Wang, Y., Chen, Q., Kang, C., & Xia, Q. (2016). Clustering of electricity consumption behavior dynamics toward big data applications. *IEEE Transactions on Smart Grid*, *7*(5), 2437–2447.
47. Labeeuw, W., & Deconinck, G. (2013). Residential electrical load model based on mixture model clustering and markov models. *IEEE Transactions on Industrial Informatics*, *9*(3), 1561–1569.
48. Xie, J., Girshick, R., & Farhadi, A. (2016). Unsupervised deep embedding for clustering analysis. In *International Conference on Machine Learning*, pp. 478–487.
49. Zhang, T., Zhang, G., Jie, L., Feng, X., & Yang, W. (2012). A new index and classification approach for load pattern analysis of large electricity customers. *IEEE Transactions on Power Systems*, *27*(1), 153–160.
50. Hong, T., & Fan, S. (2016). Probabilistic electric load forecasting: A tutorial review. *International Journal of Forecasting*, *32*(3), 914–938.
51. Xie, J., Hong, T., & Stroud, J. (2015). Long-term retail energy forecasting with consideration of residential customer attrition. *IEEE Transactions on Smart Grid*, *6*(5), 2245–2252.
52. Hoiles, W., & Krishnamurthy, V. (2015). Nonparametric demand forecasting and detection of energy aware consumers. *IEEE Transactions on Smart Grid*, *6*(2), 695–704.
53. Wang, P., Liu, B., & Hong, T. (2016). Electric load forecasting with recency effect: A big data approach. *International Journal of Forecasting*, *32*(3), 585–597.
54. Xie, J., Chen, Y., Hong, T., & Laing, T. D. (2018). Relative humidity for load forecasting models. *IEEE Transactions on Smart Grid*, *9*(1), 191–198.
55. Xie, J., & Hong, T. (2017). Wind speed for load forecasting models. *Sustainability*, *9*(5), 795.
56. Hong, T., Pinson, P., & Fan, S. (2014). Global energy forecasting competition 2012. *International Journal of Forecasting*, *30*(2), 357–363.
57. Charlton, N., & Singleton, C. (2014). A refined parametric model for short term load forecasting. *International Journal of Forecasting*, *30*(2), 364–368.
58. James Robert Lloyd. (2014). GEFCom2012 hierarchical load forecasting: Gradient boosting machines and Gaussian processes. *International Journal of Forecasting*, *30*(2), 369–374.
59. Nedellec, R., Cugliari, J., & Goude, Y. (2014). GEFCom2012: Electric load forecasting and backcasting with semi-parametric models. *International Journal of forecasting*, *30*(2), 375–381.

60. Taieb, S. B., & Hyndman, R. J. (2014). A gradient boosting approach to the Kaggle load forecasting competition. *International Journal of Forecasting*, *30*(2), 382–394.
61. Hong, T., Wang, P., & White, L. (2015). Weather station selection for electric load forecasting. *International Journal of Forecasting*, *31*(2), 286–295.
62. Høverstad, B. A., Tidemann, A., Langseth, H., & Öztürk, P. (2015). Short-term load forecasting with seasonal decomposition using evolution for parameter tuning. *IEEE Transactions on Smart Grid*, *6*(4), 1904–1913.
63. Fan, S., & Hyndman, R. J. (2012). Short-term load forecasting based on a semi-parametric additive model. *IEEE Transactions on Power Systems*, *27*(1), 134–141.
64. Goude, Y., Nedellec, R., & Kong, N. (2014). Local short and middle term electricity load forecasting with semi-parametric additive models. *IEEE Transactions on Smart Grid*, *5*(1), 440–446.
65. Ding, N., Bésanger, Y., & Wurtz, F. (2015). Next-day MV/LV substation load forecaster using time series method. *Electric Power Systems Research*, *119*, 345–354.
66. Ding, N., Benoit, C., Foggia, G., Bésanger, Y., & Wurtz, F. (2016). Neural network-based model design for short-term load forecast in distribution systems. *IEEE Transactions on Power Systems*, *31*(1), 72–81.
67. Sun, X., Luh, P. B., Cheung, K. W., Guan, W., Michel, L. D., Venkata, S.S., & Miller, M. T. (2016). An efficient approach to short-term load forecasting at the distribution level. *IEEE Transactions on Power Systems*, *31*(4), 2526–2537.
68. Borges, C. E., Penya, Y. K., & Fernandez, I. (2013). Evaluating combined load forecasting in large power systems and smart grids. *IEEE Transactions on Industrial Informatics*, *9*(3), 1570–1577.
69. Edwards, R. E., New, J., & Parker, L. E. (2012) Predicting future hourly residential electrical consumption: A machine learning case study. *Energy and Buildings*, *49*, 591–603.
70. Chitsaz, H., Shaker, H., Zareipour, H., Wood, D., & Amjady, N. (2015). Short-term electricity load forecasting of buildings in microgrids. *Energy and Buildings*, *99*, 50–60.
71. Mocanu, E., Nguyen, P. H., Gibescu, M., & Kling, W. L. (2016). Deep learning for estimating building energy consumption. *Sustainable Energy, Grids and Networks*, *6*, 91–99.
72. Shi, H., Xu, M., & Li, R. (2017). Deep learning for household load forecasting—a novel pooling deep RNN. *IEEE Transactions on Smart Grid*, *9*(5), 5271–5280.
73. Tascikaraoglu, A., & Sanandaji, B. M. (2016). Short-term residential electric load forecasting: A compressive spatio-temporal approach. *Energy and Buildings*, **111**, 380–392.
74. Yu, C.-N., Mirowski, P., & Ho, T. K. (2017) A sparse coding approach to household electricity demand forecasting in smart grids. *IEEE Transactions on Smart Grid*, *8*(2), 738–748.
75. Li, P., Zhang, B., Weng, Y., & Rajagopal, R. (2017). A sparse linear model and significance test for individual consumption prediction. *IEEE Transactions on Power Systems*, *32*(6), 4489–4500.
76. Chaouch, M. (2014). Clustering-based improvement of nonparametric functional time series forecasting: Application to intra-day household-level load curves. *IEEE Transactions on Smart Grid*, *5*(1), 411–419.
77. Hsiao, Y.-H. (2015). Household electricity demand forecast based on context information and user daily schedule analysis from meter data. *IEEE Transactions on Industrial Informatics*, *11*(1), 33–43.
78. Teeraratkul, T., O'Neill, D., & Lall, S. (2017). Shape-based approach to household electric load curve clustering and prediction. *IEEE Transactions on Smart Grid*, *9*(5), 5196–5206.
79. Yang, J., Ning, C., Deb, C., Zhang, F., Cheong, D., Lee, S. E., Sekhar, C., & Tham, K. W. (2017). k-shape clustering algorithm for building energy usage patterns analysis and forecasting model accuracy improvement. *Energy and Buildings*, *146*, 27–37.
80. Quilumba, F. L., Lee, W.-J., Huang, H., Wang, D. Y., & Szabados, R. L. (2015). Using smart meter data to improve the accuracy of intraday load forecasting considering customer behavior similarities. *IEEE Transactions on Smart Grid*, *6*(2), 911–918.
81. Wijaya, T. K., Vasirani, M., Humeau, S., & Aberer, K. (2015). Cluster-based aggregate forecasting for residential electricity demand using smart meter data. In *IEEE International Conference on Big Data*, pp. 879–887.

82. Silva, P. G. D., Ilic, D., & Karnouskos, S. (2014). The impact of smart grid prosumer grouping on forecasting accuracy and its benefits for local electricity market trading. *IEEE Transactions on Smart Grid*, *5*(1), 402–410.
83. Sevlian, R., & Rajagopal, R. (2018). A scaling law for short term load forecasting on varying levels of aggregation. *International Journal of Electrical Power & Energy Systems*, *98*, 350–361.
84. Stephen, B., Tang, X., Harvey, P. R., Galloway, S., & Jennett, K. I. (2017). Incorporating practice theory in sub-profile models for short term aggregated residential load forecasting. *IEEE Transactions on Smart Grid*, *8*(4), 1591–1598.
85. Wang, Y., Chen, Q., Sun, M., Kang, C., & Xia, Q. (2018). An ensemble forecasting method for the aggregated load with subprofiles. *IEEE Transactions on Smart Grid*, *9*(4), 3906–3908.
86. Moreno, J. J. M., Pol, A. P., Abad, A. S., & Blasco, B. C. (2013) Using the R-MAPE index as a resistant measure of forecast accuracy. *Psicothema*, *25*(4), 500–506.
87. Kim, S., & Kim, H. (2016). A new metric of absolute percentage error for intermittent demand forecasts. *International Journal of Forecasting*, *32*(3), 669–679.
88. Haben, S., Ward, J., Greetham, D. V., Singleton, C., & Grindrod, P. (2014). A new error measure for forecasts of household-level, high resolution electrical energy consumption. *International Journal of Forecasting*, *30*(2), 246–256.
89. Hong, T., Wilson, J., & Xie, J. (2014). Long term probabilistic load forecasting and normalization with hourly information. *IEEE Transactions on Smart Grid*, *5*(1), 456–462.
90. PJM. (2015). PJM Load Forecast Report January 2015 Prepared by PJM Resource Adequacy Planning Department. Technical report.
91. Hyndman, R. J., & Fan, S. (2010). Density forecasting for long-term peak electricity demand. *IEEE Transactions on Power Systems*, *25*(2), 1142–1153.
92. Xie, J., & Hong, T. (2016). Temperature scenario generation for probabilistic load forecasting. *IEEE Transactions on Smart Grid*, *9*(3), 1680–1687.
93. Dahua, G. A. N., Yi, W. A. N. G., Shuo, Y. A. N. G., & Chongqing, K. A. N. G. (2018). Embedding based quantile regression neural network for probabilistic load forecasting. *Journal of Modern Power Systems and Clean Energy*, *6*(2), 244–254.
94. Black, J., Hoffman, A., Hong, T., Roberts, J., & Wang, P. (2018). Weather data for energy analytics: From modeling outages and reliability indices to simulating distributed photovoltaic fleets. *IEEE Power and Energy Magazine*, *16*(3), 43–53.
95. Xie, J., Hong, T., Laing, T., & Kang, C. (2015). On normality assumption in residual simulation for probabilistic load forecasting. *IEEE Transactions on Smart Grid*, *8*(3), 1046–1053.
96. Bidong, L., Jakub, N., Tao, H., & Rafal, W. (2017). Probabilistic load forecasting via quantile regression averaging on sister forecasts. *IEEE Transactions on Smart Grid*, *8*(2), 730–737.
97. Wang, Y., Zhang, N., Tan, Y., Hong, T., Kirschen, D. S., & Kang, C. (2019). Combining probabilistic load forecasts. *IEEE Transactions on Smart Grid*, *10*(4), 3664–3674.
98. Xie, J., & Hong, T. (2017). Variable selection methods for probabilistic load forecasting: Empirical evidence from seven states of the united states. *IEEE Transactions on Smart Grid*, *9*(6), 6039–6046.
99. Gaillard, P., Goude, Y., & Nedellec, R. (2016). Additive models and robust aggregation for GEFCom2014 probabilistic electric load and electricity price forecasting. *International Journal of Forecasting*, *32*(3), 1038–1050.
100. Taieb, S. B., Huser, R., Hyndman, R. J., & Genton, M. G. (2016). Forecasting uncertainty in electricity smart meter data by boosting additive quantile regression. *IEEE Transactions on Smart Grid*, *7*(5), 2448–2455.
101. Arora, S., & Taylor, J. W. (2016). Forecasting electricity smart meter data using conditional kernel density estimation. *Omega*, *59*, 47–59.
102. Zhang, P., Xiaoyu, W., Wang, X., & Bi, S. (2015). Short-term load forecasting based on big data technologies. *CSEE Journal of Power and Energy Systems*, *1*(3), 59–67.
103. Humeau, S., Wijaya, T. K., Vasirani, M., & Aberer, K. (2013). Electricity load forecasting for residential customers: Exploiting aggregation and correlation between households. In *Sustainable Internet and ICT for Sustainability (SustainIT)*, pp. 1–6.

104. Zhong, S., & Tam, K.-S. (2015). Hierarchical classification of load profiles based on their characteristic attributes in frequency domain. *IEEE Transactions on Power Systems*, *30*(5), 2434–2441.
105. Wang, Y., Chen, Q., Kang, C., Xia, Q., & Luo, M. (2016). Sparse and redundant representation-based smart meter data compression and pattern extraction. *IEEE Transactions on Power Systems*, *32*(3), 2142–2151.
106. Tong, X., Li, R., Li, F., & Kang, C. (2016). Cross-domain feature selection and coding for household energy behavior. *Energy*, *107*, 9–16.
107. Vercamer, D., Steurtewagen, B., Van den Poel, D., & Vermeulen, F. (2015). Predicting consumer load profiles using commercial and open data. *IEEE Transactions on Power Systems*, *31*(5), 3693–3701.
108. Kavousian, A., Rajagopal, R., & Fischer, M. (2013). Determinants of residential electricity consumption: Using smart meter data to examine the effect of climate, building characteristics, appliance stock, and occupants' behavior. *Energy*, *55*, 184–194.
109. McLoughlin, F., Duffy, A., & Conlon, M. (2012). Characterising domestic electricity consumption patterns by dwelling and occupant socio-economic variables: An irish case study. *Energy and Buildings*, *48*, 240–248.
110. Han, Y., Sha, X., Grover-Silva, E., & Michiardi, P. (2014). On the impact of socio-economic factors on power load forecasting. In *IEEE International Conference on Big Data*, pp. 742–747.
111. Granell, R., Axon, C. J., & Wallom, D. C. (2015). Clustering disaggregated load profiles using a dirichlet process mixture model. *Energy Conversion and Management*, *92*, 507–516.
112. McLoughlin, F., Duffy, A., & Conlon, M. (2015). A clustering approach to domestic electricity load profile characterisation using smart metering data. *Applied energy*, *141*, 190–199.
113. Beckel, C., Sadamori, L., Staake, T., & Santini, S. (2014). Revealing household characteristics from smart meter data. *Energy*, *78*, 397–410.
114. Hopf, K., Sodenkamp, M., Kozlovkiy, I., & Staake, T. (2016). Feature extraction and filtering for household classification based on smart electricity meter data. *Computer Science-Research and Development*, *31*(3), 141–148.
115. Sodenkamp, M., Kozlovskiy, I., & Staake, T. (2016). Supervised classification with interdependent variables to support targeted energy efficiency measures in the residential sector. *Decision Analytics*, *3*(1), 1.
116. Wang, Y., Chen, Q., Gan, D., Yang, J., Kirschen, D. S., & Kang, C. (2018). Deep learning-based socio-demographic information identification from smart meter data. *IEEE Transactions on Smart Grid*, *10*(3), 2593–2602.
117. Albert, A., & Rajagopal, R. (2013). Smart meter driven segmentation: What your consumption says about you. *IEEE Transactions on Power Systems*, *28*(4), 4019–4030.
118. Kwac, J., Flora, J., & Rajagopal, R. (2014). Household energy consumption segmentation using hourly data. *IEEE Transactions on Smart Grid*, *5*(1), 420–430.
119. Bai, Y., Zhong, H., & Xia, Q. (2016). Real-time demand response potential evaluation: A smart meter driven method. In *IEEE Power and Energy Society General Meeting*, pp. 1–5.
120. Jindal, A., Kumar, N., & Singh, M. (2016). A data analytical approach using support vector machine for demand response management in smart grid. In *IEEE Power and Energy Society General Meeting*, pp. 1–5.
121. Dyson, M. E., Borgeson, S. D., Tabone, M. D., & Callaway., D. S. (2014). Using smart meter data to estimate demand response potential, with application to solar energy integration. *Energy Policy*, *73*, 607–619.
122. Albert, A., & Rajagopal, R. (2015). Thermal profiling of residential energy use. *IEEE Transactions on Power Systems*, *30*(2), 602–611.
123. Albert, A., & Rajagopal, R. (2016). Finding the right consumers for thermal demand-response: An experimental evaluation. *IEEE Transactions on Smart Grid*, *9*(2), 564–572.
124. Mahmoudi-Kohan, N, Moghaddam, M. P., Sheikh-El-Eslami, M. K., & Shayesteh, E. (2010). A three-stage strategy for optimal price offering by a retailer based on clustering techniques. *International Journal of Electrical Power & Energy Systems*, *32*(10), 1135–1142.

125. Joseph, S., & Erakkath Abdu, J. (2018). Real-time retail price determination in smart grid from real-time load profiles. *International Transactions on Electrical Energy Systems*.
126. Mahmoudi-Kohan, N., Moghaddam, M. P., & Sheikh-El-Eslami, M. K. (2010). An annual framework for clustering-based pricing for an electricity retailer. *Electric Power Systems Research*, *80*(9), 1042–1048.
127. Maigha & Crow, M. L. (2014). Clustering-based methodology for optimal residential time of use design structure. In *North American Power Symposium (NAPS)*, pp. 1–6.
128. Li, R., Wang, Z., Chenghong, G., Li, F., & Hao, W. (2016). A novel time-of-use tariff design based on gaussian mixture model. *Applied Energy*, *162*, 1530–1536.
129. Wijaya, T. K., Vasirani, M., & Aberer, K. (2014). When bias matters: An economic assessment of demand response baselines for residential customers. *IEEE Transactions on Smart Grid*, *5*(4), 1755–1763.
130. Weng, Y., & Rajagopal, R. (2015). Probabilistic baseline estimation via gaussian process. In *IEEE Power & Energy Society General Meeting*, pp. 1–5.
131. Zhang, Y., Chen, W., Rui, X., & Black, J. (2016). A cluster-based method for calculating baselines for residential loads. *IEEE Transactions on Smart Grid*, *7*(5), 2368–2377.
132. Hatton, L., Charpentier, P., & Matzner-Løber, E. (2016). Statistical estimation of the residential baseline. *IEEE Transactions on Power Systems*, *31*(3), 1752–1759.
133. Irish Social Science Data Archive. (2012). Commission for energy regulation (cer) smart metering project. http://www.ucd.ie/issda/data/commissionforenergyregulationcer/.
134. Luan, W., Peng, J., Maras, M., Lo, J., & Harapnuk, B. (2015). Smart meter data analytics for distribution network connectivity verification. *IEEE Transactions on Smart Grid*, *6*(4), 1964–1971.
135. Peppanen, J., Grijalva, S., Reno, M. J., & Broderick, R. J. (2016). Distribution system low-voltage circuit topology estimation using smart metering data. In *IEEE/PES Transmission and Distribution Conference and Exposition*, pp. 1–5.
136. Weng, Y., Liao, Y., & Rajagopal, R. (2016). Distributed energy resources topology identification via graphical modeling. *IEEE Transactions on Power Systems*, *32*(4), 2682–2694.
137. Liao, Y., Weng, Y., & Rajagopal, R. (2016). Urban distribution grid topology reconstruction via lasso. In *IEEE Power and Energy Society General Meeting (PESGM)*, pp. 1–5.
138. Pappu, S. J., Bhatt, N., Pasumarthy, R., & Rajeswaran, A. (2017). Identifying topology of low voltage distribution networks based on smart meter data. *IEEE Transactions on Smart Grid*, *9*(5), 5113–5122.
139. Minghao, X., Li, R., & Li, F. (2016). Phase identification with incomplete data. *IEEE Transactions on Smart Grid*, *9*(4), 2777–2785.
140. Gungor, V. C., Sahin, D., Kocak, T., Ergut, S.,Buccella, C., Cecati, C., & Hancke, G. P. (2013) A survey on smart grid potential applications and communication requirements. *IEEE Transactions on Industrial Informatics*, *9*(1), 28–42.
141. Tram, H. (2008). Technical and operation considerations in using smart metering for outage management. In *IEEE/PES Transmission and Distribution Conference and Exposition*, pp. 1–3.
142. He, Y., Jenkins, N., & Jianzhong, W. (2016). Smart metering for outage management of electric power distribution networks. *Energy Procedia*, *103*, 159–164.
143. Kuroda, K., Yokoyama, R., Kobayashi, D., & Ichimura, T. (2014). An approach to outage location prediction utilizing smart metering data. In *8th Asia Modelling Symposium (AMS)*, pp. 61–66.
144. Jiang, Y., Liu, C.-C., Diedesch, M., Lee, E., & Srivastava, A. K. (2016). Outage management of distribution systems incorporating information from smart meters. *IEEE Transactions on Power Systems*, *31*(5), 4144–4154.
145. Moghaddass, R., & Wang, J. (2017). A hierarchical framework for smart grid anomaly detection using large-scale smart meter data. *IEEE Transactions on Smart Grid*, *9*(6), 5820–5830.
146. Zheng, J., Gao, D. W., & Lin, L. (2013). Smart meters in smart grid: An overview. In *IEEE Green Technologies Conference*, pp. 57–64.

147. Andrysiak, T., Saganowski, Ł., & Kiedrowski, P. (2017). Anomaly detection in smart metering infrastructure with the use of time series analysis. *Journal of Sensors, 2017*
148. Tcheou, M. P., Lovisolo, L., Ribeiro, M. V., da Silva, E. A., Rodrigues, M. A., Romano, J. M., & Diniz, P. S. (2014). The compression of electric signal waveforms for smart grids: State of the art and future trends. *IEEE Transactions on Smart Grid*, *5*(1), 291–302.
149. Unterweger, A., & Engel, D. (2015). Resumable load data compression in smart grids. *IEEE Transactions on Smart Grid*, *6*(2), 919–929.
150. Unterweger, A., Engel, D., & Ringwelski, M. (2015). The effect of data granularity on load data compression. In *DA-CH Conference on Energy Informatics*, pp. 69–80.
151. Notaristefano, A., Chicco, G., & Piglione, F. (2013). Data size reduction with symbolic aggregate approximation for electrical load pattern grouping. *IET Generation, Transmission & Distribution*, *7*(2), 108–117.
152. Tong, X., Kang, C., & Xia, Q. (2016). Smart metering load data compression based on load feature identification. *IEEE Transactions on Smart Grid*, *7*(5), 2414–2422.
153. Rottondi, C., Verticale, G., & Krauss, C. (2013). Distributed privacy-preserving aggregation of metering data in smart grids. *IEEE Journal on Selected Areas in Communications*, *31*(7), 1342–1354.
154. Sankar, L., Rajagopalan, S. R., & Mohajer, S. (2013). Smart meter privacy: A theoretical framework. *IEEE Transactions on Smart Grid*, *4*(2), 837–846.
155. Savi, M., Rottondi, C., & Verticale, G. (2015). Evaluation of the precision-privacy tradeoff of data perturbation for smart metering. *IEEE Transactions on Smart Grid*, *6*(5), 2409–2416.
156. Eibl, G., & Engel, D. (2015). Influence of data granularity on smart meter privacy. *IEEE Transactions on Smart Grid*, *6*(2), 930–939.
157. Kement, C. E., Gultekin, H., Tavli, B., Girici, T., & Uludag, S. (2017). Comparative analysis of load-shaping-based privacy preservation strategies in a smart grid. *IEEE Transactions on Industrial Informatics*, *13*(6), 3226–3235.

Chapter 2
Electricity Consumer Behavior Model

Abstract Information acquisition devices such as smart meters are gaining popularity in recent years. The "cyber-physical-social" deep coupling characteristic of the power system becomes more prominent. Breakthroughs are needed to analyze the electricity consumer. In this situation, combining physical-driven and data-driven approaches is a significant trend. This chapter tries to decompose electricity consumer behavior into five basic aspects from the sociological perspective: behavior subject, behavior environment, behavior means, behavior result, and behavior utility. On this basis, the concept of the electricity consumer behavior model (ECBM) is proposed. The characteristics of ECBM are also analyzed. Finally, the research framework for ECBM is established.

2.1 Introduction

With the increasing integration of renewable energy and the advancement of the electricity market, the broad interaction between consumers and systems is an important part of the future smart grid. As required by the increasing integration of renewable energy, the power system should provide more flexibility to stabilize its fluctuation. However, the consumers in traditional power system often "consume the electricity passively", and never actively participate in the interaction with the power system, so the flexibility of the power system has yet to be further explored. In addition, the opening of the electricity retail market objectively requires electricity retailers to provide consumer-centric services to improve their competitiveness.

Fortunately, smart grid provides the all-around physical, information and market supports for the broad interaction between the consumers and systems: (1) Physical aspect: with the integration of distributed energy resources (DERs) such as renewable energy and storage, the traditional electricity consumers turn into the "prosumers", and can reasonably control the electric equipment and energy storage to realize the optimal utility. These DERs and control device lay the physical basis for the interaction between consumers and systems. (2) Information aspect: the advanced metering infrastructure (AMI) which consists of smart meter, communication network and data management system, plays a vital role in collecting the smart meter data and

Y. Wang et al., *Smart Meter Data Analytics*,
https://doi.org/10.1007/978-981-15-2624-4_2

realizing the bidirectional flow between the energy flow and information flow [1]. It provides an information communication basis for the interaction between consumers and systems. (3) Market aspect: the open electricity retail market will cultivate various business models. The consumer service will be conducted from the aspects of electricity price design, consumer agent, and demand response [2]. It provides a market basis for the interaction between consumers and systems.

The power system is increasingly becoming a complex system with high "power-cyber-social" coupling [3]. Since the modeling for power system from the purely physical perspective is not enough to fully depict the whole picture of the power system, the full consideration should be given to the impacts of environmental, economic, and social factors and human behaviors for the entire power system. The study on the power system with "cyber-physical" coupling characteristic has attracted broad attentions [4]. It focuses on the impact of cybersecurity and big data technology on the power system and provides a cyber perspective on the power system. However, there are currently very few studies on the modeling of the social aspect of the power system with "cyber-physical-social" deep coupling characteristic. Notably, the modeling for "consumers" in the power system is insufficient since now.

As an essential part of the whole power system, the electrical power load has been widely concerned and studied such as composite load modeling and load forecasting, which provides the basis for the planning, operation and stability analysis of the whole power system. The study on load also focuses on its electrical or power characteristics, by either conducting the composite load modeling (such as building a ZIP model) for the network computing of the power system, or conducting the sensitivity analysis and forecasting of several relevant factors of the load for the planning and operation of the power system. We know that the load is generated by the use of electrical appliances by electricity consumers. The traditional studies on power systems mainly focus on the load, rather than electricity consumers. These works fail to give full consideration to the impact of electricity consumer behavior on the power system. That is to say, the modeling of the demand side (such as composite load modeling from the physical perspective) only considers the electrical characteristics of the load rather than analyzing the massive consumers. That is to say, there have been few analyses on electricity consumer behavior.

With the further development of the smart grid, there are extensive studies on demand response, energy efficiency management, smart meter big data analytics, etc. Some studies build optimization models from the physical perspective [5, 6]; while the others focus on the data-driven analysis and electricity price design for the electrical power consumption patterns of consumers by clustering, etc. [7]. There are also analytical studies on the power consumption behavior of consumers.

Researchers all around the world have conducted a significant number of studies in terms of smart meter data analytics. These works have broad applications such as demand response, electricity price design, system operation, etc. However, current studies are often conducted by focusing on one specific application, which is similar to the "process-oriented" programming, lack of recognition of systematization of electricity consumer behavior, and has no "object-oriented" overall design. That is to say, the current studies neither have accurately analyzed the exact meaning of electricity consumer behavior, nor have built the "consumer behavior" model "systematically", and the recognition of the consumer behavior is not improved to the "system" or "model" level like that in "cyber-physical system".

The study and application of behavioristics and sociology in various industries are increasingly concerned. The Nature Research specially develops the online forum for the researchers to discuss and share the study of behavioristics and sociology and their applications in various industries where the energy industry is one of them [8]. Therefore, modeling and analysis of the demand side can be conducted from the sociology and behavioristics perspectives. The consumer in the power system is a complex subsystem, which lacks analytical models for study. Thus, the model-driven approach may not be suitable for the electricity consumer behavior modeling. Nevertheless, the big data in the power systems provide a new data-driven solution for consumer behavior analysis.

This chapter decomposes the basic components of electricity consumer behavior from the sociological perspective and proposes the concept of electricity consumer behavior model (ECBM) by analyzing the internal logical relationship among these basic components. ECBM is then transformed into a series of consumer characteristic attribute identification problems and their relationship analysis problems. The data-driven research framework of ECBM is established by conducting prospective and fundamental research in terms of consumer portrait, load structure, load profile, load forecasting, and consumer aggregation. The following Chaps. 3–12 in this book can be viewed as the approaches to electricity consumer behavior modeling.

2.2 Basic Concept of ECBM

The concept of the consumer behavior model has been widely used in the fields such as supply chain management [9], software or web design [10], consumer portrait [11], and intelligent recommendation [12] to realize the personalized consumer service. The electricity consumer is one specific consumer in the power system. ECBM can be viewed as an intersection between the consumer behavior model and the power system.

2.2.1 Definition

The word "behavior" has various meanings, and may be interpreted in different ways in different research fields. The electricity consumer behavior described in this chapter is interpreted from the sociological and psychological perspectives: the electricity consumer behavior refers to the electricity consumer's power consumption activities and attitudes under the impact of external environments. The power consumption activities are **dominant behaviors** that can be measured or perceived by the sensors such as smart meter; the power consumption attitudes (such as the

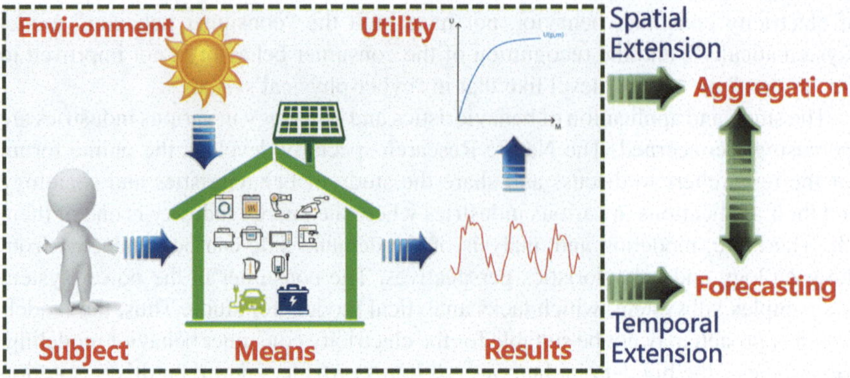

Fig. 2.1 Basic components of electricity customer behavior

attitudes to be involved in the demand response program) are **recessive behaviors** that cannot be easily observed directly, such as the way of thinking.

In the field of sociology, human behavior generally consists of behavior subject, behavior environment, behavior means, behavior results, and behavior object. Similarly, Fig. 2.1 shows the basic components and extensions of electricity consumer behavior. In the power system, the electricity consumers have their own utility functions, and their power consumption behaviors aim at pursuing greater utility. Therefore, for the electricity consumers, the behavior object in their behavior components is replaced by the behavior utility.

The basic components of the electricity consumer behavior mainly include five parts:

1. **behavior subject**: it refers to the electricity consumers themselves who have the ability of cognition and thinking, the specific social and economic information and other attributes;
2. **behavior environment**: it refers to an external environment affecting the electricity consumer behavior, such as power network, meteorological factor, electricity price, day type, etc.;
3. **behavior means**: it refers to the means to be adopted by the electricity consumer to achieve a target, including the use or control of all household appliances, electric vehicles, distributed energy storage, distributed renewable energy, etc.;
4. **behavior results**: it refers to the load profiles or specific power consumption patterns finally generated by the electricity consumer, i.e. power exchanged with the power grid;
5. **behavior utility**: it refers to the utility that the electricity consumers bring to themselves through the power consumption, including the power cost (disutility), the comfort utility, the utility for achievements of other specific targets, etc.

The above five components have close internal logical relationship: the behavior subject (electricity consumer) adopts a certain behavior means (using the electrical appliances or equipment) according to its own attribute and the behavior environment (external factors) at that time to generate the behavior result (forming the electricity consumption), and realize the highest behavior utility (such as making profit). The five components take a progressive relationship from the intrinsic electricity consumer behavior to the presentative behavior and a successive relationship from the recessive behavior to the dominant behavior. It should be pointed out that the electricity consumer behavior has a different concept from the consumer's power consumption behavior. The consumer's power consumption behavior only describes the power characteristics of the consumer's power utilization and is a consumer's dominant behavior. That is to say, the consumer's power consumption behavior is an important part of electricity consumer behavior.

For a single electricity consumer behavior, the spatial extension can be conducted, i.e. **aggregated behavior**, which refers to the collection of multiple similar consumers according to a consumer characteristic to form several consumer groups having a similar characteristic; and the time extension can also be conducted, i.e. **foreseeable behavior**, which refers to a changing trend of the consumer behavior in a period of time in the future. The power consumption behavior forecasting (load forecasting) is the most common extension.

On this basis, *the ECBM can be defined as an abstract and standard expression of electricity consumer behavior that reveals and describes the intrinsic characteristics of the behavior subject, behavior environment, behavior means, behavior result, behavior utility, foreseeable behavior, and aggregated behavior and their relationships based on diversified information using optimization modeling, data analytics, and other approaches.* The consumer smart meter data analytics for a specific application is similar to the "process-oriented" programming, which provides a specific solution for a specific application. However, the ECBM is similar to "object-oriented" overall design, which involves five basic components and two derivative components regarding the specific object of the consumer behavior, and the behavior model describes the relationship between five behavior components and two derivative components.

2.2.2 Connotation

The ECBM has the connotation covering the following aspects according to its definition:

ECBM is based on diversified data: the popularization of smart meter provides the basis for the wider and more fine-grained data collection at the demand side, including the consumer's smart meter data, electric vehicle charging and discharging data, meteorological data, electricity price data, etc. The electricity consumer has a certain ability of cognition and thinking, and the consumer can be regarded as the most complex system in the world. For the modeling of physical components in the power system, the priori physical model is provided, and then the parameters of the model can be estimated. However, human behavior modeling is different, which is usually based on a lot of observed experience. Thus the modeling of human behavior should be conducted based on the diversified data, rather than several simple physical parameters.

ECBM takes the optimization modeling and data analytics as the main approaches: For example, the consumer's power consumption optimization model under a certain external environment can be built based on a certain assumption for the utility function. Thus the consumer's power consumption behavior can be analyzed. For another example, there is no existing model to describe how the consumer's social and economic attributes affect the consumer's load profile or how the consumer's load profile can reflect the consumer's social and economic attributes. This could be deemed as a high-dimensional and nonlinear mapping relationship. In this situation, the advanced data analytics approaches such as deep learning can be applied to describe the relationship between the consumer's social and economic attributes and their load profiles.

ECBM describes the intrinsic characteristics of behavior components and their relationships: a model generally includes objective, variables, and relationships. As the consumer behavior has five basic components and temporal and spatial extensions, the ECBM should be a collection of a series of submodels, and each submodel describes the relationship among the consumer behavior components and has its own objective, variables, and relationships. For example, taking the consumer's load profile as the variable and with the target of identifying the consumer's social and economic information, the consumer portrait identification submodel can be used to build a high-dimensional and nonlinear relationship between these two. For another example, taking the external environment and load profile as the variable and with the target of stripping the distributed PV and energy storage, the load disaggregation submodel for the consumer's distributed PV and energy storage is used to build the relationship among the final net load profile and external environmental factor.

2.2.3 Denotation

The ECBM has different forms of denotation according to the consumer's basic types and the submodels.

The basic types of consumer include the residential consumer, commercial consumer, industrial consumer, building consumer, etc. Sometimes, the load aggregator can be regarded as a type of consumer as they interact with the power system on behalf of a group of consumers. Different types of consumers mean different types of behavior subject. Therefore, the attributes to describe its basic characteristics are also different. For example, for the residential customers, their portrait can be described through such attributes as age, retirement, type of work and social class. However, these attributes cannot be applicable to the building customers with the "portraits" described by the number of floors, age of the building, installation of energy management system and other attributes.

According to the submodels, the ECBM has complicated compositions and internal interactions. Therefore, it is difficult to build only one complete relationship to describe the relationships among the five basic components and the spatial and temporal-scale extensions. It needs a series of submodels to describe the mapping relationship between the two or more components. For example, the mapping relationship between the behavior subject and behavior result could be complicated; the relationship between the behavior means and behavior result is a simple additive relationship; and the relationship between the behavior environment and the behavior means could also be complicated, but can be described with PV panel energy conversion model with regard to the distributed PV. Consumer behavior has numerous submodels, which will be detailed in the research framework.

2.2.4 Relationship with Other Models

(1) From consumer behavior model to ECBM

The consumer behavior model has been widely used in the fields such as personalized recommendation system, social network, human-computer interaction design, etc. Its fundamental purpose is to realize personalized consumer service to improve market competitiveness and increase profit. For example, in the marketing field, we can try to build the consumer portrait, describe certain key characteristics of the consumer, classify the consumers, then provide different types of services according to the characteristics of each type of consumers, and promote the specific goods, etc. For another example, in the advertising push field, the consumer's purchasing behavior can be modeled according to their website browsing history and path, thereby realizing personalized advertising. From the perspective of service provider, the essence of building the consumer behavior model is to find out the possible relationship between the different "actions" (such as goods purchasing and web browsing) of the consumer, and to infer the future potential demand or preference of the consumer, thereby realizing the efficient personalized service. From the perspective of the consumer, the service enjoyed by the consumer may be convergent, or there are very diversified services, but the consumers cannot efficiently find out the service that best meets themselves, thereby facing the "information overload" problem. The modeling of consumer behavior is expected to realize the active recommendation and provision of the service.

The transformation from the "passively meeting the demand by the power system" to the "active demand response of the electricity consumer" is one of the important characteristics for the development of the smart grid and Energy Internet. Attributed to the opening and flourishing of the electricity sales market, the numerous market participants including electricity retailers and load aggregators, provide the diversified products to the consumers, such as different types of electricity price packages and diversified demand response contracts. The service products received by the electricity consumer are diversified and complicated, and also have "information

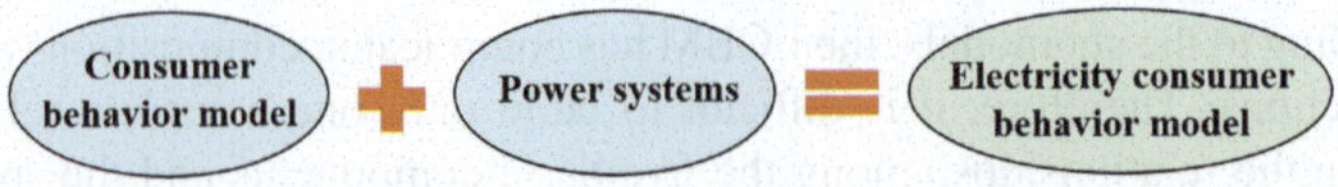

Fig. 2.2 From consumer behavior model to ECBM

overload" problem. Thus, it is better to build an ECBM for each consumer, including the consumer portrait, load structure, load pattern, load trend, and even power consumption attitude, then reducing the consumer's range of service selection in terms of electricity price package, demand response and goods recommendation in the retail electricity market, and conducting the personalized recommendation or actively providing the corresponding services. In addition, the power system has massive consumers; therefore, only by building the ECBM can the behavior of electricity consumer be abstracted to a certain extent, thereby improving service efficiency. Therefore, the ECBM is an application and expansion of the consumer behavior model in the power system, as shown in Fig. 2.2.

(2) From composite load model to ECBM

The electricity consumer plays a crucial role in the smart grid and the Energy Internet. It is not sufficient to comprehensively model the whole power system by only focusing on the physical characteristics of the power system. The modeling of electricity consumer behavior should be fully considered in order to mine its interaction characteristics. Although the study on demand response has involved consumer behavior and interaction, it mainly focuses on the arrangement of consumer's electric appliance and other more microscopic physical models. We need to model the consumer behavior more comprehensively, and especially analyze from the sociological and psychological perspectives so as to truly realize the value creation of the power system with the electricity consumer as its core.

For the whole power system, the synchronous generator set, power network, load, power, and electronic equipment are the most important and basic components. The generator, excitation system, prime mover speed governing system, and composite load modeling in the synchronous generator set are very complicated, and their parameters form the "four parameters" needing to be mainly identified by the traditional power system. Identification of the "four parameters" provides the support to the safe, stable and economic operation of the traditional power system. Figure 2.3 shows the basic components of parameter identification of the power system. On the basis of original traditional "identification of four parameters", the ECBM is added, which focuses more on the power consumption behavior of the consumer at the demand side, and tries to find the underlying basic laws of the consumer throughout the power consumption process. Extension from the composite load modeling to the consumer behavior modeling at the demand side is a transformation in perspective and thinking and is a brand-new component of the power system model. The composite load modeling and consumer behavior modeling constitute the two sides of modeling at the demand side.

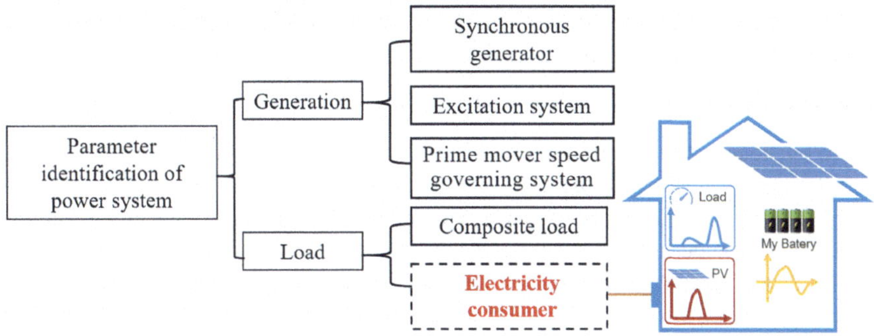

Fig. 2.3 From consumer behavior model to electricity consumer behavior model

2.3 Basic Characteristics of Electricity Consumer Behavior

The electricity consumer behavior has the following basic characteristics: near-optimality of utility, initiative, foreseeability, diversity, uncertainty, high-dimensional complexity, cluster characteristics, and weak observability, all of which will also become the basis for ECBM. They will be respectively elaborated in the following:

(1) Near-optimality of utility

The consumer, as the person having the ability of cognition and thinking, finally has the power consumption behavior due to the impact of the external environment, and meets their daily or specific demands by using or controlling certain electrical equipment, thereby maximizing the utility. In the consumer's demand response and home energy management system, its internal setting realizes the lowest power cost by reasonable arrangement and use of electrical equipment to meet the consumer comfort. The consumer cannot conduct precise modeling to their power consumption behavior and obtain the optimum like software programming but tends to increase their power utility and reduce their power cost.

(2) Initiative

The consumers do not only consume the power supplied by the power system passively, but also have certain subjective initiative, and actively changes their power consumption behavior according to the changes in the external environments, to realize the near-optimality of utility. The programs currently conducted, such as demand response and energy efficiency management, require fully mobilizing the consumer's subjective initiative, and transforming the traditional "passive load" into the "active load".

(3) Diversity

Different consumers have different utility functions and own different electric appliance. In addition, the external environments suffered by consumers in different areas are also different. Thus, the behavior results of different behavior subjects in different time periods and under the different environments are diverse, including the diversity of different consumers and the diversity of the same consumer at different periods.

(4) Foreseeability

Due to the near-optimality of the utility of the consumer, consumer behavior has certain inherent laws. When certain laws are detected, various behaviors of the consumer can be forecasted to a certain extent. For example, the load profile of the consumer in a certain time period in the future can be forecasted according to historical load profiles of the consumer. The basic patterns of consumer's future consumption can also be speculated through their social and economic information. The foreseeability of consumer behavior comes from the stability of the same consumer behavior and the similar laws of different consumer behaviors.

(5) Uncertainty

Consumer behavior not only has foreseeability but also has uncertainty. In essence, the consumer's power consumption behavior is the result of superimposing a series of random events on the basis of their long-term work and living habit. Therefore, there is inevitable uncertainty in the ECBM. Uncertainty may either come from the consumer's random behavior caused by purely random events or come from the model deviation caused by the consumer's regular behavior that is not identified. For the short term, there may be a difference in ECBM in the different periods within a day, working days, and weekend. For the long term, the ECBM will change with the change in lifestyle, upgrading of consumption level, and improvement of intelligent level of electric appliance. Therefore, the ECBM cannot be built without the depiction of its uncertainty.

(6) High-dimensional complexity

The ECBM involves a series of basic attributes of the consumer. As the natural human attribute and social attribute have high complexity, human behavior has multiple complex sides. Several simple attributes cannot be used to depict the ECBM in all dimensions. The ECBM will certainly have the high-dimensional complexity. "There are no two identical leaves in the world, not to mention the two identical people". Each consumer will be an instance in the high-dimensional ECBM space. Moreover, the consumer's power consumption behavior is closely related to their production and life, and human behavior has high subjectivity. Therefore, compared with the objective law, the consumer behavior model usually has no existing analytic mathematical expression but has complicated non-analytic and nonlinear association relationship.

(7) Cluster characteristics

The human production activity has social nature, so the electricity consumer behavior shows certain cluster characteristics. That is to say, the ECBM of different consumer individuals independently forms a series of groups in terms of attribute space or its subspace. The behavior of consumer tends to be the same in each group and has significant differences in different groups. The consumer's cluster characteristics provide the clue for the clustering analysis and aggregation modeling of the consumer model.

(8) Weak observability

The electricity consumer behavior is complex and changeable. The information interaction between the power system and electricity consumer is usually conducted through the smart meter, thereby realizing the direct observation of the load profile and other dominant behaviors. Its internal power consumption behaviors, including the power consumption behavior of single electrical appliance, output of distributed PV, response behavior of distributed energy storage, consumer attitude and other recessive behaviors, cannot be directly observed. Accordingly, the power system integrates more diversified and more fine-grained data to meet the challenge brought by this weak observability.

2.4 Mathematical Expression of ECBM

The ECBM is a model to describe the intrinsic characteristics and relationship among the main components and its extensions of the electricity consumer behavior. The main components of the electricity consumer behavior should be mathematically defined to describe the ECBM in a standard manner. The relevant mathematical notations are summarized in Table 2.1.

The electricity consumer behavior subject, i.e. the consumer, can use a series of (such as J) attributes to describe and thus forming the relatively complete consumer portrait. Accordingly, the consumer attribute space $\mathbb{C}$ is defined, The attribute set in this space is $\mathbf{C} = [c_1, c_2, \ldots c_j, \ldots c_J]$, where each element c_j in the attribute set $\mathbf{C}$ represents a consumer attribute, such as consumer type, age, social class,

Table 2.1 Mathematical notations for electricity consumer behavior model

Mathematical symbol	Physical connotation
$\mathbb{C}/\mathbf{C}$	Consumer attribute space/set
c_j	The jth attribute of the consumer
$\mathbb{E}/\mathbf{E}$	Environmental factor space/set
e_k	The kth environmental factor
I/i	Consumer set/index
T/t	Time set/index
A/a	Appliance set/index
P	Active power
O	Total utility of consumer
g_i	Utility function of the jth consumer
S_n	The nth consumer group

children, interests and preferences, and other information. The consumer attributes have various expressions forms, including the continuous variable, discrete variable, fuzzy variable, characteristic matrix, probabilistic expression of quantile, interval, or probability distribution. For example, the social and economic information of the consumer, such as age and retirement, shall be expressed with the continuous or discrete variable; the consumer's acceptance for the smart home installed can be expressed with the fuzzy number; the power consumption uncertainty of future consumer shall be expressed in probabilistic form. As the consumer portrait is time-varying in a long time scale, including the change in age and occupation, we use $\mathbf{C}_i^t = [c_{i,1}^t, c_{i,2}^t, \ldots c_{i,j}^t, \ldots c_{i,J}^t]$ to indicate the complete portrait of the ith consumer at the time t.

The electricity consumer behavior environment is the external factors stimulating or affecting the electricity consumer behavior, which are also diversified. Similarly, the behavior environment factor space $\mathbb{E}$ is defined. The environmental factor set in this space is $\mathbf{E} = [e_1, e_2, \ldots e_k, \ldots e_K]$, where each element e_k in the environmental factor set $\mathbf{E}$ represents an environmental factor, such as the power network topology, external temperature, illumination intensity, and electricity price. We use $\mathbf{E}_i^t = [e_{i,1}^t, e_{i,2}^t, \ldots e_{i,k}^t, \ldots e_{i,K}^t]$ to indicate the environment of the ith consumer at the time t.

The electricity consumer behavior means is the electrical equipment that the consumer uses the electricity to improve their own utility, including the household appliances, distributed energy storage, and distributed PV. The set of consumer's electrical equipment is defined as A, and the operating state of the ath electrical equipment is directly decided by the power $P_{i,a}^t$ consumed or generated by it. The electricity consumer behavior result is the final power exchanged with the power grid, which is defined as $P_{i,t}$.

The electricity consumer behavior utility (i.e. O_i) varies by the consumer attribute, external environment and state of electrical equipment, Therefore, the behavior utility is a function related to $\mathbf{C}_i$, $\mathbf{E}_i$ and $P_{i,a}^t$, which is defined as g_i.

So far, the five components of the ith electricity consumer behavior are respectively expressed as: behavior subject $\mathbf{C}_i^t$, behavior environment $\mathbf{E}_i^t$, behavior means $P_{i,a}^t$, behavior result P_i^t, and behavior utility g_i^t. It's worth noting that, all basic components of the consumer behavior are time-varying. The behavior subject attribute and behavior utility function often change slowly, and can be approximately deemed as no change over a period of time; and the behavior environment changes fast, causing the changing behavior means and behavior result.

The electricity consumer has the near-optimality of utility, so the behavior subject $\mathbf{C}_i^t$ realizes the maximum behavior utility g_i^t by adopting the behavior means $P_{i,a}^t$ under the behavior environment $\mathbf{E}_i^t$. The behavior subject $\mathbf{C}_i$, behavior environment $\mathbf{E}_i$ and behavior means $P_{i,a}^t$ are coupled through the utility function g_i^t:

$$\arg\max_P O_i = \arg\max_P \sum_t g_i^t(P_{i,a}^t)\big|_{\mathbf{C}_i^t, \mathbf{E}_i^t} \tag{2.1}$$

As the consumer does not completely pursue the utility optimization in a rational manner, but the "near-optimality of utility" to a certain extent, the consumer's behavior means $P_{i,a}^t$ may be affected by the consumer habit and other various factors, and thus shows uncertainties. That is to say, whether the consumer uses or how the consumer uses equipment could be regarded as a random variable having a certain expectation, which is also the direct cause that the power consumption has high uncertainty.

Without considering transmission network loss, there is a simple linear additive relationship between the behavior means $P_{i,a}^t$ and behavior result P_i^t. That is to say, the final behavior result or behavior mode is equal to the sum of consumption of all kinds of electrical equipment:

$$P_i^t = \sum_{a \in A} P_{i,a}^t \tag{2.2}$$

Except for the five basic components of the electricity consumer behavior, the aggregation behavior, extension of the consumer behavior in space, essentially refers to dividing the consumer group according to a characteristic of the consumer, i.e. dividing the consumer set I into N subgroups, where each consumer belongs to one of the N subgroups:

$$\begin{aligned} \max_{S_1, S_2, \ldots, S_N} \quad & \prod_{n=1}^{N} \prod_{i \in S_n} Prob(F_i^t | i \in S_n) \\ \text{s.t.} \quad & S_1 \cup S_2 \cup \cdots \cup S_N = I \\ & S_1 \cap S_2 \cap \cdots \cap S_N = \emptyset \end{aligned} \tag{2.3}$$

where, F_i^t denotes one characteristic which is used for dividing the consumer groups, such as consumer age, composition of consumer's electrical equipment, and shape of load profile. The objective function refers to the maximum probability of the observed characteristic when the consumer is divided into a specific group; the two constraints indicate that each consumer can only be divided into one group. The consumer group can be divided using clustering algorithm.

The foreseeable behavior, extension of the consumer in time, is generally for the behavior means and behavior result, i.e. future change trend of power $P_{i,a}^{t+h}$ or total exchange power P_i^{t+h} of specific electrical equipment in a time period in the future. Essentially, the foreseeing of future consumer behavior refers to uncovering the relationship $f_{i,a}$ or f_i within the historical data, and forecasting the future power consumption behavior according to the historical behavior:

$$\begin{aligned} \hat{P}_{i,a}^{t+h} &= f_{i,a}(\mathbf{C}_i^t, \mathbf{E}_i^t, \hat{\mathbf{E}}_i^{t+h}, P_{i,a}^t, t) \\ \hat{P}_i^{t+h} &= f_i(\mathbf{C}_i^t, \mathbf{E}_i^t, \hat{\mathbf{E}}_i^{t+h}, P_i^t, t) \end{aligned} \tag{2.4}$$

where the superscript of t refers to the historical value and current value of variables; $\hat{\mathbf{E}}_i^{t+h}$ indicates the forecasting value of the future behavior environment; $\hat{P}_{i,a}^{t+h}$ and

$\hat{P}_i^{t+h}$ respectively indicate the forecasting results of power of electrical equipment or total exchange power in the future, which can be the point forecasting result for describing the future trend, and can also be the probabilistic forecasting result including more uncertain information.

Equations (2.1)–(2.4) respectively show the coupling relationship among the five basic components of the consumer behavior and the two extension behaviors. They constitute the basic equations of ECBM. It should be pointed out that the abstract expression of the above equations is concise, but its specific relationship is very complicated, which is mainly reflected in the following three aspects: (1) In Eq. (2.1), it is not easy to obtain the consumer utility function g_i, and the near-optimality instead of optimality of the consumer's utility makes the consumer behavior has great complexity and uncertainty. Therefore, the relationship among the behavior subject $\mathbf{C}_i$, behavior environment $\mathbf{E}_i$, and behavior means $P_{i,a}^t$ is complicated and has great uncertainty. (2) In Eq. (2.2), it is easy to obtain the final exchange power according to the summation of the powers of all kinds of electrical equipment, but conversely, it is difficult to decompose it. (3) In Eq. (2.3), the attribute or characteristic F_i^t used for consumer classification should be carefully extracted or selected, and the optimization problem needs to be transformed into the clustering problem and other problems. (4) In Eq. (2.4), the input feature selection, forecasting model selection, and the model training process are also very complicated.

2.5 Research Paradigm of ECBM

The ECBM is composed of a series of submodels that describe the intrinsic characteristics or their relationship among consumer behavior components. Each submodel can be abstracted to the form of $Y = h(X)$, which tries to identify one behavior attribute Y of the consumer, given another consumer behavior information X. $h(\cdot)$ is the function to be trained. That is to say, the ECBM is established by identifying the consumer behavior attribute Y. This section will introduce the research framework of the ECBM, including the basic research paradigm and research contents.

Figure 2.4 gives a basic research paradigm of the ECBM, which mainly includes three modules, i.e. data collection, consumer behavior model, and consumer interaction. In the three modules, the data collection is the basis, the consumer behavior model is the core, and the interaction between consumers and systems is the purpose. The three modules are progressive successively to form a closed loop, thereby realizing the continuous updating and optimization of the ECBM.

Specifically, in the data collection module, various data related to the consumer's characteristics shall be widely collected. There are two ways to collect the data: (1) active collection, such as smart meter data, meteorological data, and electricity price data; and (2) consumer feedback, including the direct feedback data (for example, whether interested in a demand response program) and indirect feedback data (for example, consumer's power consumption at the different electricity prices).

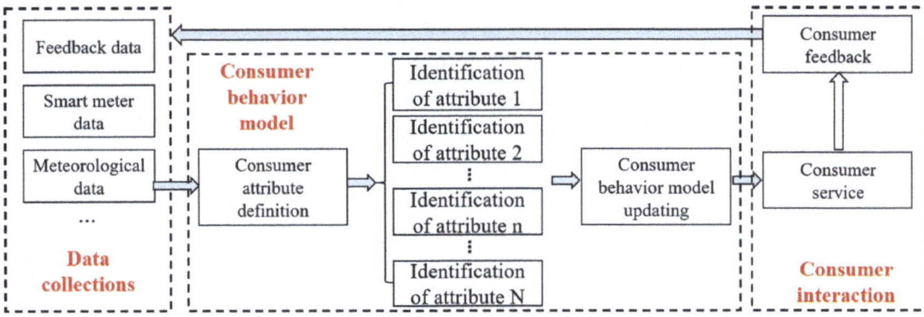

Fig. 2.4 Research paradigm of electricity customer behavior modeling

The consumer model module mainly includes three steps: the consumer attribute definition, consumer attribute identification, and ECBM updating:

(1) Firstly, different consumer attributes need to be defined from different aspects according to the diversified requirements of power system for the consumer, such as implementation of demand response, electricity price design, and recommendation of personalized electrical appliance and other commodities. Generally, the attributes cab be sorted out from the aspects of endogenous attributes, behavior attributes, and preference attributes. The details are discussed in the next sections.

(2) Secondly, the attributes need to be identified. This step is the key of the whole ECBM, which needs to determine the expression forms of different attributes and the identification method of each attribute. For example, the expression form of the electricity consumer uncertainty is presented as a series of quantiles, and the identification method is the probabilistic quantile regression method.

(3) Finally, the ECBM needs to be updated, i.e. updating the set formed by all attributed values. The updates include directly substituting the original result with the latest result, or comprehensively considering the attribute value calculated latest and historical attribute value with the weight decay.

2.6 Research Framework of ECBM

In the research paradigm of ECBM, the consumer behavior model is the core which is established mainly based on the consumer attribute definition. The electricity consumer attribute shall have the following four characteristics:

(1) The attribute should be defined for real applications: The consumer attribute is the standard expression for describing the electricity consumer characteristics. The consumer is complicated and shall be comprehensively depicted with massive attributes/However, the purpose of establishing the consumer behavior model is to realize the personalized service for the consumer and the optimization and interaction between the consumer and the power grid. Therefore, the consumer attribute should be screened, and the important attributes that have great application potential in

the power system shall be reserved. For example, the socioeconomic information of the consumer shall be detected and applied in terms of voice service, electrical appliance promotion; for another example, the identification of power consumption pattern provides the basis for the time-of-use electricity pricing.

(2) The attribute may drift: The consumer attributes are not unalterable but may change over time. For the attribute drift, the consumer attribute shall be modified in real-time or regularly, such as the attribute modification method based on the weight decay. To timely establish and update the variable consumer behavior model, various consumer behavior data including the smart meter data, meteorological data, electricity price data, and questionnaire data shall be reacquired by scrolling or periodically. On this basis, the core relationship and parameter of the consumer behavior model are updated or modified.

(3) The attribute should be consistent: The internal consistency shall be ensured among the consumer attribute sets. Different attributes are used for depicting the different aspects of the consumer. The obtained attribute values cannot contradict each other but should verify each other and depict the consumer and their power consumption characteristics as fully as possible.

(4) The attribute can be evaluated: Different attribute values have different forms of expression, but these forms of expression shall be able to be evaluated to guide the data acquisition and attribute identification. For example, the probability model can be evaluated through the quantile loss, and the classified discrete value can be evaluated through the accuracy or classification entropy. All attributes shall be expressed with the specialized value, and have corresponding evaluation indexes, including the qualitative evaluation and quantitative evaluation.

According to the above basic characteristics, Fig. 2.5 concludes several consumer attributes tp reflect the electricity consumer behavior from the perspectives of endogenous attributes, consumption attributes, and preference attributes by taking the residential customer as an example.

Figure 2.6 concludes the multi-dimensional research framework of ECBM and its analysis method according to the components of consumer behavior.

For the behavior subject, the consumer portrait can be described, including the consumer's basic attributes such as sex and age, occupation and salary, social class and state of the house, and the consumer preference attribute such as demand response willingness, and power consumption preference. Figure 2.7 gives the average weekly load curve of three consumers and their corresponding socioeconomic information. This kind of relationship can help to obtain the socio-economic information of some consumers from the load profiles conveniently and more intuitively. The power consumption of the retired consumer #1018 at the working hours is also maintained at the higher level, while that of the consumers #1020 and #1032 that have not retired is relatively low at the working hours except for the weekends, both of which accord with the working states of the three consumers. The consumer #1032 has a small number of bedrooms, and their power level is also relatively low. The consumer #1018 having children in the family still has higher power level at late night, which may be because that the house of this consumer is bungalow (similar to villa), all family members live together, and other members of some families still keep the

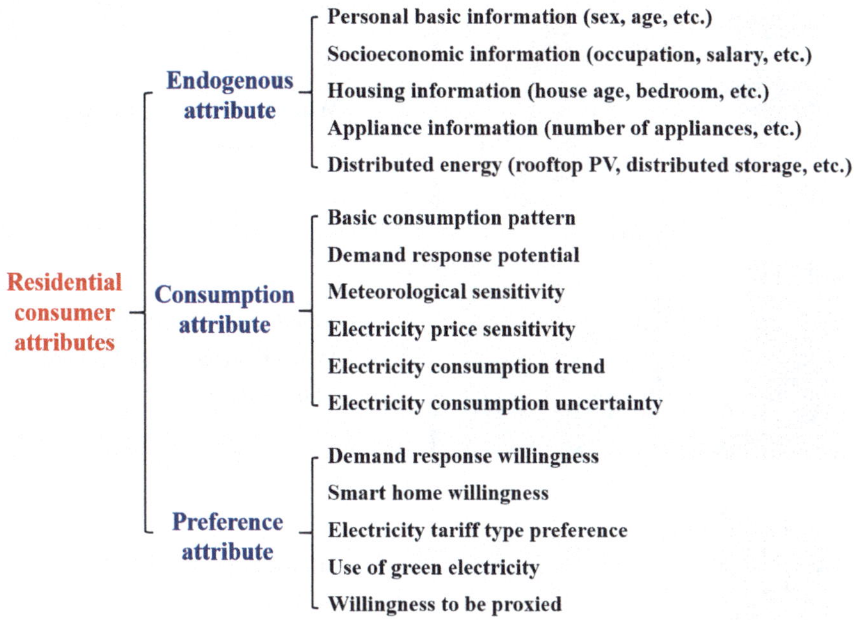

Fig. 2.5 Attributes classification for residential consumer modeling

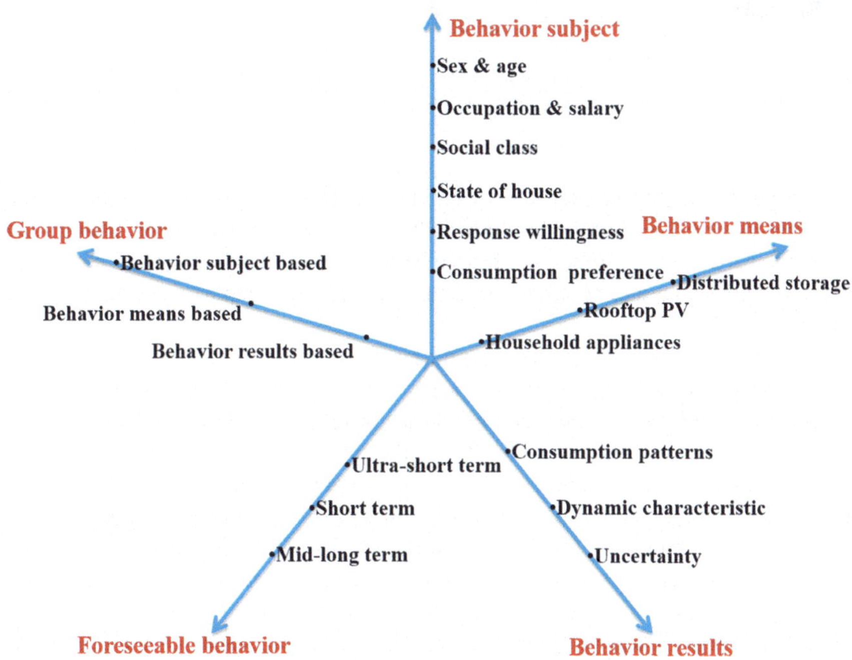

Fig. 2.6 Multi-dimensional analysis for electricity customer modeling

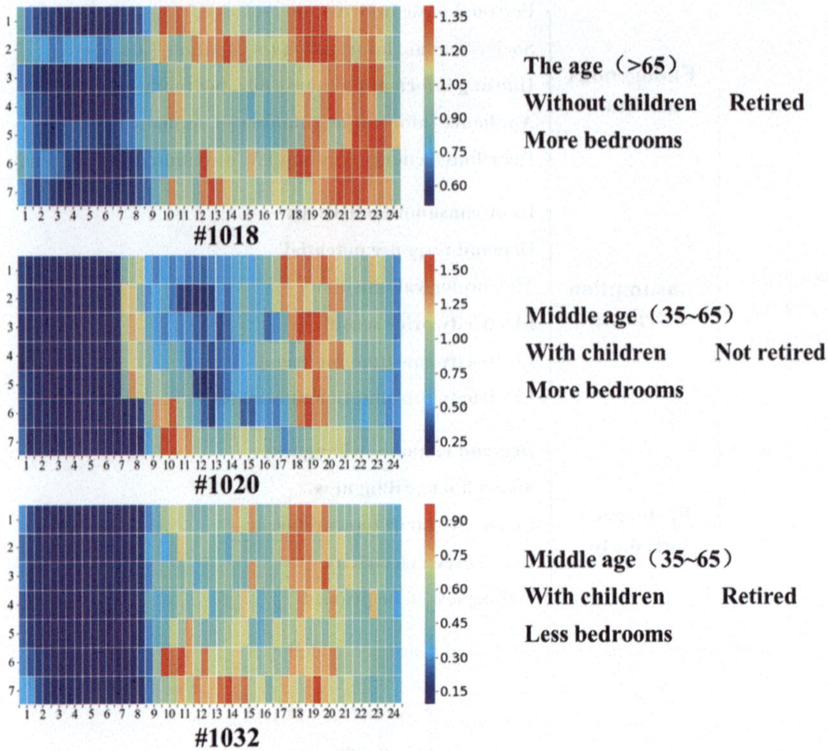

Fig. 2.7 Illustration of the correspondence between load profiles and characteristics of consumers

active power at night to look after the children at night. The three consumers keep the active power from 6:00 to 8:00 PM, which accords with the power habit of the general family. Chapter 10 builds a bridge between the consumer's power consumption and their social and economic information with the deep convolutional neural network.

The behavior means refers to interpreting the structural analysis of consumer's power consumption behavior in two aspects. One is to directly decompose the operating state of one or some equipment according to the total load profile. The non-intrusive Load Monitoring (NILM) is the important approach to conduct the structural analysis of the power consumption behavior of the residential customer and even building customer, which decomposes the load into several power curves of single electric equipment with more fine-grained smart meter data. The study on NILM can be traced back to the 1970s, but the current relevant studies do not fully consider the impact of access to distributed renewable energy and energy storage. The other interpretation is to analyze the different components of the consumer. For example, the consumer's power consumption behavior structure is analyzed as the meteorological sensitive component, electricity price-sensitive component, and

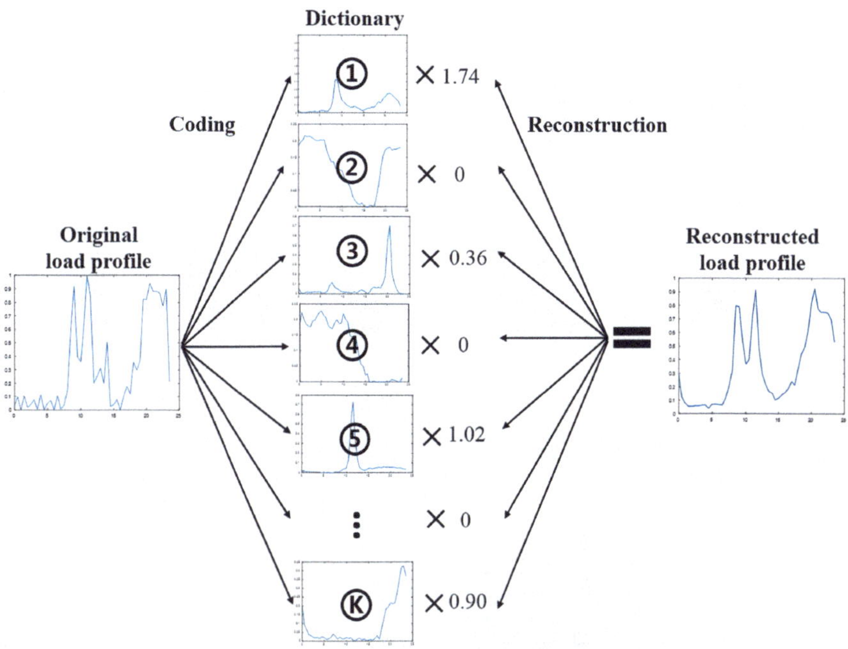

Fig. 2.8 Illustration of sparse coding-based partial usage pattern extraction

basic power component, etc., or analyzed as the seasonal component, weekly component, daily component, etc., or analyzed as the low-frequency stable component, high-frequency random component, etc.

The behavior result can be used for identifying the various indexes, such as consumer's basic power consumption patterns, dynamic characteristics, and uncertainty of power consumption. The consumers' power consumption pattern can be extracted by clustering of the load curve. Chapter 8 re-examines the consumer's load profile from the sparse perspective, believing that the consumer's load profile is essentially the superposition of several power consumption behaviors, as shown in Fig. 2.8. Then, the consumer behavior mode extracting problem is modeled as a sparse coding problem, which can effectively identify the consumer's partial usage pattern, as well as compress the massive smart meter data.

For the foreseeable behavior, the estimation of future power consumption behavior may have different time scales, such as ultra-short-term, short term, and medium and long term. The consumer load forecasting is a typical foreseeing for the behavior result to describe the uncertainty of consumer's power consumption behavior in the future. Currently, researchers around the world conduct more and more probabilistic load forecasting studies facing a single consumer. For example, Chap. 12 proposes a quantile long and short- term memory network model to conduct the probabilistic forecasting for the single consumer. Figure 2.9 gives a typical illustration for proba-

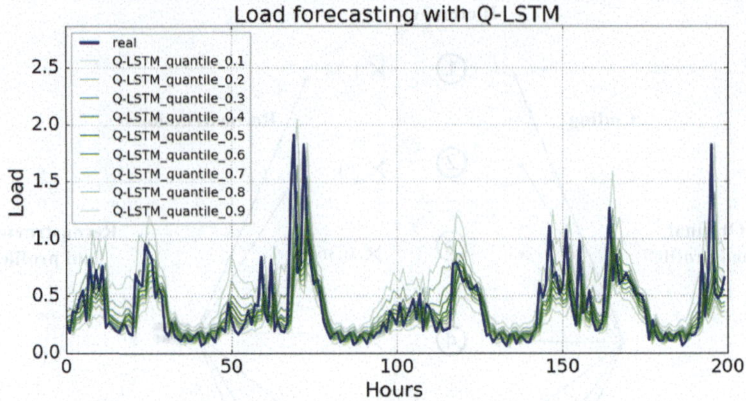

Fig. 2.9 Illustration of individual probabilistic load forecasting

Fig. 2.10 Illustration of consumer segmentation

bilistic load forecasting of ultra-short time residential customer, which describes its future uncertainty by a series of quantiles.

For the aggregation behavior, the consumers are grouped according to the different standards, i.e. a consumer behavior characteristic, such as identifying the group according to the consumer's basic attribute, use of electrical Fig. 2.10.

2.7 Conclusions

This chapter proposes the basic concept of ECBM, decomposes the basic components of the consumer behavior, including the behavior subject, behavior environment, behavior means, behavior result, and behavior utility, then further extends to the aggregation behavior and foreseeable behavior. On this basis, the theoretical research framework of ECBM is proposed through several illustrations. This chapter is expected to provide the reference for the study of ECBM, build the data-driven consumer-centric research and application, and further promote the interaction between consumers and systems in the context of Energy Internet.

References

1. Wang, Y., Chen, Q., Kang, C., Zhang, M., Wang, K., & Zhao, Y. (2015). Load profiling and its application to demand response: A review. *Tsinghua Science and Technology*, *20*(2), 117–129.
2. Wang, Q., Zhang, C., Ding, Y., Xydis, G., Wang, J., & Østergaard, J. (2015). Review of real-time electricity markets for integrating distributed energy resources and demand response. *Applied Energy*, *138*, 695–706.
3. Xue, Y., & Xinghuo, Y. (2017). Beyond smart grid-cyber-physical-social system in energy future [point of view]. *Proceedings of the IEEE*, *105*(12), 2290–2292.
4. Xin, S., Guo, Q., Sun, H., Chen, C., Wang, J., & Zhang, B. (2017). Information-energy flow computation and cyber-physical sensitivity analysis for power systems. *IEEE Journal on Emerging and Selected Topics in Circuits and Systems*, *7*(2), 329–341.
5. Palensky, P., & Dietrich, D. (2011). Demand side management: Demand response, intelligent energy systems, and smart loads. *IEEE transactions on Industrial Informatics*, *7*(3), 381–388.
6. Siano, P. (2014). Demand response and smart grids-a survey. *Renewable and Sustainable Energy Reviews*, *30*, 461–478.
7. Yang, J., Zhao, J., Wen, F., & Dong, Z. (2018). A model of customizing electricity retail prices based on load profile clustering analysis. *IEEE Transactions on Smart Grid*, *10*(3), 3374–3386.
8. Behavioural and social sciences at nature research. https://socialsciences.nature.com/.
9. Harland, C. M. (1996). Supply chain management: Relationships, chains and networks. *British Journal of Management*, *7*, S63–S80.
10. Koufaris, M., Kambil, A., & LaBarbera, P. A. (2001). Consumer behavior in web-based commerce: An empirical study. *International Journal of Electronic Commerce*, *6*(2), 115–138.
11. Kooti, F., Lerman, K., Aiello, L. M., Grbovic, M., Djuric, N., & Radosavljevic V. (2016). Portrait of an online shopper: Understanding and predicting consumer behavior. In *Proceedings of the 9th ACM International Conference on Web Search and Data Mining*, pp. 205–214. ACM
12. Koufaris, M. (2002). Applying the technology acceptance model and flow theory to online consumer behavior. *Information Systems Research*, *13*(2), 205–223.

Chapter 3
Smart Meter Data Compression

Abstract The huge amount of household load data requires highly efficient data compression techniques to reduce the great burden on data transmission, storage, processing, application, etc. This chapter proposes the generalized extreme value distribution characteristic for household load data and then utilizes it to identify load features, including load states and load events. Finally, a highly efficient lossy data compression format is designed to store key information of load features. The proposed feature-based load data compression method can support highly efficient load data compression with little reconstruction error and simultaneously provide load feature information directly for applications. A case study based on the Irish Smart Metering Trial Data validates the high performance of this new approach, including in-depth comparisons with the state-of-the-art load data compression methods.

3.1 Introduction

Smart meters typically capture the domestic loads accumulated over a 30 min period, offering a previously unknown degree of insight into the behavior in an individual dwelling as an aggregation of appliance loads [1]. With the rollout of smart meters, there is an explosive increase in smart metering load data. The yearly volume of load profile data for the 1.658 million households in Ireland (statistics obtained from Central Statistics Office) could amount to 216 GB. Compared with Ireland, in which the number of households is relatively small, the volume of load profile data generated by the 230 million smart meters installed by the State Grid Corporation of China is estimated to be 29 TB each year. It should be noted that all encapsulating identifiers and length fields are omitted to treat different data formats equally. Hence, the real volume of load profile data is larger.

The accompanying hundreds of millions of load profile data recorded by smart meters have also caused "big data" problems covering data transmission, storage, processing, and application, etc. Smart meters are typically connected with narrow-band powerline communication (PLC) links and upload load data to the aggregator installed in the transformer. Owing to the limited bandwidth, the reliability of data transmittance will decrease with increasing data volume [2]. The storage requirement

Y. Wang et al., *Smart Meter Data Analytics*,
https://doi.org/10.1007/978-981-15-2624-4_3

and processing time would also increase with increasing data. However, the volume of smart meter data exasperates these applications. Compressing load profile data allows for a substantial reduction in data volume, thus providing a highly efficient framework to transmit, store, and process these load profile data.

Data compression can be divided into either lossy compression or lossless compression. Lossy compression typically reduces bits through identifying unnecessary information in the data and removing it, whereas lossless compression usu-ally reduces bits through eliminating statistical redundancy. Lossy compression drop nonessential detail and retain information key to the data's applications from the data source; thus it can be mainly applied to accelerate similarity search, which supports important load data mining applications such as load profiling [3, 4] and customer segmentation [5–7]. The similarity between two load profiles is typically measured with distance index like Euclidean distance. The similarity within compressed load data through lossy data compression can be calculated more efficiently compared to lossless data compression because distance calculation between part information is faster than the complete information.

In terms of load profile data compression, a resumable load data compression (RLDC) method is proposed in [2]. This method is mainly based on the differential coding method. In this method, for a load profile, the first load value is recorded completely, and the following data are the value differences between consecutive load values. Most consecutive values of load profiles in households exhibit little value difference; thus, the difference can be stored by fewer bits, thereby conserving storage. This method can accomplish resumable data compression with improved compression efficiency by orders of magnitude compared with transmission encodings that are currently used for electricity metering. However, because of the differential coding technique, the compressed data record the difference between consecutive load values rather than the original load values or symbols marking the load level, thus making it inconvenient and inefficient for direct processing by data mining methods. Reference [8] exploits the effects of using the symbolic aggregate approximation (SAX) method [9] to do lossy data compression. By symbolizing the average load value in a fixed time window, this method provides high compression efficiency, and the compressed data can be easily processed by data mining methods. However, the compressed data lose some of the high-frequency signals; hence, the data reconstruction precision for this lossy method is not high.

Above all, there is an urgent requirement to design a smart meter data compression method that can provide high compression efficiency, high reconstruction precision, and a simple data compression format for applications. Here, we propose a feature-based load data compression method (FLDC). The method is a lossy smart metering load data compression method which is designed to fulfill the above requirement. In the method, the general extreme value (GEV) distribution characteristic of household load data is validated and utilized to identify load features such as load states and load events for low-resolution load data. The identified load features are stored in the proposed highly efficient data compression format, which can support highly efficient load data compression with little compression error and simultaneously provide load feature information directly for application. With the method presented in this chapter, this compressed data volume will be only 1.8% of the original data volume, reducing Irish smart meter data from 216 GB to 3.88 GB and China's smart meter data from 29 TB to 0.52 TB (assuming data properties similar to the test data).

3.2 Household Load Profile Characteristics

Smart meters installed in households typically record electric power consumption data in an interval of 30 min [1]. These data compose household load profiles showing certain characteristics, including small value difference and generalized extreme value distribution (GEV) characteristics, which are illustrated as follows.

3.2.1 Small Consecutive Value Difference

As demonstrated in [2], an important characteristic of household load profiles is that the value difference between two consecutive load values in a day is small compared with the peak load for the load data sampling in 1 s intervals. This characteristic suggests that the household load remain in one low state for most time intervals. We confirm that this characteristic also exists for the load data at a granularity of 30 min, and the character becomes increasingly more significant as the load decreases to a very low level.

The analyzed household load data are from Electric Ireland and the Sustainable Energy Authority of Ireland (SEAI) [10]. SEAI released online fully anonymous datasets from smart metering trials for electricity customers. The smart metering trials were conducted in 2009 and 2010, with more than 5,000 Irish homes and businesses participating. The participating households were all carefully recruited to ensure that they were representative of the national population and that their load profiles were also representative of the national profile. The Irish Smart Metering Trial Data collected are composed of the smart metering data, which recorded daily consumer energy consumption in 30-min intervals. To evaluate what percent of consecutive load value exhibits little difference, a cumulative probability analysis of the consecutive value difference rate is done for household load in the data set. The result for the typical household #1008 is shown in Fig. 3.1. The consecutive value difference rate $r_{n,t}$ at interval t of day n is calculated as follows:

$$r_{n,t} = \frac{P_{n,t} - P_{n,t-1}}{P_{n,max}} \tag{3.1}$$

where $P_{n,t}$ is the load at interval t of day n, $P_{n,t-1}$ is the load at interval $t-1$, and $P_{n,max}$ is the peak load on day n.

The cumulative probability versus consecutive value difference rate plot illustrates that 70% of the load values exhibit a consecutive value difference rate smaller than

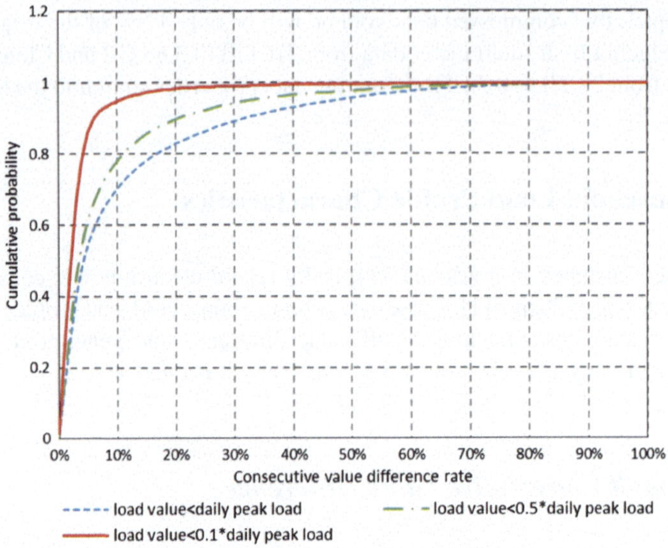

Fig. 3.1 Cumulative probability versus consecutive value difference rate for household #1008

10%, which suggests that most value differences are smaller than 10% of the daily peak load. This small difference allows household load data to be compressed because most load values would be the same if the 10% value difference is ignored. If we count only the load values below 50% of the daily peak load, the probability will increase to 78%. As the load values decrease below 10% of the daily peak load, the probability increases to 95%.

The cumulative probability plot illustrates that when the household load is at a very low level, it is stable and exhibits little change. As the load level improves, the household load becomes unstable and shows a large change rate.

3.2.2 Generalized Extreme Value Distribution

Reference [1] proposes employing a linear mixture of Gaussian distributions to enumerate and validates that the mixture distribution performs well in capturing domestic load characteristic. The mixture distribution is a linear combination of parametric distributions; thus, it can be inferred that the fitting degree also increases as the degree of freedom of the model increases. However, the complexity of the model also increases and requires more training time as a result of an increasing degree of freedom. Also, the overfitting problem may occur due to the high degree of freedom.

Hence, to illustrate the distribution characteristic of household load profiles, we conduct extensive analysis and comparison on the load of all the 4225 households in the Irish Smart Metering Trial Data for the variety of possible unimodal distributions

Table 3.1 Four best distribution fitting results of household #1008

Distribution name	Generalized extreme value	Exponential	Generalized Pareto	*t* location scale
Parameter name	μ, σ, k	μ	μ, σ, k	μ, σ, n
Parameter value	0.43, 0.25, 0.39	0.73	−0.07, 0.78	1.05, 0.15
AIC	22902	31003	35145	35362

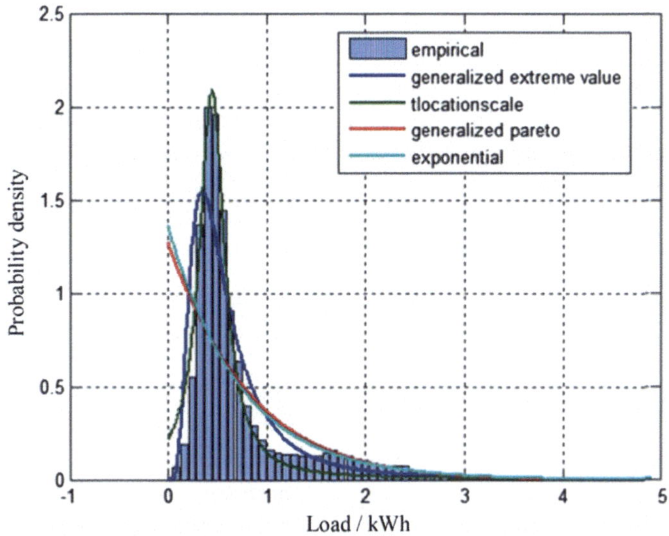

Fig. 3.2 Four best distributions for the household load of household #1008. The GEV distribution provided the best fit. The bin interval for the empirical histogram equals 0.0769 kWh

and finally choose the best four distributions, including the GEV distribution, the exponential distribution, the generalized Pareto distribution and the t-distribution. Here we show the result of the typical household #1008 in detail in Table 3.1 and Fig. 3.2, and illustrate the summary result of all the households in Table 3.2. As shown in Fig. 3.2, the GEV distribution significantly outperforms the other distributions. The Akaike information criterion (AIC) is often used to evaluate a distribution fit. A lower AIC value means a better distribution fit. According to the AIC values shown in Tables 3.1 and 3.2, the AIC values of the GEV distribution are the lowest; thus the GEV distribution performs best in the distribution fits of the household load data.

Our distribution fitting result show that the GEV distribution also fits smart meter data well. The GEV distribution is often used to model the probability of extreme events. It performs well because domestic events of consumption electricity highly resemble extreme events. For most times of day, residents do not require the use of high-power appliances such as ovens, washers, dryers, air conditioners, and electric water heater; thus, the load remains at a low level. When residents switch on high-

Table 3.2 Statistics of AIC value for the distribution fits of all the 4225 households

AIC	Generalized extreme value	Exponential	Generalized Pareto	*t* location scale
Mean	−2058	9903	2929	14550
Median	3833	13350	11551	19924
25% quantile	−20885	−8832	−12477	−5560
75% quantile	22859	31208	30215	40461

power appliances, the load will soon increase to a high level but typically will not maintain it for a long time. This behavior pattern of domestic electricity consumption leads to the load typically remaining at a low level for most of the time and a high level for rare occasions.

The GEV distribution combines the three possible types of limiting distributions for extreme values into a single form. The distribution function is

$$\begin{aligned} F(x) &= \exp[-\{1 + k\frac{(x-u)}{\sigma}\}^{-\frac{1}{k}}],\ k \neq= 0 \\ &= \exp[-\exp\{-\frac{(x-u)}{\sigma}\}],\ k = 0 \end{aligned} \quad (3.2)$$

with x bounded by $u - \sigma/k$ from below if $k > 0$ and from above if $k < 0$. Here, u and σ are location and scale parameters, respectively, and the shape parameter k determines which extreme value distribution is represented: Fréchet, Weibull, and Gumbel correspond to $k > 0$, $k < 0$, and $k = 0$.

The Fréchet type has a lower bound below which the probability density equals 0, whereas the Weibull type has an upper bound above which the probability density equals 0. The Gumbel type has no restriction on value [11]. Most households consume electricity, so their load typically shows a zero lower bound; hence, the best-fitted GEV distributions of their load profile data typically belong to the Fréchet type ($k > 0$).

3.2.3 *Effects on Load Data Compression*

The effects of these characteristics on load data compression can be summarized as follows.

3.2.3.1 Small Consecutive Value Difference Shows that Low-Level Load Is More Stable

The small consecutive value difference illustrates that the household load rarely changes between two consecutive time intervals. As shown in Fig. 3.1, when the

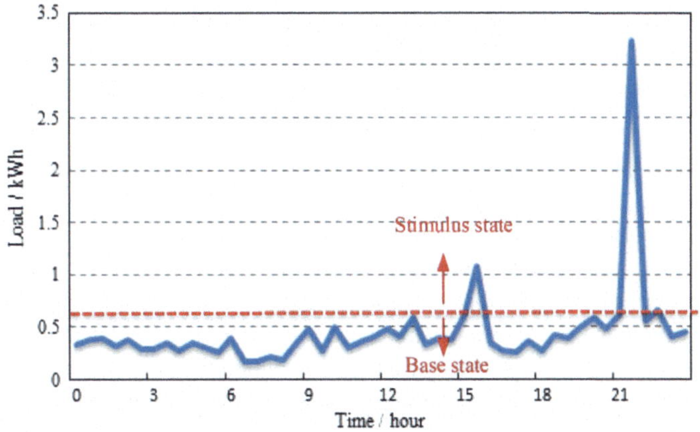

Fig. 3.3 Boundary separating the base state and stimulus state for the load profile of household #1008

load level decreases, this characteristic will strengthen. This means that when the household load decreases, the consecutive value difference is smaller, and the load would become more stable. When the household load steps into a high level, as the consecutive value difference increases, the load would become more unstable.

To differentiate stable and unstable load levels, Fig. 3.3 plots a state boundary, below and above which the load is defined as "base state" and "stimulus state", respectively. As shown in Fig. 3.3, there is a state boundary set to differentiate the base state and stimulus state. The load below the boundary is in the base state; otherwise, the load is in the stimulus state.

Base state: In this state, the load level and consecutive value difference are both low.

Stimulus state: In this state, the load level and consecutive value difference are both high.

Load event: The phenomenon of the household load deviating from the base state, experiencing several stimulus states accompanied with large consecutive value difference and finally returning to the base state is defined as a "load event". The load event can be detected by searching for the transition from the base state to the stimulus state.

The household load typically remains in the base state, which is often accompanied by small value differences between adjacent sampling load values. Load events are often caused by the switching of high-power appliances, such as air conditioners, microwave ovens, washers, and dryers. As the load event finishes, the load will return to the base state and remain nearly unchanged again. The state boundary and corresponding load events construct the key features of household load profiles and hence are our identification target.

For data compression, because the base state is a stable state that exhibits little value difference and the load events rarely occur, the compression efficiency can be improved significantly by recording the time and load of typical load events. The remaining data are all base state loads. This process would not yield much compression error because the consecutive value difference in the base state is low.

3.2.3.2 GEV Distribution Can Be Used to Decide Load State Boundary

For GEV distribution, the value is distributed densely at a low level-i.e., the base state–and loosely at a high level-i.e., the stimulus state. Hence, we can adopt the quantile at which the cumulative distribution function equals to a predetermined probability as the state boundary. According to the fitted GEV cumulative distribution function through maximum likelihood estimation (MLE), when the confidence probability is ascertained, the state boundary separating the base state from the stimulus state can be calculated.

3.3 Feature-Based Load Data Compression

The feature-based load data compression method consists of 6 steps (denoted as A to F) shown in Fig. 3.4. The input is a household load time series denoted as $x_t (t = 1, 2, \ldots, N)$, where x represents the household load, t represents the time interval, and N is the length of the series. The output is a compressed binary data representation. The first five steps from A to E comprise load feature identification and clustering, through which typical load features including base states and load events are coded in step F to represent the original load time series x_t, providing powerful compression and reconstruction performance.

3.3.1 Distribution Fit

Household loads obey the GEV distribution; hence, the first step is modeling household load time series x_t by a distribution fit through the MLE algorithm. Because the household load characteristics differ by season, the distribution fit is made for

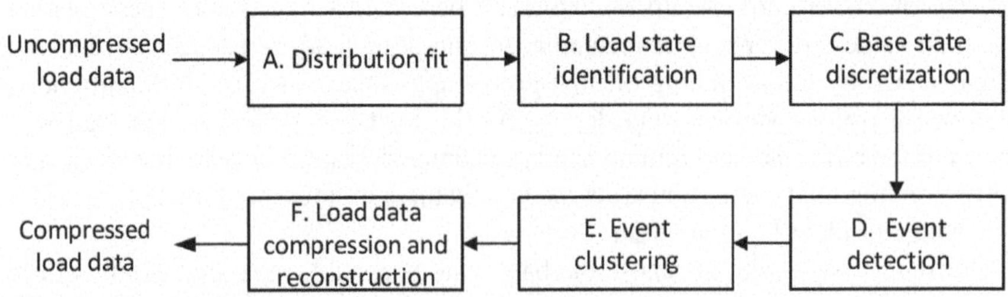

Fig. 3.4 Framework of FLDC

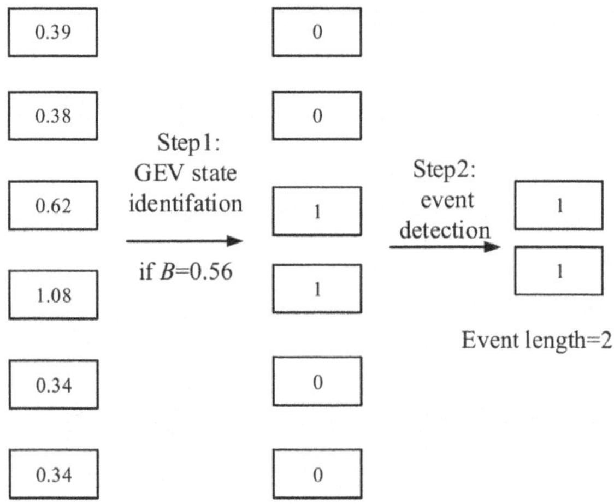

Fig. 3.5 GEV state identification and event detection for the household load

load data in spring (Mar. to May), summer (Jun. to Aug.), autumn (Sep. to Nov.) and winter (Dec. to Feb.). Given a confidence probability α and cumulative probability density function $F(x)$, the load state boundary B is calculated as follows:

$$B = x \quad \text{if } F(x) = \alpha \tag{3.3}$$

3.3.2 *Load State Identification*

Through the GEV distribution fit and boundary calculation, a load state matrix $S = [S_1, S_2, \ldots, S_N]$ composed of 0 (base state) and 1 (stimulus state) is generated by determining whether each load value in the original load profile data is below the boundary B, as shown in step 1 of Fig. 3.5.

$$S_t = \begin{cases} 0 & \text{if } x_t \leq B \\ 1 & \text{if } x_t > B \end{cases} \tag{3.4}$$

3.3.3 *Base State Discretization*

The load data in the base state are discretized by predetermined breakpoints according to the fitted GEV distribution. As shown in Fig. 3.6, breakpoints are a sequence of quantiles $C = [c_0, c_1, \ldots, c_d]$ such that the area under the fitted GEV probability density function $f(x)$ from c_{i-1} and $c_i = \alpha/d(i = 1, 2, \ldots, d)$, where α is the confidence probability, d is the discretization interval number, and $c_0 = u\sigma/k$, $c_d = B$.

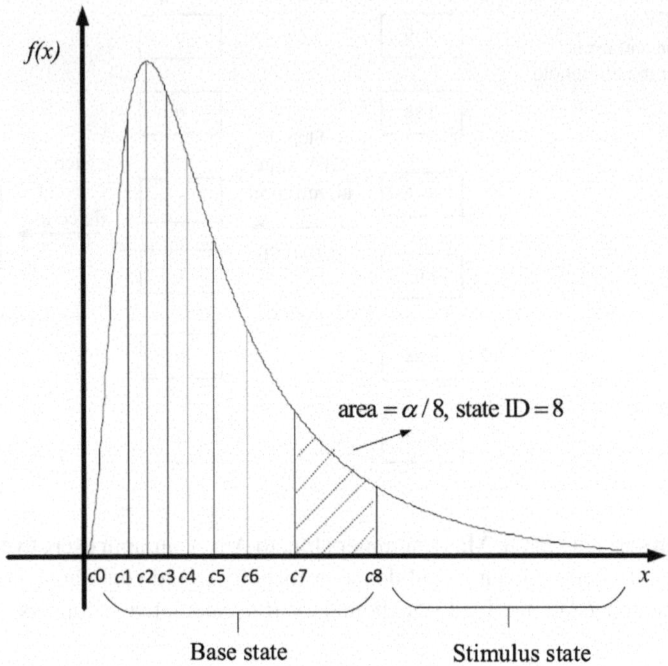

Fig. 3.6 Base state discretization

For any load series in the base state whose average value is falling into the interval between c_{i-1} and c_i, the series is coded by sub-base state ID i and expected value $E(i)$ between c_{i-1} and c_i:

$$\text{sub-base state } ID(x) = i \quad if \quad c_{i-1} < x \leq c_i \tag{3.5}$$

$$E_i = \int_{c_{i-1}}^{c_i} x \cdot f(x)dx \tag{3.6}$$

As shown in Fig. 3.6, the base state is separated into 8 sub-base states, with 9 breakpoints from c_0 to c_8. c_8 is the state boundary B; hence, below c_8, the load is in the base state or the stimulus state. The area under the GEV probability density function between two consecutive breakpoints equals $\alpha/8$. It can be seen that the original base state is coded by one number 0, and after discretization, it is divided into 8 sub-base states; thus, the coding resolution of the base state is significantly improved.

3.3.4 Event Detection

As shown in Step 2 of Fig. 3.5, event detection is performed after scanning all nonzero segments in the state matrix S. A load event occurs when the load deviates from the base state and moves into the stimulus state. Before the load returns to the base state, the load may experience several stimulus states. Hence, the event detection algorithm is composed of two steps, which are 0–1, 1–0 edge detection, and event load slicing.

(1) 0–1 Edge Detection: When the state changes from 0 to 1, a load event starts. Hence, the load event start time ts is calculated as follows:

$$ts = t + 1 \ \text{ if } \ S_{t+1} - S_t = 1 \tag{3.7}$$

(2) 1–0 Edge Detection: Increase t by 1 iteratively until the state changes from 1 to 0, at which point a load event ends. Hence, the load event end time te is calculated as follows:

$$te = t - 1 \ \text{ if } \ S_t - S_{t-1} = -1 \tag{3.8}$$

(3) Event Load Profile Slicing: The event load profile (ELP) is sliced from x_t according to each matched detected start time ts and end time te.

$$\text{ELP} = [x_{ts} \cdots x_{te}] \tag{3.9}$$

The number of stimulus state intervals is defined as the length of the ELP:

$$\text{length(ELP)} = te - ts + 1 \tag{3.10}$$

3.3.5 Event Clustering

After all load events are detected, the sliced load event profiles are used to construct a load event segment pool, on which load event clustering is based.

The length of the ELP represents the operation time of high-power appliances. As shown in Fig. 3.7, the first step is to classify load events according to their lengths. In addition to the length of the load event, the profile shape and load level are also important metrics for clustering. Here, the Euclidean distance is used as the distance between two ELPs with the same length. Based on the Euclidean distance, the hierarchical clustering algorithm [12] is applied to cluster load events into M groups in which the load events share a similar load event profile. The group ID is coded with integers from 1 to M. Finally, profiles in the same group are averaged to shape the representative profile, which is combined with the group ID.

3.3.6 Load Data Compression and Reconstruction

Here, we propose a special data format for data compression, as shown in Fig. 3.8. This data structure is event-based, with every 16 bits recording one load event.

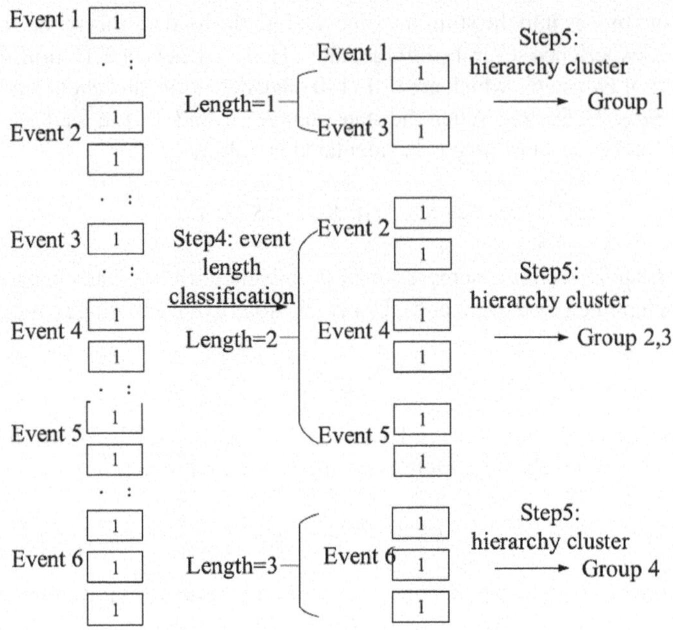

Fig. 3.7 Load event clustering decomposed into two steps: event classification based on lengths and hierarchical clustering based on Euclidean distance between ELPs

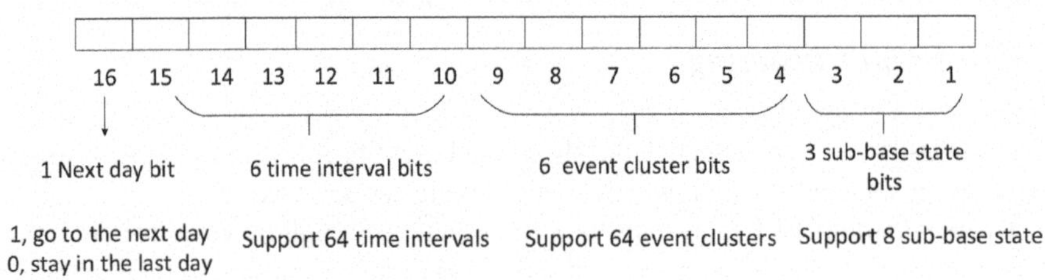

Fig. 3.8 Data format for compression

The 16 bits are equal to 2 bytes, which is easy for CPUs to process. Of the 16 bits, the first is named the next day bit so that the day on which the load event occurs can be determined. If this bit equals 0, the event occurs on the same day. If this bit equals 1, the event occurs on the following day. Following the next day bit, there are 6-time interval bits, which record the time when the load event starts. The maximum time interval provided by six bits is 64. The next 6 bits are responsible for coding the event group ID. With 6 bits, the data compression format can support no more than 64 event clusters. The final 3 sub-base state bits can support recoding of no more than 8 sub-base states.

This data compression format improves the compression efficiency significantly because all of the load values in the base state are recoded by the integer sub-base state

ID, and the event is represented by the integer event group ID. For household loads, events rarely occur, which is beneficial for significantly improving data compression.

The data reconstruction is divided into two steps: event reconstruction and base state reconstruction. In the event reconstruction process, the representative load profile of the event group is used to reconstruct the original event load profile. The start time and event group ID of any identified load event are recorded in the data compression format, as shown in Fig. 3.8. The baseload data before load events are replaced by the expected values corresponding to the sub-base state IDs, which are recorded in the last three bits of the data compression format.

3.4 Data Compression Performance Evaluation

The evaluation of data compression performance can be described in two aspects: compression efficiency–i.e., the extent to which data can be reduced through the compression method–and reconstruction precision–i.e., the difference between the reconstructed data and uncompressed data. In this section, the performance of FLDC is evaluated in these respects with an extensive comparison to the state-of-the-art data compression methods using the Irish Smart Metering Trial Data. Before discussing the results, we describe the related data formats for the smart metering data, evaluation index, and dataset.

3.4.1 Related Data Formats

In this section, the performance of FLDC is evaluated in these respects with an extensive comparison to the state-of-the-art data compression methods using the Irish Smart Metering Trial Data. Before discussing the results, we describe the related data formats for the smart metering data, evaluation index, and dataset.

PAA: In this method, time-series data with n dimensions is divided into w equally sized "frames". The mean value of the data falling within a frame is calculated, and a vector of these values constructs the compressed data. The compression efficiency of this method is decided by the w parameter. The smaller w is, the higher the compression efficiency will be. However, the reconstruction precision decreases with decreasing w. The original dimension of the daily load profiles is 48; here, we set $w = 6$, which compromises the compression efficiency and precision. The final compressed data depict the average load level at a granularity of 3 h.

SAX: This method first transforms the data into the Piecewise Aggregate Approximation (PAA) representation and then symbolizes the PAA representation into a discrete string. The compressed data are strings coded by ASCII, which can reduce the data size further than PAA. However, the string represents only the interval in which the data value falls; hence, the compressed data through SAX often have low performance on data reconstruction.

Haar DWT: This method is based on a three-level discrete wavelet transform, with which the approximate signal in level 3 is retained as the compressed load data. The wavelet adopted is the Haar wavelet, which has a square shape and provides a strong capability to reduce noise resulting from switching of high-power appliances.

RLDC: This method is based on the differential coding method. Most consecutive values of load profiles in house-holds exhibit little value difference; thus, the difference can be stored by fewer bits, thereby conserving storage. This method can improve compression efficiency by order of magnitude without any compression error compared to the uncompressed unsigned integer data format.

3.4.2 Evaluation Index

In terms of compression efficiency, one common index is the average value size in bits. This index evaluates the number of bits required to store one load value. The lower the index is, the higher the compression efficiency will be. For uncompressed double-precision float data, the bit number per value is constant at 64. For the uncompressed unsigned integer data described in IEC 61334-6, which is also referred to as A-XDR encoding, the bit number per value equals 16 if we use 16 bits to store an integer. The other evaluation index is the compression ratio, which is defined here as the uncompressed data volume divided by the double-precision floating point data volume.

In terms of reconstruction precision—because in most time intervals, the customer load remains low compared with the peak load–a micro-error would lead to large percent error. Even if the absolute error between the data before and after compression was small, a percent error evaluation method, such as MAPE, would also produce an extremely large percent error. Here, we propose a new precision evaluation metric defined as the mean peak percent error (MPPE), which uses the daily peak load as the denominator to calculate the percent error:

$$MPPE = \frac{1}{T}\sum_{t=1}^{T}\frac{\text{absolute error at period } t}{\text{daily peak load}} \times 100\% \tag{3.11}$$

where T equals the number of overall time intervals. For data compression, it is important to evaluate the accuracy of both the reconstructed time and load level of load events; thus there is no requirement to use the new error metrics proposed by [13] which reduces the so-called "double penalty" effect, incurred by forecasts whose features are displaced in space or time.

3.4.3 Dataset

The dataset for compression performance evaluation is taken from the Irish Smart Metering Trial Data from SEAI. The smart metering data are recorded in 30-min intervals; hence, the uncompressed daily load profile comprises 48 double-precision floating point data, which equals $48 \times 8 = 384$ Bytes. FLDC's compression performance is validated on 536 continuous daily load profiles of 4225 households, with the overall data volume amounting to 829.32 MB.

3.4.4 Compression Efficiency Evaluation Results

There are 4225 households evaluated overall, of which 20 household load compression efficiencies are plotted in Fig. 3.9. For the 20 household load data, the average value size in bits given by FLDC is 1.24, which surpasses most approaches significantly and is only a bit lower than SAX, which has the highest compression efficiency. It is noted that FLDC falls behind SAX by 0.25 bits per value. However, the latter method loses the capability of high reconstruction precision, which will be discussed in the next part.

As shown in Table 3.3, the overall evaluation result shows that the mean compression ratio of the 4225 households through FLDC reaches a high level of 55.71, which is near that of SAX.

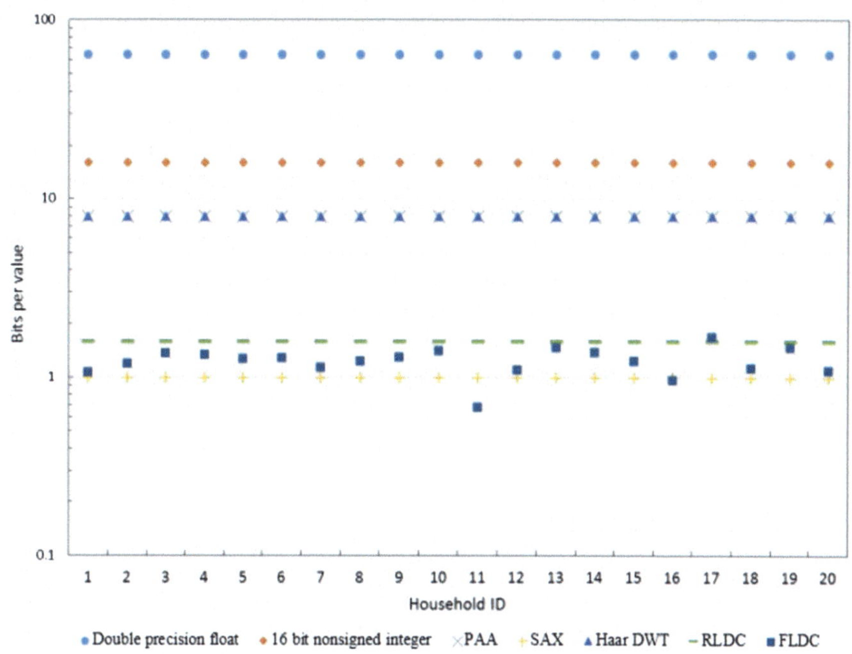

Fig. 3.9 Average value size in bits for 20 households from the Irish smart meter data

Table 3.3 Average Compression Efficiency and Reconstruction Precision for 4255 Households

	Average bits per value	Average compression ratio	MPPE (%)
Double precision float	64	1	0
16 bit unsigned integer	16	4	0
PAA	8	8	10.48
SAX	1	64	11.42
Harr DWT	8	8	10.48
RLDC [6]	1.6	40	0
FLDC	1.27	55.71	5.57

3.4.5 Reconstruction Precision Evaluation Results

Figure 3.10 shows the load reconstruction profiles of FLDC compared with PAA, SAX, DWT, and RLDC for households #1009, #1015, and #1018. Figure 3.11 shows the data reconstruction precision of FLDC for 20 households compared to the existing methods. With the exception of RLDC, it can be seen that FLDC outperforms the other methods significantly. Because RLDC does not yield any compression error, the reconstruction precision is 100%, and the reconstruction profile is the same as the uncompressed load.

The existing methods—PAA, SAX, and DWT—cannot capture the load event with high time and load level resolution, whereas FLDC restores the load event profile nearly without error. Figure 3.10c shows that the start time interval of the first load event in a day for household #1018 obtained by PAA, SAX, and DWT is 4:00 a.m. whereas the real start time interval equals 7:00 a.m.

As shown in the MPPE column of Table 3.3, the average MPPE of FLDC for all 4225 residents equals 5.57%, which indicates that the average reconstruction error is only 5.57% of the daily peak load for the 4225 households. Although it provides high compression efficiency, SAX loses the capability of high reconstruction precision and hence has the highest MPPE, which is 11.42%. PAA and DWT have the same MPPE, both equal 10.18%.

3.4.6 Performance Map

Figure 3.12 shows a performance map in which the state-of-the-art methods are located according to their performance in terms of reconstruction precision (1-MPPE) and compression ratio. It can be seen that any of SAX, RLDC, and FLDC cannot beat each other methods in both dimensions of performance, and they all significantly outperform PAA and DWT in the dimension of compression ratio. The compression ratio of SAX is the highest, but its reconstruction precision is the lowest.

Fig. 3.10 Data reconstruction for households (a) #1009; (b) #1015; (c) #1018

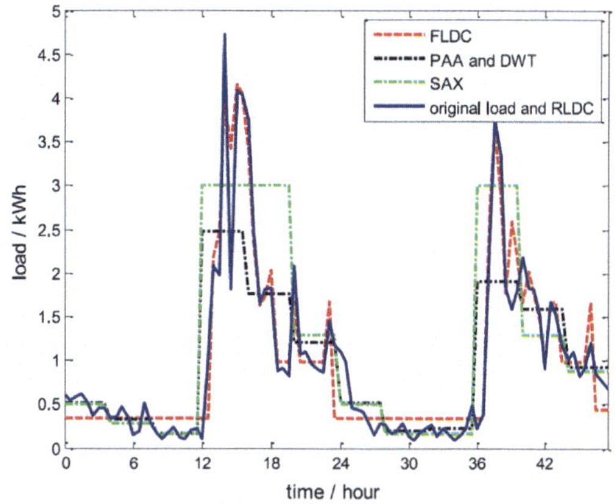

(a) Household #1009

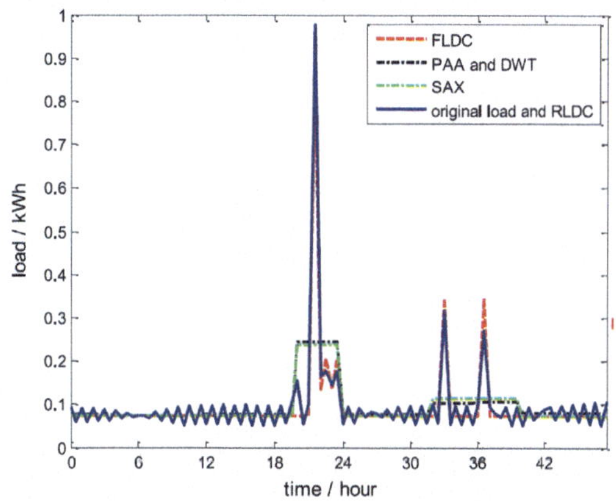

(b) Household #1015

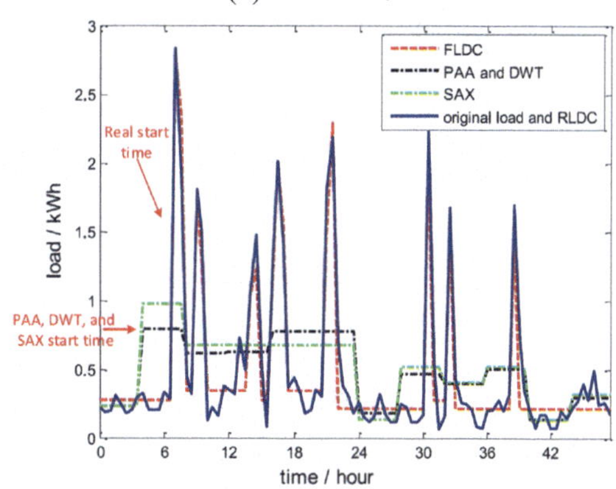

(c) Household #1018

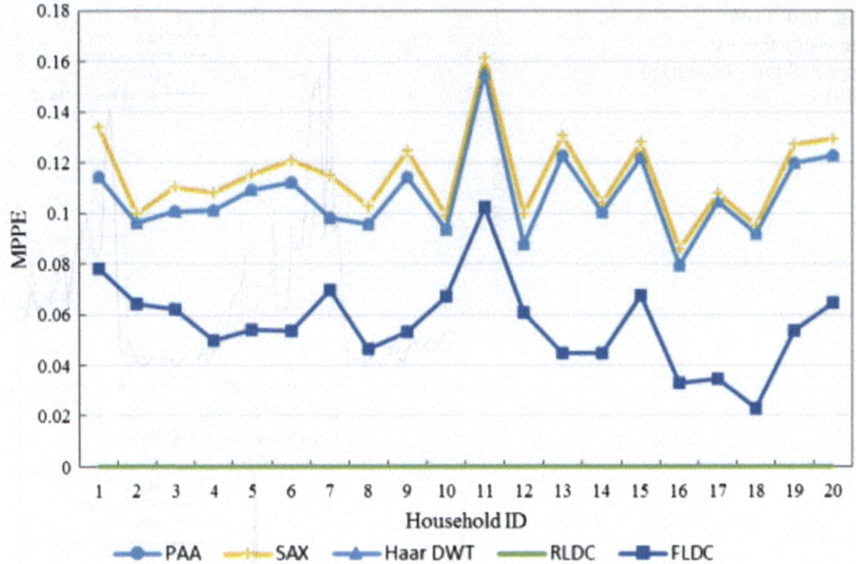

Fig. 3.11 Data reconstruction precision for 20 households. The MPPE of RLDC equals 0. PAA and DWT have the same reconstruction precision. With the exception of RLDC, FLDC has the lowest MPPE

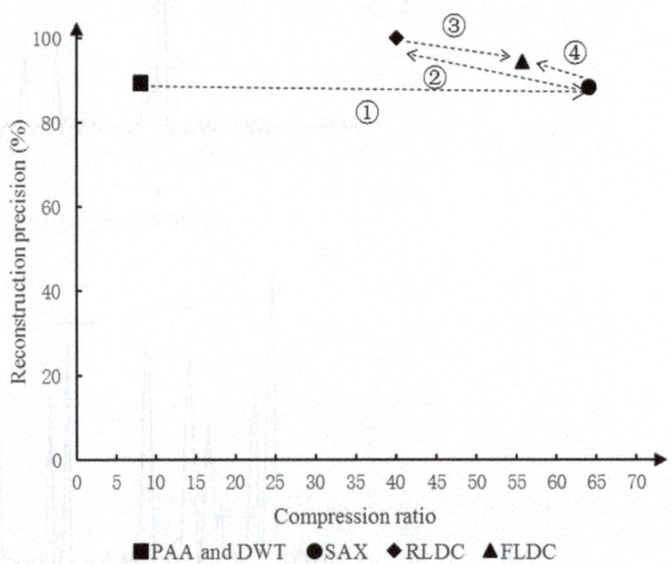

Fig. 3.12 The reconstruction precision versus compression ratio for data compression methods. 1 From PAA and DWT to SAX: compression ratio increases by 800%, reconstruction precision decreases by 0.94%; 2 From SAX to RLDC: compression ratio decreases by 37.5%, reconstruction precision increases by 11.42%; 3 From RLDC to FLDC: compression ratio increases by 39.3%, reconstruction precision decreases by 5.57%; 4 From SAX to FLDC: compression ratio decreases by 13.0%, reconstruction precision increases by 5.85%

From PAA and DWT to SAX and RLDC, there is a huge improvement in compression ratio. However, when the compression ratio has been as high as 40–64, it becomes difficult to improve them without reducing the reconstruction precision.

Compared with SAX, FLDC improves the reconstruction precision from 88.58 to 94.43% while sacrificing only 13.0% compression ratio. From RLDC to FLDC, the compression ratio is improved by 39.3% at the expense of 5.57% of reconstruction precision. Although a 39.3% compression ratio increase is much smaller than the 800% compression ratio increases from PAA and DWT to SAX, it is still significant progress. Actually, FLDC realizes a better compromise of compression ratio and reconstruction precision and yields a large improvement in compression ratio with little loss of reconstruction precision.

3.5 Conclusions

This chapter proposes a smart metering load data compression method based on load feature identification. This feature-based load data compression identifies load features from the uncompressed load data and restores load features rather than original data values. According to the GEV distribution characteristic, load features are classified into two types: base states and load events. The base state load is then discretized into several sub-base states, which improves the coding resolution. The load events are clustered into load event groups in which the load events share a representative load event profile. Finally, we design an event-based data compression format, within which every 16 bits record one load event, and the baseload before the event starts. Owing to the GEV distribution characteristic of household load, the base state load rarely changes, and load events rarely occur, thus giving FLDC the capability of high compression ratio with little compression error while simultaneously providing feature information.

The advantages of FLDC include the following:

(1) Applied to the Irish smart meter data, the data compression ratio is as high as 55.71, with an average reconstruction error equaling 5.57% of the daily peak load;

(2) The data compression and reconstruction are simple and efficient, enabling both online and offline application;

(3) The compressed data directly show load feature information including the base state and load event type.

References

1. Stephen, B., & Galloway, S. J. (2012). Domestic load characterization through smart meter advance stratification. *IEEE Transactions on Smart Grid*, *3*(3), 1571–1572.
2. Unterweger, A., & Engel, D. (2015). Resumable load data compression in smart grids. *IEEE Transactions on Smart Grid*, *6*(2), 919–929.

3. Piao, M., Shon, H. S., Lee, J. Y., & Ryu, K. H. (2014). Subspace projection method based clustering analysis in load profiling. *IEEE Transactions on Power Systems*, *29*(6), 2628–2635.
4. Wang, Y., Chen, Q., Kang, C., Zhang, M., Wang, K., & Zhao, Y. (2015). Load profiling and its application to demand response: A review. *Tsinghua Science and Technology*, *20*(2), 117–129.
5. Tsekouras, G. J., Hatziargyriou, N. D., & Dialynas, E. N. (2007). Two-stage pattern recognition of load curves for classification of electricity customers. *IEEE Transactions on Power Systems*, *22*(3), 1120–1128.
6. Chicco, G., Ionel, O. M., & Porumb, R. (2013). Electrical load pattern grouping based on centroid model with ant colony clustering. *IEEE Transactions on Power Systems*, *28*(2), 1706–1715.
7. Espinoza, M., Joye, C., Belmans, R., & Demoor, B. (2005). Short-term load forecasting, profile identification, and customer segmentation: A methodology based on periodic time series. *IEEE Transactions on Power Systems*, *20*(3), 1622–1630.
8. Notaristefano, A., Chicco, G., & Piglione, F. (2013). Data size reduction with symbolic aggregate approximation for electrical load pattern grouping. *IET Generation Transmission & Distribution*, *7*(2), 108–117.
9. Lin, J., Keogh, E., Lonardi, S., & Chiu, B. (2003). A symbolic representation of time series, with implications for streaming algorithms. In *ACM SIGMOD Workshop on Research Issues in Data Mining and Knowledge Discovery*.
10. Commission for Energy Regulation (CER). (2012). CER Smart Metering Project - Electricity Customer Behaviour Trial, 2009–2010. Irish Social Science Data Archive. SN: 0012-00.
11. Walden, A. T., & Prescott, P. (1983). Maximum likeiihood estimation of the parameters of the three-parameter generalized extreme-value distribution from censored samples. *Journal of Statistical Computation and Simulation*, *16*(3–4), 241–250.
12. Johnson, S. C. (1967). Hierarchical clustering schemes. *Psychometrika*, *32*(3), 241–254.
13. Haben, S., Ward, J., Vukadinovic Greetham, D., Singleton, C., & Grindrod, P. (2014). A new error measure for forecasts of household-level, high resolution electrical energy consumption. *International Journal of Forecasting*, *30*(2), 246–256.

Chapter 4
Electricity Theft Detection

Abstract As the problem of electricity thefts via tampering with smart meters continues to increase, the abnormal behaviors of thefts become more diversified and more difficult to detect. Thus, a data analytics method for detecting various types of electricity thefts is required. However, the existing methods either require a labeled dataset or additional system information which is difficult to obtain in reality or have poor detection accuracy. In this chapter, we combine two novel data mining techniques to solve the problem. One technique is the Maximum Information Coefficient (MIC), which can find the correlations between the non-technical loss (NTL) and a certain electricity behavior of the consumer. MIC can be used to precisely detect thefts that appear normal in shapes. The other technique is the clustering technique by fast search and find of density peaks (CFSFDP). CFSFDP finds the abnormal users among thousands of load profiles, making it quite suitable for detecting electricity thefts with arbitrary shapes. Next, a framework for combining the advantages of the two techniques is proposed. Numerical experiments on the Irish smart meter dataset are conducted to show the good performance of the combined method.

4.1 Introduction

Fraudulent users can tamper with smart meter data using digital tools or cyber-attacks. Thus, the form of electricity thefts is very different from the form in the past, which relies mostly on physically bypassing or destructing mechanical meters [1]. Cases of organized energy theft spreading tampering tools and methods against smart meters that caused a severe loss for power utilities were reported by the U.S. Federal Bureau of Investigation [2] and Fujian Daily [3] in China. In total, the non-technical loss (NTL) due to consumer fraud in the electrical grid in the U.S. was estimated to be \$6 billion/year [4]. Because the traditional detection methods of sending technical staff or Video Surveillance are quite time-consuming and labor-intensive, electricity theft detection methods that take advantage of the information flow in power system are urgently needed to solve the problem of the "Billion-Dollar Bug".

Y. Wang et al., *Smart Meter Data Analytics*,
https://doi.org/10.1007/978-981-15-2624-4_4

The existing non-hardware electricity theft detection methods can be classified into three categories: artificial intelligence-based (AI-based), state-based, and game theory-based [5]. The AI-based methods use machine learning techniques, such as classification and clustering to analyze the load profiles of consumers to find the abnormal users because the consumption patterns of fraudulent users are believed to differ from those of benign users. Classification methods [6–8] usually require a labeled dataset to train the classifier, whereas clustering methods [9–11] are unsupervised and can be applied to an unlabeled dataset. The state-based methods [12, 13] use additional measurements, such as power, voltage, and current in the distribution network to detect electricity thefts. Because fraudulent users are incapable of tempering with the network measurements, conflicts will arise between the system states and smart meter records. Although high detection accuracy can be achieved, these methods require the network topology and additional meters. The game theory-based methods [14, 15] assume that there is a game between fraudulent users and power utilities and that different distributions of fraudulent users' and benign users' consumption can be derived from the game equilibrium. Detection can be conducted according to the difference between the distributions. Because the game theory-based methods focus on theoretical analysis with strong assumptions, they are beyond the scope of this chapter.

In fact, the existing methods have some issues that must be addressed further. For AI-based methods, due to the difficulty in building a labeled dataset of electricity thefts, the application of classification methods is limited. Because the clustering methods are unsupervised, tampered load profiles with normal shapes can not be detected, resulting in low detection accuracy. For the state-based methods, the measurement data and system information are much more difficult to obtain. In real applications, the consumption patterns, which are the focus of AI-based methods and the state consistency which is the focus of state-based methods should both be considered and utilized.

In this chapter, a real and general scene in which an observer meter is installed for every area containing a group of users is considered. The recorded data of the observer meter are the sum of the electricity consumptions of the area during a certain time interval. The data are available to most of the distribution system operators (DSOs) or electricity retailers. We attempt to combine the advantages of AI- and state-based methods to propose a detecting framework that adapts to the least parameters or system information to ensure general application and achieves good accuracy without any labeled training set. In particular, the maximum information coefficient (MIC) [16] is used to detect the association between NTL and the tampered load profiles with minimal additional system information. Next, CFSFDP is applied to catch thieves whose load profiles are more random and arbitrary according to their abnormal density features. We ensemble the two techniques by combining the suspicion ranks to cover most types of electricity thefts.

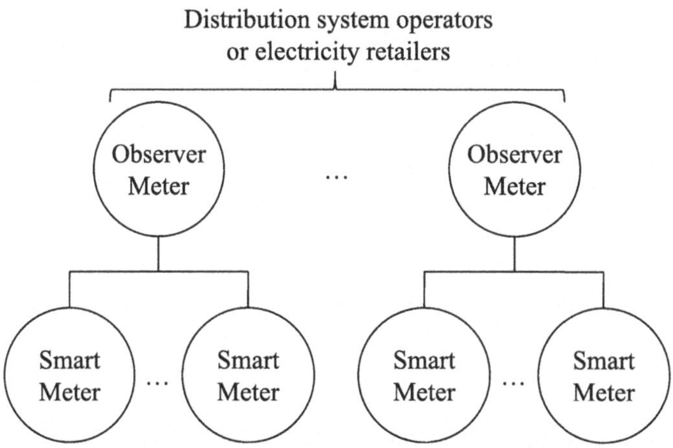

Fig. 4.1 Observer meters for areas and smart meters for customers

4.2 Problem Statement

4.2.1 Observer Meters

Our method is applicable to the scene of Fig. 4.1, where an observer meter is installed in an area with a group of customers. An observer meter is more secure than a normal smart meter is, making it almost impossible for fraudulent users to tamper with the meter. We believe that DSOs and electricity retailers have access to observer meter data.

4.2.2 False Data Injection

Electricity thieves tend to reduce the quantity of their billed electricity. Thus an FDI that has certain impacts on the tampered load profiles is used to simulate the tampering behaviors of the electricity thieves. We use six FDI types similar to those mentioned in [10] that have time-variant modifications on load profiles. Table 4.1 shows our FDI definitions and Fig. 4.2 gives an example of the tampered load profiles. In Table 4.1, x_t is the ground true power consumption during time interval t, and $\tilde{x}_t$ is the tampered data recorded by the smart meter. There are many other FDI types in the literature [5, 17]. However, a characteristic can be generalized according to their definitions and examples: an FDI type either keeps the features and fluctuations of the original curve or creates new patterns. This is the same for other sophisticated FDI types, so our method can handle them as well.

Table 4.1 Six FDI types [10]

Types	Modification
FDI1	$\tilde{x}_t \leftarrow \alpha x_t$ where $0.2 < \alpha < 0.8$ is randomly generated
FDI2	$\tilde{x}_t \leftarrow \begin{cases} x_t, & \text{if } x_t \le \gamma \\ \gamma, & \text{if } x_t > \gamma \end{cases}$ where γ is a randomly defined cut-off point, and $\gamma < \max \mathbf{x}$
FDI3	$\tilde{x}_t \leftarrow \max\{x_t - \gamma, 0\}$ where γ is a randomly defined cut-off point, and $\gamma < \max \mathbf{x}$
FDI4	$\tilde{x}_t \leftarrow f(t) \cdot x_t$ where $f(t) = \begin{cases} 0, & \text{if} t_1 < t < t_2 \\ 1, & \text{otherwise} \end{cases}$, $t_1 - t_2$ is a randomly defined time period longer than 4 h
FDI5	$\tilde{x}_t \leftarrow \alpha_t x_t$ where $0.2 < \alpha_t < 0.8$ is randomly generated
FDI6	$\tilde{x}_t \leftarrow \alpha_t \bar{\mathbf{x}}$ where $0.2 < \alpha_t < 0.8$ is randomly generated, $\bar{\mathbf{x}}$ is the average consumption of the load profile

The index i in $x_{i,t}$, $\tilde{x}_{i,t}$ and $\mathbf{x}_i$ is omitted here for simplicity

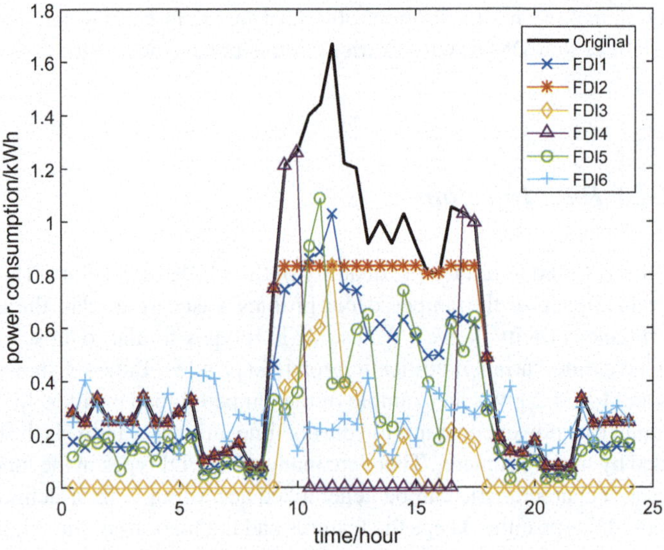

Fig. 4.2 An example of the FDI types

4.2.3 A State-Based Method of Correlation

The NTL of an area e_t can be calculated by subtracting the observer meter data E_t from the sum of the smart meter data $\tilde{x}_{i,t}$ in the area:

$$e_t = E_t - \sum_{i \in \mathscr{A}} \tilde{x}_{i,t} \tag{4.1}$$

where $\mathscr{A}$ is the set of the labels of all meters in the ares. Let $\mathscr{F}$ denote the set of the labels of tampered meters in the area, and $\mathscr{B} = \mathscr{A} / \mathscr{F}$ be the set of the labels of benign meters. Equation (4.1) can be represented as:

$$e_t = \sum_{i \in \mathscr{F}} (x_{i,t} - \tilde{x}_{i,t}) \tag{4.2}$$

where $x_{i,t}$ is the ground truth electricity consumption by consumer i. According to the analysis in Sect. 4.2.2, if the tampered data $\tilde{x}_{i,t}$ has a positive correlation with the ground truth data $x_{i,t}$, then the NTL value of $(x_{i,t} - \tilde{x}_{i,t})$ caused by user i is also correlated with $\tilde{x}_{i,t}$. Because e_t is composed of several $(x_{i,t} - \tilde{x}_{i,t})$, the correlation between vector $\mathbf{e}$ and $\tilde{\mathbf{x}}_i$ when $i \in \mathscr{F}$ should be stronger than the correlation when $i \in \mathscr{B}$:

$$Corr(\mathbf{e}, \tilde{\mathbf{x}}_i)\Big|_{i \in \mathscr{F}} > Corr(\mathbf{e}, \tilde{\mathbf{x}}_i)\Big|_{i \in \mathscr{B}} \tag{4.3}$$

where $Corr(\cdot, \cdot)$ is a proper correlation measurement for two vectors. Figure 4.3 shows a real electricity theft case in Shenzhen [18], where $\mathbf{e}$ and $\tilde{\mathbf{x}}_i$ have a high correlation. In FDI1, the correlation is linear and certain; however, in many other situations, the correlation is rather fuzzy. Note that Eq. (4.3) may not hold for some FDI types (e.g., FDI6 which produces a totally random curve); however, we can filter out a large part of electricity thefts by using Eq. (4.3). The selection of measurement $Corr(\cdot, \cdot)$ that can precisely reveal the fuzzy relationship between NTL and tampered load profiles is of vital importance.

4.3 Methodology and Detection Framework

The overall detection methodology is based on the two novel data mining techniques, i.e., MIC and CFSFDP. MIC utilizes the analysis in Sect. 4.2.3 to detect associations between the area NTL and tampered load profiles. CFSFDP is used to determine the load profiles with abnormal shapes. According to the suspicion ranks given by the two methods, a combined rank is given to take the advantages of both methods.

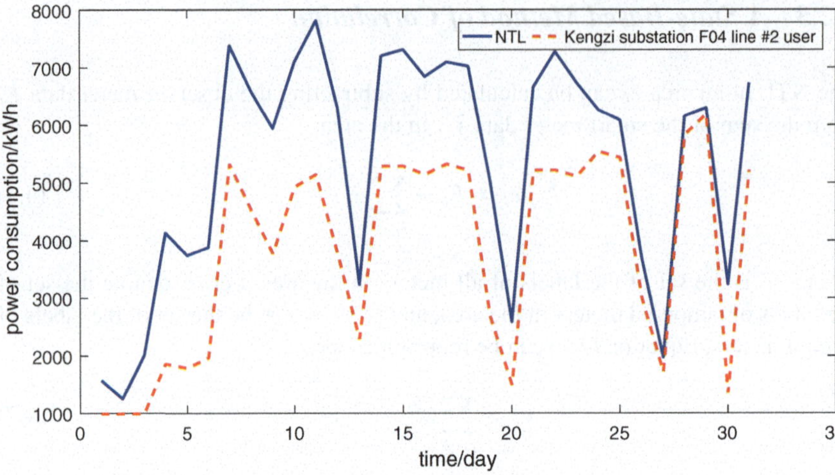

Fig. 4.3 A real case of NTL and power consumption of the suspected user [18]

4.3.1 Maximum Information Coefficient

In statistics, the Pearson correlation coefficient (PCC) is an effective measurement for the correlation between two vectors. The PCC has a value between $+1$ and -1. If two vectors have a strict linear correlation, then the absolute value of PCC is 1. If two vectors are irrelevant, then the value is 0. However, the PCC cannot detect more sophisticated associations, such as quadratic or cubic and time-variant relations. The mutual information (MI) of two variables is used as a good measurement of relevance because it detects all types of associations. MIC is based on the calculation of MI and has been proven to have a better performance than MI in many occasions [16].

Given a finite set $\mathscr{D}$ of ordered pairs, the x-values of $\mathscr{D}$ can be partitioned into a bins and the y-values of $\mathscr{D}$ can be partitioned into b bins. This creates an a-by-b grid G in the finite 2D space. Let $\mathscr{D}|_G$ be the distribution induced by the points in $\mathscr{D}$ on the cells of G. For $\mathscr{D} \subset \mathbb{R}^2$ and $a, b \in \mathbb{N}^*$, define

$$I^*(\mathscr{D}, a, b) = \max_G I(\mathscr{D}|_G) \tag{4.4}$$

where the maximum is over all grids G with a columns and b rows, and $I(\mathscr{D}|_G)$ is the MI of $\mathscr{D}|_G$. The characteristic matrix $\mathbf{M}(\mathscr{D})$ is defined as

$$\mathbf{M}(\mathscr{D})_{a,b} = \frac{I^*(\mathscr{D}, a, b)}{\log \min\{a, b\}} \tag{4.5}$$

The MIC of a finite set $\mathscr{D}$ with sample size $|\mathscr{D}|$ and grid size less than $B(n)$ is given by

$$MIC(\mathscr{D}) = \max_{ab<B(|\mathscr{D}|)} \{\mathbf{M}(\mathscr{D})_{a,b}\} \tag{4.6}$$

We use $B(|\mathscr{D}|) = |\mathscr{D}|^{0.6}$ in this chapter because it is found to work well in practice. The value of MIC falls in the range of [0, 1], and a larger value indicates a stronger association.

The $MIC(\cdot)$ is applied as the $Corr(\cdot, \cdot)$ in Eq. (4.3) to detect electricity thefts whose consumption behaviors have strong relevance to the NTL in the area.

4.3.2 CFSFDP-Based Unsupervised Detection

To tackle the FDI types that cannot be detected by the method of correlation, we use clustering to find the outliers in the numerous load profiles. Density-based clustering methods have been widely adopted in anomaly detecting. CFSFDP [19] is a newly-proposed method that has been proven to be very powerful in large dataset clustering and outlier detection.

In CFSFDP, two values are defined for the pth load profile: its local density ρ_p and its distance δ_p from other load profiles of higher density. Both values depend on the distances d_{pq} between the data points. Equation (4.7) gives the definition of ρ_p:

$$\rho_p = \sum_q \chi(d_{p,q} - d_c) \tag{4.7}$$

where d_c is the cut-off distance and $\chi(\cdot)$ is the kernel function. The cut-off kernel is $\chi(x) = \begin{cases} 1, & \text{if } x < 0 \\ 0, & \text{otherwise} \end{cases}$. Because the local density ρ_p is discrete in Eq. (4.7), a Gaussian kernel is occasionally used to estimate ρ_p, as shown in Eq. (4.8), to avoid conflicts:

$$\rho_p = \sum_{q \neq p} \exp\left[-\left(\frac{d_{p,q}}{d_c}\right)^2\right] \tag{4.8}$$

The definition of δ_p is shown in Eq. (4.9):

$$\delta_p = \min_{q:\rho_q > \rho_p} d_{p,q} \tag{4.9}$$

For those load profiles with the highest local density, δ_p is conventionally written as

$$\delta_p = \max_q d_{p,q} \tag{4.10}$$

Although the cut-off distance d_c is exogenous in the definitions, it can be automatically chosen by a rule of thumb suggested in [19]. Figure 4.4 shows an example of 28 data points among which #26∼28 are abnormal. The abnormal data points usually deviate from the normal majority, thus they only have a few neighborhood points

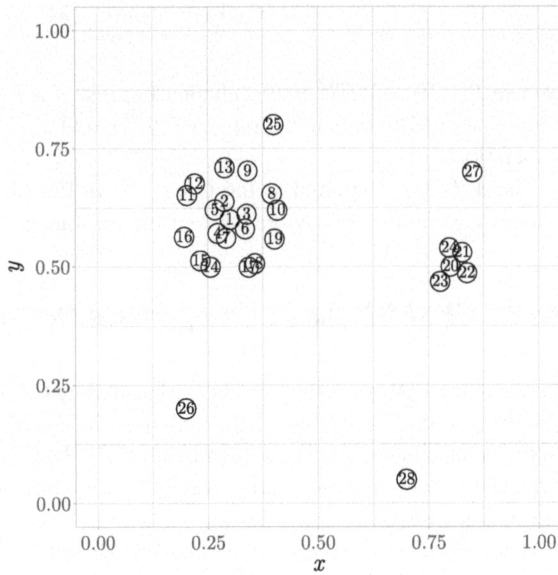

Fig. 4.4 An example distribution of data points

and their distance to the high-density area is larger than the normal points. From the definitions above, the spatial distribution of the abnormal points results in a small ρ_p and a large δ_p (Fig. 4.5). We define the degree of abnormality ζ_p in Eq. (4.11):

$$\zeta_p = \frac{\delta_p}{\rho_p + 1} \tag{4.11}$$

Compared with k-means and other partition-based clustering methods, density-based clustering can consider clusters with an arbitrary shape without any parameter selection. Moreover, the algorithm of CFSFDP is so simple that once the local density ρ_p of all the load profiles is calculated, δ_p and ζ_p can be easily obtained without any iteration. Load profiles with strange or arbitrary shapes are very likely to have a high value of ζ_p. Thus we can find out the abnormal load profiles according to their ζ_p values, which are very helpful in detecting electricity thefts that MIC cannot consider.

4.3.3 Combined Detecting Framework

Figure 4.6 shows the framework of how to utilize MIC and CFSFFDP in electricity theft detecting and how to combine the results of the two independent but complementary methods.

For an area with n consumers and m-day recorded data series, a time series of NTL is first calculated using Eq. (4.1). Next, we normalize each load profile $\tilde{\mathbf{x}}_p$ by dividing it with $\max_t \tilde{\mathbf{x}}_p$ and then reconstruct the smart meter dataset into a normalized load

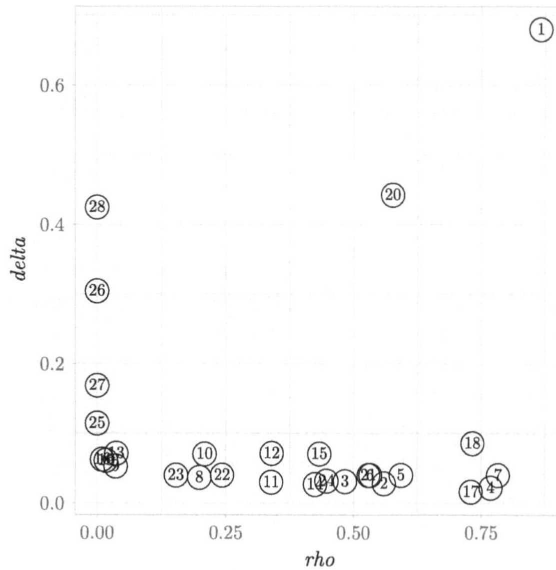

Fig. 4.5 Scatter plot of (ρ_p, δ_p) of the example data points

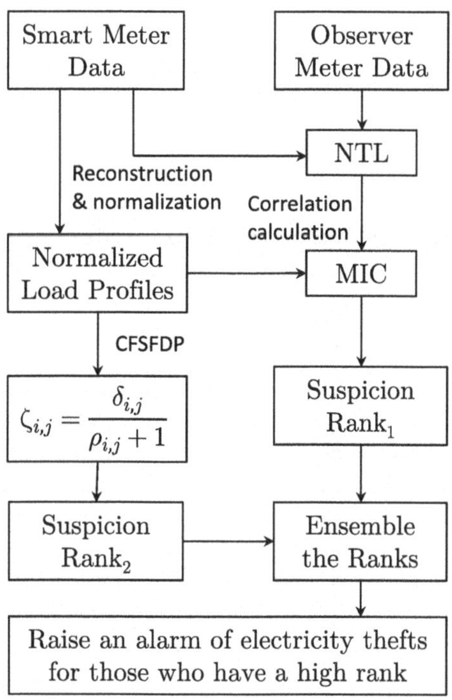

Fig. 4.6 The detection framework of the MIC-CFSFDP combined method

profile dataset with $n \times m$ vectors. This procedure retains the shape of each load curve to the greatest extent and helps the clustering method focus on the detection of arbitrary load shapes. Let $\mathbf{u}_{i,j}$ denote the normalized vector of the ith consumer's load profile on the jth day and $\mathbf{e}_j$ denote the NTL loss vector of the area on the jth day. For every i and j, $MIC(\mathbf{u}_{i,j}, \mathbf{e}_j)$ is calculated according to the equations in Sect. 4.3.1. Moreover, $\rho_{i,j}$ and $\delta_{i,j}$ are calculated using CFSFDP, and the degree of abnormality $\zeta_{i,j}$ for vector $\mathbf{u}_{i,j}$ is obtained.

For consumer i with m MIC or ζ values, a k-means clustering method with $k = 2$ is used to detect the MIC or ζ values of suspicious days by classifying the m days into 2 groups. The mean of the MIC or ζ values that belong to the more suspicious group is taken as the suspicion degree for consumer i. Thus, the two suspicion ranks of the n consumers can be extracted by inter-comparing the $n \times m$ MIC or ζ values.

The idea of combining the two ranks is based on the famous Rank Product (RP) method [20], which is frequently used in Biostatistics. In this chapter, we use the arithmetic mean and the geometric mean of the two ranks to combine the methods, as in Eq. (4.12).

$$\begin{aligned} Rank_{\text{Arith}} &= \frac{Rank_1 + Rank_2}{2} \\ \text{or } Rank_{\text{Geo}} &= \sqrt{Rank_1 \times Rank_2} \end{aligned} \tag{4.12}$$

Finally, a consumer is considered committing electricity theft if his combined *Rank* is high.

4.4 Numerical Experiments

4.4.1 Dataset

We use the smart meter dataset from Irish CER Smart Metering Project [21] that contains the load profiles of over 5000 Irish residential users and small & medium-sized enterprises (SMEs) for more than 500 days. Because all users have completed the pre-trial or post-trial surveys, the original data are considered ground truth. We use the load profiles of all 391 SMEs in the dataset from July 15 to August 13, 2009. Thus, we have $391 \times 30 = 11\,730$ load profiles in total, and each load profile consists of 48 points, with a time interval of half an hour. The 391 SMEs are randomly and evenly divided into several areas with observer meters. For each area, several users are randomly chosen as fraudulent users, and certain types of FDI are used to tamper with their load profiles. Fifteen of the 30 load profiles of each fraudulent user are tampered with.

4.4.2 Comparisons and Evaluation Criteria

To demonstrate the effectiveness of our proposed method, we use other correlation analysis and unsupervised outlier detection methods for comparison:

- Pearson correlation coefficient (PCC): a famous statistic method for bivariate correlation measurement.
- Kraskov's estimator for mutual information [22]: an improved method for estimating the MI of two continuous samples.
- Fuzzy C-Means (FCM): an unsupervised fuzzy clustering method. The number of cluster centers is chosen to range from 4 to 12 in this chapter.
- Density-based Local Outlier Factor (LOF) [23]: a commonly used method of density-based outlier detection.

To obtain comprehensive evaluation results in the unbalanced dataset, we use the AUC (Area Under Curve) and MAP (Mean Average Precision) values mentioned in [7]. The two evaluation criteria have been widely adopted in classification tasks. The AUC is defined as the area under the receiver operating characteristic (ROC) curve, which is the trace of the false positive rate and the true positive rate. Define the set of fraudulent users $\mathscr{F}$ as the positive class and benign users $\mathscr{B}$ as the negative class. The suspicion *Rank* is in ascending order according to the suspicion degree of the users. AUC can be calculated using *Rank* as in Eq. (4.13):

$$\text{AUC} = \frac{\sum_{i \in \mathscr{F}} Rank_i - \frac{1}{2}|\mathscr{F}|(|\mathscr{F}|+1)}{|\mathscr{F}| \times |\mathscr{B}|} \tag{4.13}$$

Let Y_k denote the number of electricity thieves who rank at top k, and define the precision $P@k = \frac{Y_k}{k}$. Given a certain number of N, MAP@N is the mean of $P@k$ defined in Eq. (4.14):

$$\text{MAP}@N = \frac{\sum_{i=1}^{r} P@k_i}{r} \tag{4.14}$$

where r is the number of electricity thieves who rank in the top N and k_i is the position of the ith electricity thieves. We use MAP@20 in this chapter. In the random guess (RG), the true positive rate equals the false positive rate; thus, the AUC for RG is always 0.5, and the MAP for RG is $|\mathscr{F}|/(|\mathscr{F}|+|\mathscr{B}|)$ which is the proportion of electricity thieves among all users. We consider these values to be the benchmarks.

Note that all the numerical experiments in this chapter are repeated for 100 randomly generated scenarios to avoid contingency among the results. The values of AUC and MAP are calculated using the mean value to show the average performance.

4.4.3 Numerical Results

In this subsection, we divide the users into 10 areas and randomly choose 5 electricity thieves for each area. Thus, there are approximately 39 users in each area, and the ratio of fraudulent users is 12.8%.

Figure 4.7 shows the comparison results of the methods. Tables 4.2 and 4.3 shows the detailed values of AUC and MAP@20 of the correlation-based methods and the unsupervised clustering-based methods for the six FDI types. The type MIX indicates that the 5 electricity thieves randomly choose one of the six types. We believe that different fraudulent users might choose different FDI types. The results for the detection of single FDI type show the advantage of each method under certain situations, while the results for type MIX are of significance in practice. In CFSFDP, the cut-off kernel is used because it is faster than the Gaussian kernel and because we have a large dataset in which conflicts do not occur. In the application of FCM, there are 9 different results due to the number of cluster centers, and we only present the best among them. MI denotes the Kraskov's estimator for mutual information, and Arith and Geo are abbreviations for arithmetic and geometric mean, respectively. The best results among the 8 methods are in bold for each FDI type in Tables 4.2 and 4.3.

The results demonstrate that the correlation-based methods exhibit excellent performance in detecting FDI1. The blue lines in Fig. 4.7 show that MIC has a more balanced performance in both AUC and MAP@20. MIC also shows its superiority in detecting type MIX. The correlation-based method performs poorly in detecting FDI5 and FDI6 because the tampered load profiles become quite random, and the correlation no longer exists. The unsupervised clustering methods, especially CFSFDP and LOF, have quite high values of AUC in detecting FDI4, FDI5, and FDI6; however, they have zero performance in FDI1 because after normalization the tampered load profiles appear exactly the same as the original load profiles. FCM has poor performance in types, except for FDI6; thus FCM may not be a good tool for electricity theft detection. Furthermore, during the numerical experiments, we notice that the performance of FCM is heavily affected by the number of cluster centers, and it is quite unpractical to tune the number in a wider range. From the black lines in Fig. 4.7, CFSFDP is found to have the best performance in detecting FDI5, FDI6, and type MIX among all the clustering methods. The MAP@20 of CFSFDP is much higher than that of LOF for these types.

The combined methods have taken the advantages of both MIC and CFSFDP. For FDI1, for which MIC specializes in, the performance of our combined methods is not as good as that of MIC. However, our method achieves a rather high AUC of 0.766 in detecting FDI1. For FDI5 and FDI6, for which CFSFDP specializes in, our methods also have high values of AUC and MAP@20. The combined methods achieved improvements in the remaining types. The MIC-CFSFDP combined methods maintain the excellent performance of the original two methods in their own specialized situations while achieving significant improvements in the remaining situations, resulting in the best detection accuracy in type MIX and high and steady

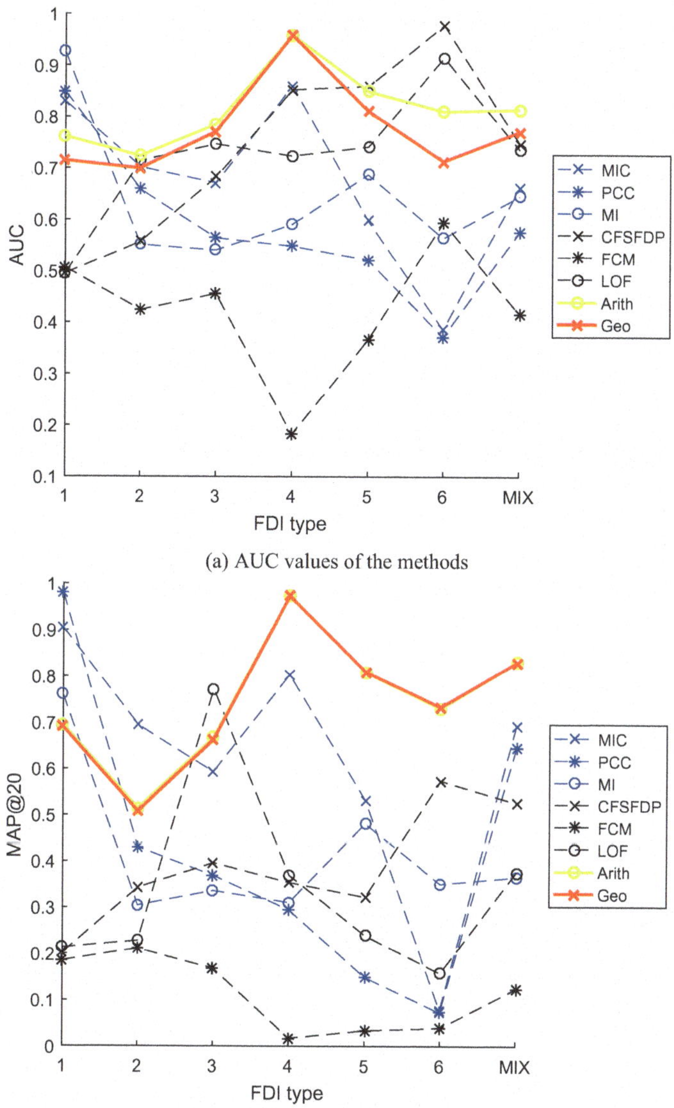

(a) AUC values of the methods

(b) MAP@20 values of the methods

Fig. 4.7 The evaluation results of the original and combined methods

detection accuracy for FDI1 to FDI6. The AUC value for type MIX increased from 0.748 to 0.816 (approximately 10%), and the MAP@20 value for type MIX increased from 0.693 to 0.831 (approximately 20%). The results for Arith and Geo are similar in most cases, and Arith performs slightly better in AUC. It is worthwhile to mention that weight factors in type MIX alter the detection accuracy. Although we assume

Table 4.2 Average evaluation results of the methods

Type	AUC(%)							
	Correlation			Unsupervised clustering			Combined	
	MIC	PCC	MI	CFSFDP	FCM	LOF	Arith	Geo
FDI1	83.1	84.9	**92.7**	49.5	50.6	49.5	76.6	71.5
FDI2	70.3	66.0	55.2	55.7	42.4	71.4	**72.5**	70.0
FDI3	67.2	56.5	54.1	68.3	45.5	74.7	**78.7**	77.2
FDI4	86.1	55.9	59.1	85.3	18.2	72.3	**96.0**	95.9
FDI5	59.9	52.2	68.8	**86.0**	36.6	74.1	85.1	81.2
FDI6	38.6	37.2	56.5	**97.9**	59.5	91.6	81.2	71.4
MIX	**66.2**	57.6	64.6	**74.8**	41.6	73.6	**81.6**	77.2

Table 4.3 Average evaluation results of the methods

Type	MAP@20(%)							
	Correlation			Unsupervised clustering			Combined	
	MIC	PCC	MI	CFSFDP	FCM	LOF	Arith	Geo
FDI1	90.6	**98.1**	76.2	20.2	18.5	21.4	69.6	69.1
FDI2	**69.5**	43.1	30.5	34.3	21.2	22.8	51.5	50.8
FDI3	59.4	36.9	33.7	39.6	16.8	77.2	**66.8**	66.3
FDI4	80.4	29.5	31.1	35.4	1.7	37.0	**97.5**	97.4
FDI5	53.3	15.0	48.2	32.3	3.4	23.9	**81.0**	**81.0**
FDI6	7.8	7.3	35.1	57.4	3.9	15.8	73.1	**73.4**
MIX	**69.3**	64.5	36.6	**52.6**	12.4	37.6	**83.1**	**83.1**

identical weights for the FDI types, the combined methods achieve improvements in accuracy for other non-extreme weight factors.

Figure 4.8 shows the standard deviations σ of AUC and MAP@20 in the 100 randomly generated scenes of type MIX for each method. σ of AUC is approximately 4% for all the methods, and Arith has a minimum σ_{AUC} of 3.08%. $\sigma_{\text{MAP@20}}$ is distributed between 9 and 17%. $\sigma_{\text{MAP@20}}$ of Arith and Geo are 9.16 and 9.13%, respectively, and are smaller than those of all the other methods. The combined methods improve both the accuracy and the stability of the original methods.

Figure 4.9 presents the average time consumption of the six methods for one detection of the whole 11 730 load profiles. For FCM, we only show the results of 4 and 12 cluster centers. The test was done on an Intel Core i7-7900X@4.30GHz desktop computer with 32GB RAM. Among these methods, Kraskov's estimator for MI has the most time-consuming. The combining process only requires simple calculation and sorting, and its time consumption is less than 1 s.

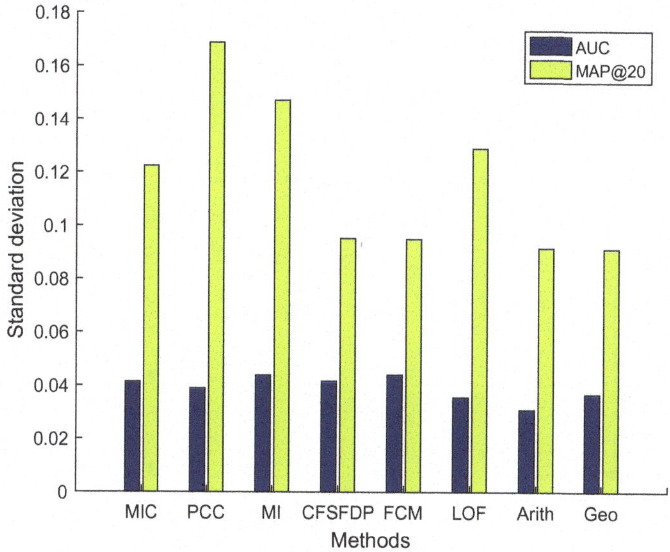

Fig. 4.8 Standard deviations of the evaluation results

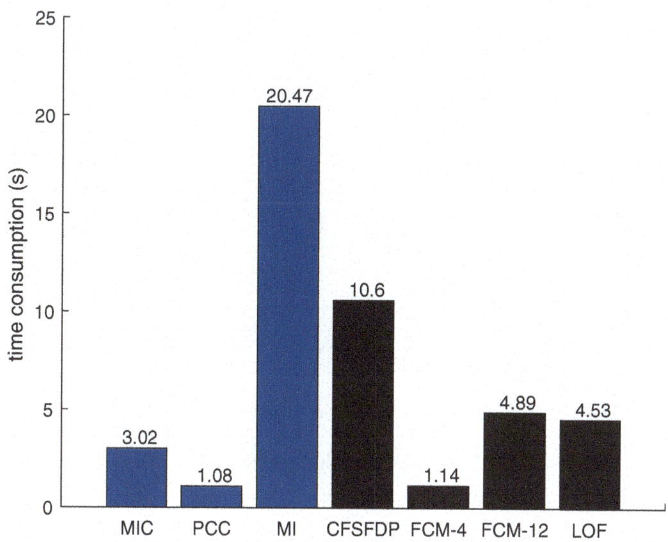

Fig. 4.9 Time consumption of the correlation and clustering-based methods

4.4.4 Sensitivity Analysis

When applying the electricity detection methods in real-world conditions, the number of electricity consumers or electricity thieves per area varies over a wide range,

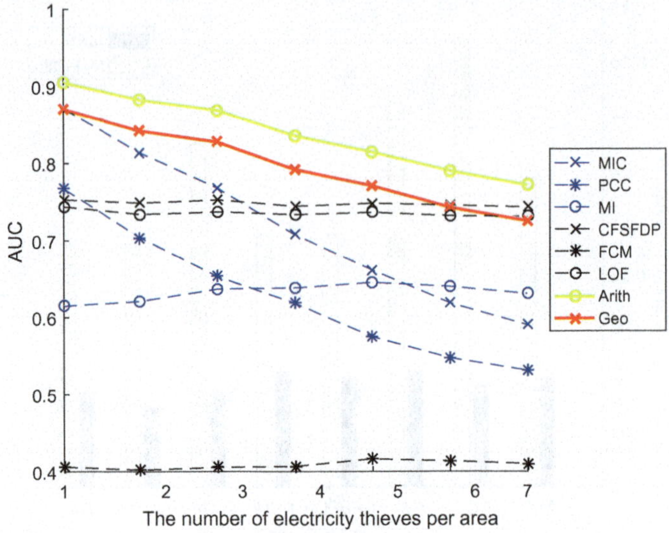

(a) AUC values of the methods

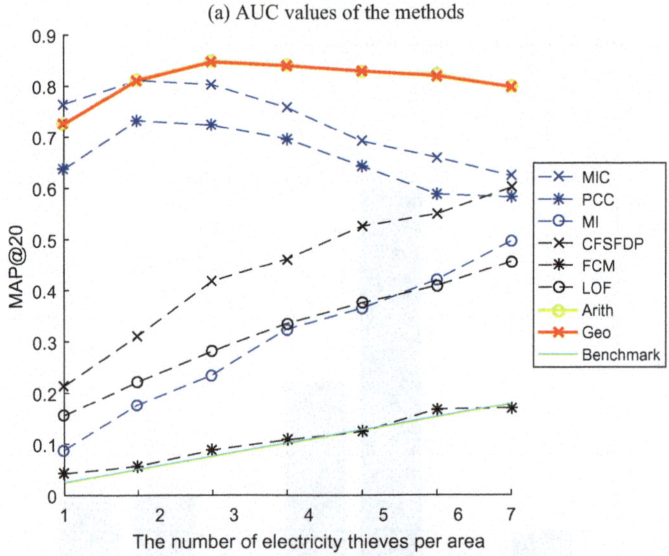

(b) MAP@20 values of the methods

Fig. 4.10 Performance of the methods with different numbers of electricity thieves per area

resulting in different detection accuracy and stability. In this subsection, we attempt to analyze the sensitivity in the two aspects. First, we hold the number of electricity consumers per area to 39 and change the number of electricity thieves per area from 1 to 7. Seven electricity thieves per area represent approximately 18% of all users; this is a very severe condition. Next, we hold the number of electricity thieves per

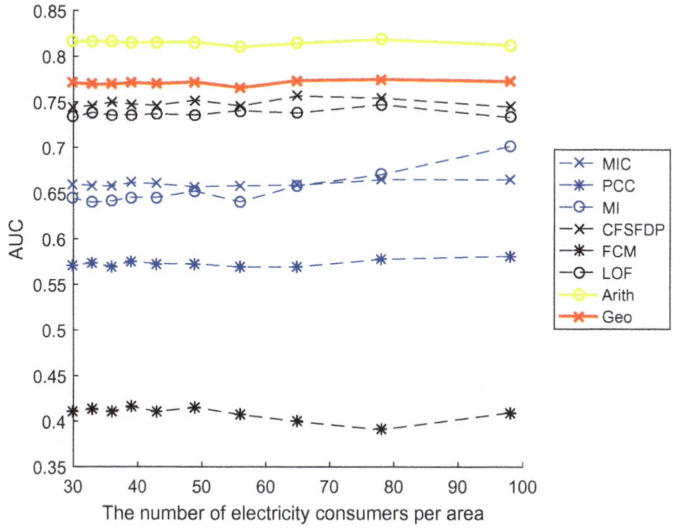

(a) AUC values of the methods

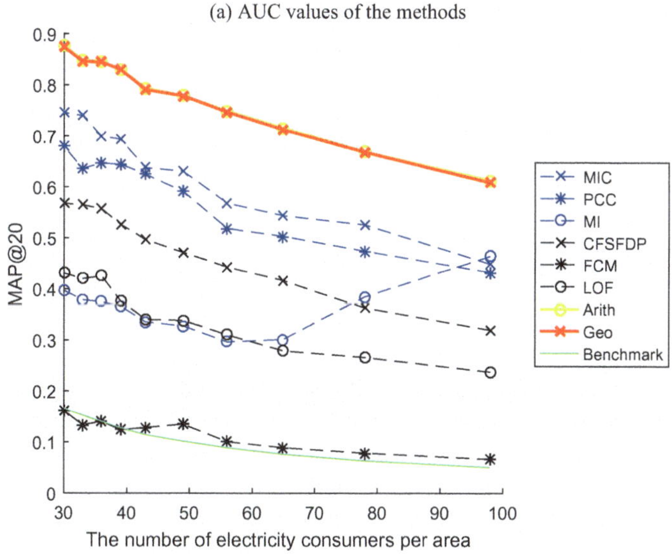

(b) MAP@20 values of the methods

Fig. 4.11 Performance of the methods with different numbers of electricity consumers per area

area to 5 and change the number of electricity consumers per area from 30 to 98 (which is achieved by dividing the 391 users into 4 to 13 areas). Figures 4.10 and 4.11 show the evaluation results for the two aspects of sensitivity analysis. Due to space limitations, we only present the results for type MIX.

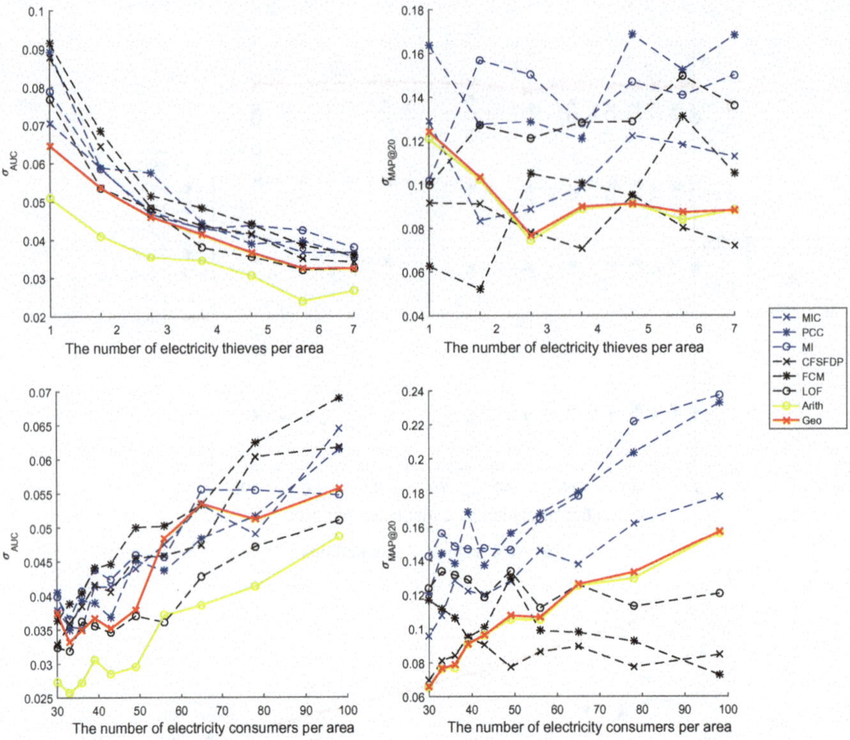

Fig. 4.12 Standard deviations of the methods with different numbers of electricity thieves and electricity consumers per area

As the number of electricity thieves per area changes, we can see from the AUC values that MIC and PCC perform well under the conditions of fewer electricity thieves and that MI is more robust in this aspect. However, MIC and PCC perform better in MAP@20 than MI. MIC can detect electricity thieves more precisely under these conditions. CFSFDP always performs the best of the three unsupervised clustering methods. The combined method of Arith maintains excellent performance for both AUC and MAP@20.

As the number of electricity consumers per area increases, most of the methods give a stable performance against the benchmark value. MIC is the best overall of the correlation-based methods, and CFSFDP is the best among the clustering-based methods. The combined methods achieve improvements against other methods in all conditions.

Figure 4.12 shows the change in standard deviations during the two aspects of sensitivity analysis. σ_{AUC} shows a certain trend as the number of electricity thieves or electricity consumers increases. As the electricity theft problem becomes more severe, σ_{AUC} decreases slightly, whereas $\sigma_{\text{MAP@20}}$ changes in a more disordered way. $\sigma_{\text{MAP@20}}$ of most methods have an upward trend as the number of electricity consumers per area increases. Although the combined methods do not always have the smallest standard deviation, the change of σ is over a rather small range, which is adequate for the methods in the practical application.

4.5 Conclusions

This chapter proposes a combined method for detecting electricity thefts against AMI in the Energy Internet. We first analyze the basic structure of the observer meters and the smart meters. Next, a correlation-based detection method using MIC is given to quantify the association between the tampered load profiles and the NTL. Considering the FDI types that have little association with the original data, an unsupervised CFSFDP-based method is proposed to detect outliers in the smart meter dataset. To improve the detection accuracy and stability, we ensemble the two techniques by combining the suspicion ranks. The numerical results show that the combined method achieves good and steady performance for all FDI types in various conditions.

References

1. Jiang, R., Lu, R., Wang, Y., Luo, J., Shen, C., & Shen, X. S. (2014). Energy-theft detection issues for advanced metering infrastructure in smart grid. *Tsinghua Science and Technology, 19*(2), 105–120.
2. Federal Bureau of Investigation. (2012). Cyber intelligence section: smart grid electric meters altered to steal electricity.
3. Fujian Daily. (2013). The first high-tech smart meter electricity theft case in China reported solved.
4. McDaniel, P., & McLaughlin, S. (2009). Security and privacy challenges in the smart grid. *IEEE Security & Privacy, 7*(3), 75–77.
5. Jokar, P., Arianpoo, N., & Leung, V. C. M. (2016). Electricity theft detection in AMI using customers' consumption patterns. *IEEE Transactions on Smart Grid, 7*(1), 216–226.
6. Nizar, A. H., Dong, Z., & Wang, Y. (2008). Power utility nontechnical loss analysis with extreme learning machine method. *IEEE Transactions on Power Systems, 23*(3), 946–955.
7. Zheng, Z., Yatao, Y., Niu, X., Dai, H.-N., & Zhou, Y. (2018). Wide & deep convolutional neural networks for electricity-theft detection to secure smart grids. *IEEE Transactions on Industrial Informatics, 14*(4), 1606–1615.
8. Ahmad, T., Chen, H., Wang, J., & Guo, Y. (2018). Review of various modeling techniques for the detection of electricity theft in smart grid environment. *Renewable and Sustainable Energy Reviews, 82*, 2916–2933.
9. Passos, L. A. Jr., Oba Ramos, C. C., Rodrigues, D., Pereira, D. R., de Souza, A. N., Pontara da Costa, K. A., & Papa, J. P. (2016). Unsupervised non-technical losses identification through optimum-path forest. *Electric Power Systems Research, 140*, 413–423.
10. Zanetti, M., Jamhour, E., Pellenz, M., Penna, M., Zambenedetti, V., & Chueiri, I. (2017). A tunable fraud detection system for advanced metering infrastructure using short-lived patterns. *IEEE Transactions on Smart Grid, 10*(1), 830–840.
11. Sun, M., Konstantelos, I., & Strbac, G. (2016). C-vine copula mixture model for clustering of residential electrical load pattern data. *IEEE Transactions on Power Systems, 32*(3), 2382–2393.

12. Aranha Neto, E. A. C., & Coelho, J. (2013). Probabilistic methodology for technical and non-technical losses estimation in distribution system. *Electric Power Systems Research*, *97*, 93–99.
13. Leite, J. B., & Mantovani, J. R. S. (2016). Detecting and locating non-technical losses in modern distribution networks. *IEEE Transactions on Smart Grid*, *9*(2), 1023–1032.
14. Cárdenas, A., Amin, S., Schwartz, G., Dong, R., & Sastry, S. (2012). A game theory model for electricity theft detection and privacy-aware control in AMI systems. In *50th Annual Allerton Conference on Communication, Control, and Computing (Allerton), 2012* (pp. 1830–1837). Monticello: IEEE.
15. Amin, S., Schwartz, G. A., Cardenas, A. A., & Sastry, S. S. (2015). Game-theoretic models of electricity theft detection in smart utility networks: Providing new capabilities with advanced metering infrastructure. *IEEE Control Systems*, *35*(1), 66–81.
16. Reshef, D. N., Reshef, Y. A., Finucane, H. K., Grossman, S. R., McVean, G., Turnbaugh, P. J., Lander, E. S., Mitzenmacher, M., & Sabeti, P. C. (2011). Detecting novel associations in large data sets. *Science*, *334*(6062), 1518–1524.
17. Han, W., & Xiao, Y. (2016). Combating TNTL: Non-technical loss fraud targeting time-based pricing in smart grid. In *International Conference on Cloud Computing and Security* (pp. 48–57). Berlin: Springer.
18. Yijia, T., & Hang, G. (2016). Anomaly detection of power consumption based on waveform feature recognition. In *2016 11th International Conference on Computer Science & Education (ICCSE)* (pp. 587–591). Nagoya: IEEE.
19. Rodriguez, A., & Laio, A. (2014). Clustering by fast search and find of density peaks. *Science*, *344*(6191), 1492–1496.
20. Breitling, R., Armengaud, P., Amtmann, A., & Herzyk, P. (2004). Rank products: a simple, yet powerful, new method to detect differentially regulated genes in replicated microarray experiments. *FEBS Letters*, *573*(1–3), 83–92.
21. Commission for Energy Regulation (CER). (2012). CER Smart Metering Project - electricity customer behaviour trial, 2009–2010. Irish Social Science Data Archive. SN: 0012-00.
22. Kraskov, A., Stögbauer, H., & Grassberger, P. (2004). Estimating mutual information. *Physical Review E*, *69*(6), 066138.
23. Breunig, M. M., Kriegel, H.-P., Ng, R. T., & Sander, J. (2000). LOF: identifying density-based local outliers. In *ACM sigmod record* (Vol. 29, pp. 93–104). New York: ACM.

Chapter 5
Residential Load Data Generation

Abstract Due to the technical limitations of metering and privacy concerns of customers, the large-scale and real-time collection of residential load data still remains a big challenge. To address the problem, we use the generative adversarial networks (GANs) to produce synthetic residential loads as an alternative. Different from existing load generation models, the GAN model is based on deep neural networks (DNNs). It includes a generator network that outputs synthetic loads and a discriminator network that differentiates the real or fake loads. Taking advantage of the learning ability of DNNs, we can capture hidden features of the load pattern and describe them accurately. In this chapter, we conduct an investigation of frequently-used GAN variants accounting for their performance at generating residential load. We design the architectures and training methods for different GANs and propose different metrics to evaluate the model performance comprehensively. Case studies demonstrate that the auxiliary classifier GAN (ACGAN) outperforms other models on the real load data from an Irish smart meter trial. It is practical to use the ACGAN to generate synthetic residential loads when in shortage of real data.

5.1 Introduction

The residential load data play an important role in various research and application fields. However, the large-scale and real-time collection of residential load data still remains a big challenge. First, collecting a large volume of load data is still costly due to the technical barriers of data storage and communication. Second, processing and analyzing real load data might bring potential legal risks due to the rising privacy concern of customers and the promulgation of relevant laws in recent years [1]. Generally speaking, although smart meters have been widely deployed in many areas around the world, the recorded load data still suffer from barriers to being completely utilized at present. To solve the problem of shortage of available residential load data, researches present to generate synthetic loads as an alternative [2].

Existing load generation methods can be classified into 2 categories: bottom-up and top-down. Bottom-up methods decompose the total electricity consumption in the household into loads of individual appliances. This kind of approaches mainly

Y. Wang et al., *Smart Meter Data Analytics*,
https://doi.org/10.1007/978-981-15-2624-4_5

includes two steps. First, construct the electrical model and usage model for different appliances. Next, generate load profiles of live appliances and sum them up to form the total profile. Capassp et al. present a model of residential end-users to establish the load diagram of an area in [3]. The total consumption is constructed from the relevant socioeconomic and demographic characteristics, unitary energy consumption and load profiles of individual household appliances. McKenna et al. propose a model capturing closed-loop load behavior with bottom-up load modeling techniques for the residential sector in [4]. It incorporates time-variant load models and discrete state-space representation of loads of thermal appliances. Tsagarakis et al. convert user activity profiles into load profiles in [5]. The user activity profiles, including time series of daily resident activities, are based on an appliance ownership statistics and electrical characteristics dataset. Dickert et al. present a time-series-based probabilistic load curve model for residential customers in [6]. The total loads are constructed by investigating each possible appliance, respective power consumption, frequency of use, turn-on time, operating time as well as the potential correlation between appliances. Collin et al. propose a Markov chain Monte Carlo method to generate load profiles based on the electrical characteristics of appliances [7]. Stephen et al. propose a Markov chain-based generating method derived from practice theory of human behavior [8]. To conclude, bottom-up approaches model the residential load based on the details of end-use appliances, thus are interpretable and precise. However, these approaches are usually of high computational effort and additional historical and statistics data requirements.

Top-down methods consider the residential load as a whole and fit the relationship between the load and relevant influence factors. Labeeuw et al. propose a top-down model based on a dataset of over 1300 load profiles in [9]. The load profiles are first clustered by a mixed model. Then two Markov models are used to construct the user behavior and randomness of the behavior respectively. Xu et al. compare the bottom-up model with an agent-based approach and top-down model based on neural networks in [10]. Uhrig et al. use the generalized extreme value distribution to describe the distribution of residential loads in [11]. By introducing corresponding transition matrices, the synthetic load profiles are generated directly with Markov chains. Gu et al. propose to use the Generative Adversarial Network (GAN) to generate residential load under typical user behavior groups in [12]. The GAN model can generate synthetic load profiles from random noise upon finishing training the neural networks. To conclude, top-down models are mainly data-driven and of low model complexity. They are suitable for scenarios where low computational effort is preferred to the consumption details of end-use appliances.

Different from the industrial or commercial load, the residential load has strong randomness, volatility, and is difficult to predict. Thus, conventional methods have difficulty in balancing the model complexity and the fidelity of synthetic loads. Numerical experiments in [12] indicate that the GAN model is suitable for load generation. First, GANs are of low complexity due to its general architecture and standard training process. Second, GANs are of low computational cost. Once upon finishing the training, they can be used to generate synthetic loads quickly. Third, generated loads follow a similar distribution to that of the real loads without losing diversity. Due to these fine properties, GANs have become a new research hotspot of generative models in recent years. Various variants have been derived based on the original GAN. In this chapter, we test several classic and popular GAN variants for residential load synthesis. A comprehensive investigation is conducted to find the most proper GAN variant. Different metrics are applied to evaluate model performance.

5.2 Model

5.2.1 Basic Framework

Basic GANs contain two parts: one generator and one discriminator [13]. The generator converts random noise vectors to synthetic samples that subject to the same distribution as the real samples. The discriminator tells whether the input sample is real or synthetic. Both the generator and discriminator are composed of neural networks. Their structures should be designed according to the specific task, e.g. convolution and transposed convolution layers are commonly used in image generation tasks, fully connected layers are widely used in vectorized data generation tasks. The basic structure sketch of the GAN is shown in Fig. 5.1.

Since GAN actually is constructed by neural networks, the training process is similar to traditional networks. Both the generator and the discriminator search optimal parameters by stochastic gradient descent of their loss functions. The main difference is that we need to deal with the training of two networks synchronously. In practice, two networks get trained alternately to ensure that their abilities are balanced. During the training process, the discriminator gets better at distinguishing from real and synthetic samples, while the generator gets better at generating samples that could fool the discriminator. When the game between the generator and discriminator reaches a Nash equilibrium, no one gets better through the training. The discriminator could not tell the difference between the real samples and generated samples, and then we could use the generator to produce synthetic samples.

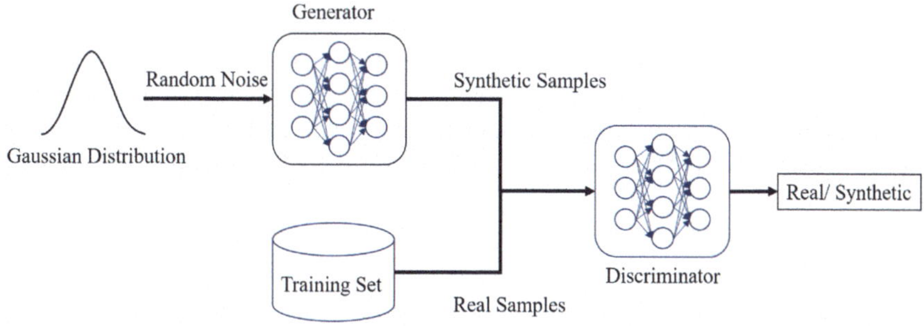

Fig. 5.1 The basic structure sketch of the GAN

Denote the trainable parameters in the generator and the discriminator as θ_G and θ_D, the mapping function of the generator and the discriminator as $G(\cdot)$ and $D(\cdot)$, the random noise vector subjected to a given distribution as $z \sim p(z)$, the synthetic and real samples as $\boldsymbol{x}_s$ and $\boldsymbol{x}_r$, the distribution of real samples as $p(\boldsymbol{x}_r)$. From the generator we have

$$\boldsymbol{x}_s = G(z; \theta_G) \tag{5.1}$$

Now that the discriminator outputs the probability that the input sample is real, the loss function of the discriminator l_D could be defined as

$$l_D = -\mathbb{E}_{z\sim p(z)} \log(1 - D(\boldsymbol{x}_s; \theta_D)) - \mathbb{E}_{\boldsymbol{x}_r\sim p(\boldsymbol{x}_r)} \log D(\boldsymbol{x}_r; \theta_D) \tag{5.2}$$

The l_D decreases during the training, indicating that the expectation of discriminator outputs of real samples tending to 1 while that of synthetic samples tending to 0. The generator aims at confusing the discriminator, it outputs synthetic samples that make the discriminator presents wrong judgments. Thus the loss function of the gererator l_G could be defined as

$$l_G = -\mathbb{E}_{z\sim p(z)} \log D(\boldsymbol{x}_s; \theta_D) \tag{5.3}$$

The l_G decreases during the training, indicating that the expectation of discriminator outputs of synthetic samples tending to 1. The θ_G and θ_D are updated by gradient descent of l_D and l_G until the two loss functions converge.

5.2.2 *General Network Architecture*

Theoretically, the architecture of the generator and discriminator could be arbitrary. However, experiments show that the training process for a randomly designed GAN is unstable. Thus, the model performance depends heavily on network architectures of the generator and discriminator. Radford proposed a well designed GAN architecture called Deep Convolutional GAN (DCGAN) in [14]. According to his suggestions and our experiments, the following guidelines are adopted in this chapter when designing the network architecture for a GAN.

- Use strided convolutional layers to down-sample in the discriminator and fractional-strided convolutional layers to up-sample in the generator.
- Use batch norm before the up-sample operation in the generator and after the down-sample operation in the discriminator except for the input.
- Use Relu activation in the generator for all layers except for the output, which uses the Tanh.
- Use leaky Relu activation in the discriminator for all layers except for the output, which uses the Sigmoid.
- Use dropout in the discriminator.

To design GANs for residential load generation, we need to take the load characteristics into consideration. First, smart meters in residents usually record the loads every 15 or 30 min. Gather recorded load points in a day, and we could form a daily load curve, which could be viewed as a 1D vector mathematically. Second, the household load is closely related to the living and working habits of family members, which often take a week as a cycle. Third, neighboring loads in a daily load curve have close relationships. Fourth, loads at similar time intervals on different weekdays have close relationships. Based on the first and second points mentioned above, we arrange daily load curves for a complete week as rows to form a 2D load matrix. The load matrix can be viewed as a one-channel image with every load data point as a pixel. According to the third and fourth points above, a load matrix is similar to an image since their pixels are both relevant to the neighboring pixels. By transforming load curves into load matrics, we could generate loads for a complete week synchronously without missing the relevance of loads on different weekdays. Also, we can use convolutional layers in the generating and discriminating networks to discover the deep features behind the load pattern.

Suppose the residential loads are sampled every 30 min, then we have 336 points in a week. Arrange them to form a one-channel 2D image with a size of $1 \times 7 \times 48$. Then the output size of the generator and input size of the discriminator are ascertained. According to the guidelines above, designing of the network architecture is given below.

5.2.2.1 Generator

Denote the length of the input noise vector as N_z, the height and width of the output image as N_h and N_w. First, we use a fully-connected layer to map the $1 \times N_z$ input to a higher dimensional space. The output size of the first layer is $1 \times N_{fc1}$. Here N_{fc1} is given by

$$N_{fc1} = 128 \times \lceil N_h/4 \rceil \times \lceil N_w/4 \rceil \tag{5.4}$$

Reshape the $1 \times N_{fc1}$ vector into a 128-channels image with the size of $\lceil N_h/4 \rceil \times \lceil N_w/4 \rceil$. The second layer is a fractional-strided convolutional layer to up-sample the first layer output. Number of input channels of this layer is set as 128, number of output channels is set as 128 too, size of the convolving kernel is set as 3×3, stride of the convolution is set as 2, number of zero-padding that will be added to both sides of each dimension in the input and output image are both set as 1. The output size of this layer can be derived by

$$H_{out} = (H_{in} - 1) \times stride - 2 \times input\_padding + (kernel[1] - 1) + output\_padding + 1 \tag{5.5}$$

$$W_{out} = (W_{in} - 1) \times stride - 2 \times input\_padding + (kernel[2] - 1) + output\_padding + 1 \tag{5.6}$$

Thus the output size of this layer is a 128-channels image with the size of $\lceil N_h/2 \rceil \times \lceil N_w/2 \rceil$. In the third layer, we apply a batch norm layer to normalize the input. The mean and standard-deviation are calculated per-dimension over the batches. The size of this layer's output remains unchanged. In the fourth layer we use the ReLU as the activation function. It is an element-wise function goes as

$$ReLU(x) = \max(0, x) \tag{5.7}$$

From the fifth to the seventh layer, we place a fractional-strided convolutional layer, a batch norm layer and an activation layer in turn. The parameters of these layers are the same as the former except that the number of output channels of the fractional-strided convolutional layer is set as 64. Then the output of the seventh layer is a 64-channels image with the size of $(N_h + 1) \times N_w$ (when N_h is odd, e.g. 7). In the eighth layer, we use a convolutional layer to regularize the output channel and size. The number of input channels is set as 64, the number of output channels is set as 1, size of the convolving kernel is set as 4×3, stride of the convolution is set as 1, number of zero-padding that will be added to both sides of each dimension in the input is set as 1. The output size of this layer can be derived by

$$H_{out} = \lfloor \frac{H_{in} + 2 \times padding - (kernel[1] - 1) - 1}{stride} + 1 \rfloor \tag{5.8}$$

$$W_{out} = \lfloor \frac{W_{in} + 2 \times padding - (kernel[2] - 1) - 1}{stride} + 1 \rfloor \tag{5.9}$$

Thus the output of this layer is a 1-channel image with the size of $N_h \times N_w$. Finally, we use the Tanh activation as the last layer to regularize the output value. Layers in the generator and their parameters are listed in Table 5.1 (denote the number of samples per batch as b).

5.2.2.2 Discriminator

The input of the discriminator is real or synthetic load images. We use convolutional layers to down-sample the original image and map it to the real/fake binary space. The network architecture of the discriminator is approximately symmetrical to the generator. The convolutional layers in the discriminator have same parameters except for the number of input and output channels. The size of the convolving kernel is set as 3×3, the stride of the convolution is set as 2, and the number of zero-padding that will be added to both sides of each dimension in the input is set as 1. In the first layer we place a convolutional layer. The number of input channels is 1; the number of output channels is set as 16. Then the output of this layer is a 16-channel image with the size of 4×24 according to 5.8 and 5.9. In the second layer we use the LeakyReLU with $negative\_slope$ 0.2 as the activation function. It is an element-wise function goes as

Table 5.1 General network architecture of the generator

No.	Layer	Parameters	Input size	Output size
1	Fully-connected	–	$b \times N_z$	$b \times 3072$
2	Fractional-strided convolutional	Kernel size: 3×3; stride: 2; padding: 1	$b \times 128 \times 2 \times 12$	$b \times 128 \times 4 \times 24$
3	Batchnorm	–	$b \times 128 \times 4 \times 24$	$b \times 128 \times 4 \times 24$
4	Activation function	ReLU	$b \times 128 \times 4 \times 24$	$b \times 128 \times 4 \times 24$
5	Fractional-strided convolutional	Kernel size: 3×3; stride: 2; padding: 1	$b \times 128 \times 4 \times 24$	$b \times 64 \times 8 \times 48$
6	Batchnorm	–	$b \times 64 \times 8 \times 48$	$b \times 64 \times 8 \times 48$
7	Activation function	ReLU	$b \times 64 \times 8 \times 48$	$b \times 64 \times 8 \times 48$
8	Convolutional	Kernel size: 4×3; stride: 1; padding: 1	$b \times 64 \times 8 \times 48$	$b \times 1 \times 7 \times 48$
9	Activation function	Tanh	$b \times 1 \times 7 \times 48$	$b \times 1 \times 7 \times 48$

$$LeakyReLU(x) = \max(0, x) + negative\_slope \times \min(0, x) \qquad (5.10)$$

In the third layer, we use a dropout2D layer that randomly zero out entire channels of the input with the probability of 0.25. The main difference between dropout2D and normal dropout is that the first abandon channels of the input while the second abandons pixels of the input. As described in [15], if adjacent pixels within feature maps are strongly correlated (as is often the case in early convolution layers) then normal dropout will not regularize the activations and will otherwise just result in an effective learning rate decrease. Under this circumstance, dropout2D will help promote independence between feature maps and should be used instead. From the forth to the seventh layer, we use a convolutional layer with the number of input channels as 16, the number of output channels as 32, a leaky-ReLU activation function layer, a dropout2D layer and a batch norm layer in turn. The output of the seventh layer is a 32-channels image with a size of 2×12. From the eighth to the eleventh layer, we use a convolutional layer with the number of input channels as 32, the number of output channels as 64, a leaky-ReLU activation function layer, a dropout2D layer and a batch norm layer in turn. The output of the eleventh layer is a 64-channels image with a size of 1×6. Reshape the output into a 1D vector with a size of 1×384. Use a fully-connected layer to map the vector into the binary space of real/fake in the twelfth layer. Finally use the Sigmoid activation to regularize the output value within [0, 1]. Layers in the discriminator and their parameters are listed in Table 5.2.

The network architecture presented in this part is general designing. When it comes to the variants of the GAN, the network architecture needs fine-tuning.

Table 5.2 General network architecture of the discriminator

No.	Layer	Parameters	Input size	Output size
1	Convolutional	Kernel size: 3 × 3; stride: 2; padding: 1	$b \times 1 \times 7 \times 48$	$b \times 16 \times 4 \times 24$
2	Activation function	LeakyReLU	$b \times 16 \times 4 \times 24$	$b \times 16 \times 4 \times 24$
3	Dropout2D	–	$b \times 16 \times 4 \times 24$	$b \times 16 \times 4 \times 24$
4	Convolutional	Kernel size: 3 × 3; stride: 2; padding: 1	$b \times 16 \times 4 \times 24$	$b \times 32 \times 2 \times 12$
5	Activation function	LeakyReLU	$b \times 32 \times 2 \times 12$	$b \times 32 \times 2 \times 12$
6	Dropout2D	–	$b \times 32 \times 2 \times 12$	$b \times 32 \times 2 \times 12$
7	Batchnorm	–	$b \times 32 \times 2 \times 12$	$b \times 32 \times 2 \times 12$
8	Convolutional	Kernel size: 3 × 3; stride: 2; padding: 1	$b \times 32 \times 2 \times 12$	$b \times 64 \times 1 \times 6$
9	Activation function	LeakyReLU	$b \times 64 \times 1 \times 6$	$b \times 64 \times 1 \times 6$
10	Dropout2D	–	$b \times 64 \times 1 \times 6$	$b \times 64 \times 1 \times 6$
11	Batchnorm	–	$b \times 64 \times 1 \times 6$	$b \times 64 \times 1 \times 6$
12	Fully-connected	–	$b \times 384$	$b \times 1$
13	Activation function	Sigmoid	$b \times 1$	$b \times 1$

5.2.3 *Unclassified Generative Models*

In this part, we introduce unclassified GAN variants for residential load generation. The structure and network architecture are inherited from Tables 5.1 and 5.2. Loss functions for the generator and discriminator are also defined in this part.

5.2.3.1 Boundary Equilibrium GAN

In order to overcome the instability and poor convergence when training the GAN, boundary equilibrium GAN (BEGAN) modifies the output of the discriminator and loss functions. The discriminator reconstructs the input instead of classifying it. The network architecture of the discriminator is presented in Table 5.3. First, the discriminator uses convolutional and fully-connected layers to down-sample the input to the feature space. Then we use fully-connected and fractional-strided convolutional layers to up-sample the features to the original space. The network architecture of the generator keeps the same as Table 5.1.

To define the loss function for the BEGAN, we first introduce l_1 distance here. For 2D images $\boldsymbol{x}_1$ and $\boldsymbol{x}_2$ with the height and width of N_h and N_w pixels, their l_1 distance can be expressed as

$$l_1(\boldsymbol{x}_1, \boldsymbol{x}_2) = \frac{\sum_{i=1}^{N_w} \sum_{j=1}^{N_h} |\boldsymbol{x}_1(i, j) - \boldsymbol{x}_2(i, j)|}{N_w * N_h} \tag{5.11}$$

Table 5.3 Network architecture of the BEGAN discriminator

No.	Layer	Parameters	Input size	Output size
1	Convolutional	Kernel size: 3 × 3; stride: 2; padding: 1	$b \times 1 \times 7 \times 48$	$b \times 64 \times 4 \times 24$
2	Activation function	LeakyReLU	$b \times 64 \times 4 \times 24$	$b \times 64 \times 4 \times 24$
3	Fully-connected	–	$b \times 6144$	$b \times 32$
4	Batchnorm	–	$b \times 32$	$b \times 32$
5	Activation function	ReLU	$b \times 32$	$b \times 32$
6	Fully-connected	–	$b \times 32$	$b \times 6144$
7	Batchnorm	–	$b \times 6144$	$b \times 6144$
8	Activation function	ReLU	$b \times 6144$	$b \times 6144$
9	Fractional-strided convolutional	Kernel size: 4 × 3; stride: 2; padding: 1	$b \times 64 \times 4 \times 24$	$b \times 1 \times 7 \times 48$

Denote the noise vectors and real load images sampled in the tth training step as z^t and $\boldsymbol{x}_r^t$. The reconstruction error of real samples can be expressed as

$$\mathscr{L}(\boldsymbol{x}_r^t) = l_1(\boldsymbol{x}_r^t, D(\boldsymbol{x}_r^t)) \tag{5.12}$$

The reconstruction error of synthetic samples can be expressed as

$$\mathscr{L}(G(z^t)) = l_1(G(z^t), D(G(z^t))) \tag{5.13}$$

Then the loss function of the discriminator and generator can be defined as

$$l_D = \mathscr{L}(\boldsymbol{x}_r^t) - k_t \mathscr{L}(G(z^t)) \tag{5.14}$$

$$l_G = \mathscr{L}(G(z^t)) \tag{5.15}$$

The weight k_t is updated by

$$k_{t+1} = k_t + \lambda(\gamma \mathscr{L}(\boldsymbol{x}_r^t) - \mathscr{L}(G(z^t))) \tag{5.16}$$

In the formula above, λ is the update step of k, $\gamma \in [0, 1]$ is the parameter that determines the diversity of synthetic samples. The larger γ is, the more diversity synthetic samples have. According to [16], we set $k_0 = 0$, $\lambda = 0.001$ and $\gamma = 0.9$. During the iteration, if k surpasses the bound [0, 1], it would be clipped.

5.2.3.2 Boundary-Seeking GAN

The boundary-seeking GAN (BGAN) retains network architectures in Tables 5.1 and 5.2. It modifies the loss function from vanilla GAN so that the generator could produce samples on the decision boundary of the current discriminator. As proposed in [17], the optimal generator is the one that can make the discriminator be 0.5 everywhere. In order to make the discriminative results of synthetic samples $D(G(z))$ near the decision boundary, the BGAN tries to minimize the distance between $D(G(z))$ and $1 - D(G(z))$. Thus, the loss function of the generator is

$$l_G = -\mathbb{E}_{z\sim p(z)}\left[\frac{1}{2}(\log D(G(z)) - \log(1 - D(G(z))))^2\right] \tag{5.17}$$

The loss function of the discriminator remains unchanged and is given in Eq. 5.2.

5.2.3.3 Wasserstein GAN

Besides the instability during the training, the vanilla GAN has some other problems, e.g. be easy to encounter the mode collapse, the loss function cannot indicate the training process, etc. The Wasserstein GAN (WGAN) solves these problems by redesigning the network architecture and the loss function. Mathematically, optimizing the vanilla GAN is equivalent to minimizing the Jensen–Shannon divergence between the distribution of real samples and generated samples [18]. However, if the two distributions do not overlap or overlap parts are negligible in high-dimensional space, their JS divergence is constant. Under such kind of circumstance, the JS divergence can neither reflect the distance nor provide meaningful gradients for training the networks. WGANs use the Wasserstein distance to measure the similarity between the real and synthetic distribution. The advantage of Wasserstein distance over JS divergence is that even if two distributions do not overlap, Wasserstein distance can still reflect their distance. Briefly, the JS divergence is discontinuous while the Wasserstein distance is smooth. When we use the gradient descent method to optimize the trainable parameters in neural networks, the JS divergence can not provide gradient at all while the Wasserstein distance can.

To implement the Wasserstein distance in practice, [18] suggested the following modifications.

- Remove the Sigmoid activation in the last layer of the discriminator.
- Remove the log in the loss functions Eqs. (5.2) and (5.3):

$$l_D = \mathbb{E}_{z\sim p(z)} D(\boldsymbol{x}_s) - \mathbb{E}_{\boldsymbol{x}_r\sim p(\boldsymbol{x}_r)} D(\boldsymbol{x}_r) \tag{5.18}$$

$$l_G = -\mathbb{E}_{z\sim p(z)} D(\boldsymbol{x}_s) \tag{5.19}$$

- Every iteration the discriminator parameters updated, clip their values into a fixed range, e.g. $[-0.01, 0.01]$.
- Use the RMSProp optimizer instead of the Adam optimizer.

However, WGAN is also hard to train in practice. In the vanilla WGAN, trainable parameters of the discriminator are clipped into a given range to satisfy Lipschitz condition. This brings two main problems. First, the parameters will be concentrated on the boundary. In other words, parameters are either maximized or minimized. As a result, the network tends to learn a simple mapping function. Second, weight clipping may cause gradient vanishing or explosion. If we set the clipping threshold a little bit smaller, the gradient would decrease exponentially after several layers, thus leads to gradient vanishing. On the contrary, if we set it a little bit larger, the gradient would increase exponentially after several layers, thus leads to gradient explosion. Author of the WGAN proposes corresponding improvement methods in [19]. The solution is that we do not need to impose Lipschitz restriction on the whole space. Instead, we only need to impose Lipschitz restriction on where the generated and real samples gather and the area between them. In practice, we add a penalty term to the loss function of the discriminator. Denote the synthetic sample as $\boldsymbol{x}_s$, the real sample as $\boldsymbol{x}_r$, a random variable ε drawn from $\mathcal{N}(0, 1)$. First, we randomly interpolate on the segment between $\boldsymbol{x}_s$ and $\boldsymbol{x}_r$

$$\hat{\boldsymbol{x}} = \varepsilon \boldsymbol{x}_r + (1 - \varepsilon)\boldsymbol{x}_s \tag{5.20}$$

Denote the distribution of $\hat{\boldsymbol{x}}$ as $p_{\hat{\boldsymbol{x}}}$, then we define the penalty term as

$$l_P = \mathbb{E}_{\hat{\boldsymbol{x}} \sim p_{\hat{\boldsymbol{x}}}}[\left\| \nabla_{\hat{\boldsymbol{x}}} D(\hat{\boldsymbol{x}}) \right\| - 1]^2 \tag{5.21}$$

The loss function of the discriminator can be expressed as

$$l_D = \mathbb{E}_{z \sim p(z)} D(\boldsymbol{x}_s) - \mathbb{E}_{\boldsymbol{x}_r \sim p(\boldsymbol{x}_r)} D(\boldsymbol{x}_r) + \lambda l_P \tag{5.22}$$

In the formula above, λ is the weight of the penalty. We set $\lambda = 10$ in this chapter since [19] found that it works well across a variety of architectures and datasets.

The network architecture of the discriminator and generator are shown in Tables 5.1 and 5.2 except the batch norm layers in the discriminator. As suggested in [19], all the batch norm layers in the discriminator are omitted since we penalize the norm of the discriminator's gradient with respect to each input independently, not the entire batch.

5.2.4 *Classified Generative Models*

Household electricity consumption is significantly affected by the living habits of family members. Since daily life usually exhibits typical categories, the residential load curves can be classified into several groups as well [20]. In this part, we will introduce models that can generate load curves of different categories.

5.2.4.1 Conditional GAN

The conditional GAN (CGAN) modifies the model input. Load curve labels are added to both the generator and the discriminator [21]. Denote the number of categories as K, the label as $\boldsymbol{y}$. First, we implement One-Hot encoding to process the label. After One-Hot encoding, $\boldsymbol{y}$ is converted into a $1 \times K$ vector filled with 0 except that the kth entry is 1, which represents the sample belongs to the kth category.

Since the generator has an additional input vector, we modify its architecture in Table 5.1. The new architecture is shown in Fig. 5.2. We replace the first layer with two parallel fully-connected layers, each of them outputs a 1D feature vector with a size of 1×1536. Then we concatenate two vectors to form a 1×3072 vector. Subsequent layers remain unchanged.

Similar to the modification of the generator, we add labels to the discriminator input. Use the One-Hot encoding to convert the label $\boldsymbol{y}$ into a K-channels image with the size of 7×48. All channels are filled with 0 except that the kth channel is 1. The new architecture is shown in Fig. 5.3. We replace the first layer in Table 5.2 with two convolutional layers, one with 1-channel input and 8-channel output while the other with K-channel input and 8-channel output. The stride step and kernel size are the same as other convolutional layers in Table 5.2. Concatenate the two outputs to form a 16-channels image. Subsequent layers remain unchanged.

Loss functions of the CGAN remain unchanged as defined in Eqs. 5.2 and 5.3.

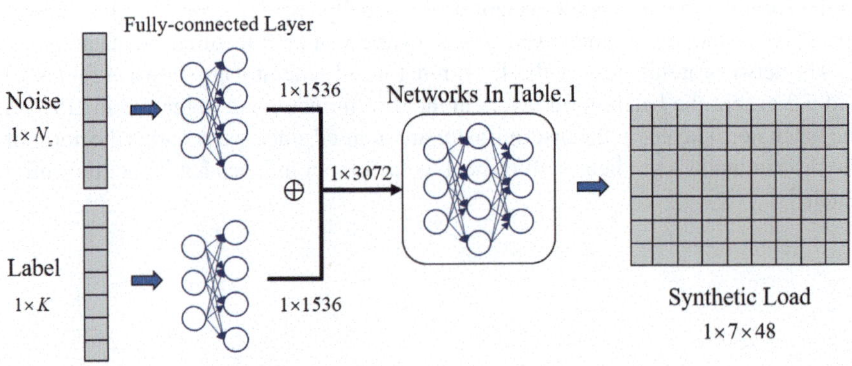

Fig. 5.2 Network architecture of the CGAN generator

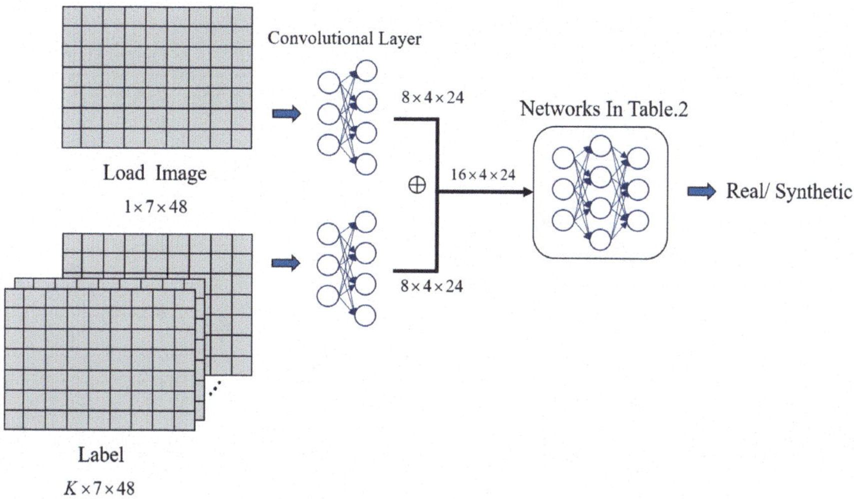

Fig. 5.3 Network architecture of the CGAN discriminator

5.2.4.2 InfoGAN

The InfoGAN is an information-theoretic extension to the original GAN [22]. By splitting the random noise input of the generator into the random part and the latent informatics part, InfoGAN can learn interpretable representation of the training data. Denote the $1 \times N_l$ latent informatics vector as $\boldsymbol{c}$, the $1 \times N_z$ noise vector as $\boldsymbol{z}$, the $1 \times K$ label vector as $\boldsymbol{y}$. The generator function can be expressed as $G(\boldsymbol{z}, \boldsymbol{c}, \boldsymbol{y})$. The discriminator outputs real or synthetic, the predicted label and inferred latent information of the input, denoted as $D_{disc}(\boldsymbol{x})$, $D_{cate}(\boldsymbol{x})$ and $D_{con}(\boldsymbol{x})$ respectively.

Most of the generator architecture in Table 5.1 is retained. The only difference is the number of neurons in the input of the first fully-connected layer, which is increased to $N_z + N_l + K$ from N_z. The discriminator retains the first eleven layers in Table 5.2. We use three parallel fully-connected layers and corresponding activation functions in the tail of discriminator to output $D_{disc}(\boldsymbol{x})$, $D_{cate}(\boldsymbol{x})$ and $D_{con}(\boldsymbol{x})$. The parameters of additional layers are shown in Table 5.4.

Except for the loss functions defined in Eqs. 5.2 and 5.3, the InfoGAN has an additional loss which is called the information loss. It is defined as the weighted sum of the classification error and reconstruction error. Denote the synthetic sample as $\boldsymbol{x}_s$, the classification error is the cross entropy between the true label $\boldsymbol{y}$ and predicted label $\hat{\boldsymbol{y}} = D_{cate}(\boldsymbol{x}_s)$. Suppose the kth element of $\boldsymbol{y}$ is 1, then we have

$$cross\_entropy(\hat{\boldsymbol{y}}, k) = -\log \frac{\exp(\hat{\boldsymbol{y}}_k)}{\sum_{i=1}^{K} \exp(\hat{\boldsymbol{y}}_i)} \tag{5.23}$$

Table 5.4 Network architecture of the InfoGAN discriminator

No.	Layer	Parameters	Input size	Output size
1–11	Table 5.2	–	$b \times 1 \times 7 \times 48$	$b \times 64 \times 1 \times 6$
12(1)	Fully-connected	–	$b \times 384$	$b \times 1$
13(1)	Activation function	Sigmoid	$b \times 1$	$b \times 1$
12(2)	Fully-connected	–	$b \times 384$	$b \times K$
13(2)	Activation function	Softmax	$b \times K$	$b \times K$
12(3)	Fully-connected	–	$b \times 384$	$b \times N_l$

The reconstruction error is the mean squared error between each element in the recovered latent information $\hat{\boldsymbol{c}} = D_{con}(\boldsymbol{x}_s)$ and the input latent information $\boldsymbol{c}$

$$mse(\hat{\boldsymbol{c}}, \boldsymbol{c}) = \frac{\sum_{i=1}^{N_l} (\hat{\boldsymbol{c}}_i - \boldsymbol{c}_i)^2}{N_l} \tag{5.24}$$

Finally, the information loss can be defined as

$$l_{info} = \lambda_{cate} cross\_entropy(\hat{\boldsymbol{y}}, k) + \lambda_{con} mse(\hat{\boldsymbol{c}}, \boldsymbol{c}) \tag{5.25}$$

Here λ_{cate} and λ_{con} are both set as 1. In every training step, the generator and discriminator first update according to l_G and l_D then they update together according to l_{info}, respectively.

5.2.4.3 Auxiliary Classify GAN

The ACGAN is the latest variant of classified generative models and has been widely used in the generation of labeled samples. The generator gets the noise vector and the label as input, outputs synthetic samples of the given type. Different from the CGAN, we apply the embedding layer to process the label instead of One-Hot encoding. The network architecture of the generator is shown in Fig. 5.4. After embedding the label into the noise space, we multiply the embedded label and noise to incorporate the randomness and type information. Subsequent layers remain unchanged as in Table 5.1. The discriminator outputs the truth probability and label prediction. It is almost the same as that of the InfoGAN. We only need to omit the 12(3) layer in Table 5.4.

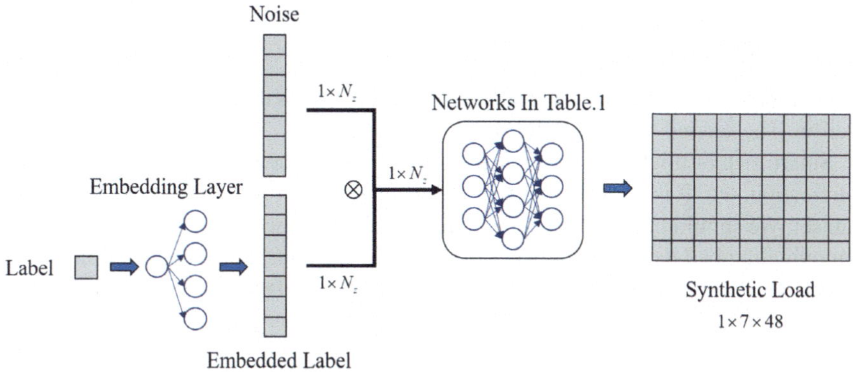

Fig. 5.4 Network architecture of the ACGAN generator

Denote the truth probability as $D_{disc}(\boldsymbol{x})$, the predicted label as $D_{cate}(\boldsymbol{x})$, the synthetic sample and label as $\boldsymbol{x}_s$ and $\boldsymbol{y}_s$, the real sample and label as $\boldsymbol{x}_r$ and $\boldsymbol{y}_r$. The objective of the generator is to make the truth probability $D_{disc}(\boldsymbol{x}_s)$ approximate 1, and the classification accuracy higher. Thus its loss function can be defined as,

$$l_G = \frac{1}{2}(-\mathbb{E}_{z\sim p(z)} \log D_{disc}(\boldsymbol{x}_s) + cross\_entropy(D_{cate}(\boldsymbol{x}_s), \boldsymbol{y}_s)) \tag{5.26}$$

The objective of the discriminator is composed of two parts. In addition to making $D_{disc}(\boldsymbol{x}_r)$ approximate 1 and $D_{disc}(\boldsymbol{x}_s)$ approximate 0, the discriminator should increase the classification accuracy for both real and synthetic samples. Thus its loss function can be defined as,

$$\begin{aligned} l_D = &\frac{1}{2}(-\mathbb{E}_{z\sim p(z)} \log(1 - D_{disc}(\boldsymbol{x}_s)) + cross\_entropy(D_{cate}(\boldsymbol{x}_s), \boldsymbol{y}_s)) + \\ &\frac{1}{2}(-\mathbb{E}_{\boldsymbol{x}_r\sim p(\boldsymbol{x}_r)} \log D_{disc}(\boldsymbol{x}_r) + cross\_entropy(D_{cate}(\boldsymbol{x}_r), \boldsymbol{y}_r)) \end{aligned} \tag{5.27}$$

5.3 Methodology

In this section, we will present the methodology to generate residential load data. It contains three steps including data preprocessing, model training, and model evaluation. Different metrics to evaluate the generation performance are also given in this part.

5.3.1 Data Preprocessing

Data preprocessing includes two stages. The first stage is data cleaning and regularization. The second is data clustering and labeling.

Smart meters might come across errors during the measurement, storage, communication, etc. It is unavoidable to have some absurd or missing data in the whole dataset. We should omit samples that contain null or negative load values first. After removing the bad data, we apply l_1 norm regularization to each sample. Denote the weekly load curve as $\boldsymbol{x}$ with the size of $1 \times N$, the regularized curve as $\hat{\boldsymbol{x}}$, then we have

$$\hat{\boldsymbol{x}} = \frac{\boldsymbol{x}}{\sum_{i=1}^{N} \boldsymbol{x}_i} \tag{5.28}$$

After regularization, the sum of all points in a load curve equals to 1. The reason of applying l_1 norm regularization is that we care more about the consumption pattern rather than its absolute value.

To generate load curves of a specific type, we need to classify the dataset before training. k-means clustering is used to label the load curves in this chapter. We use the Silhouette Coefficient (SC) and sum of the squared errors (SSE) to find the best k for the dataset.

Denote the dataset as $\{\boldsymbol{x}_1, \boldsymbol{x}_2, \ldots, \boldsymbol{x}_N\}$, the clusters as $\{\mathcal{S}_1, \mathcal{S}_2, \ldots, \mathcal{S}_k\}$, the clustering centers as $\{\boldsymbol{c}_1, \boldsymbol{c}_2, \ldots, \boldsymbol{c}_k\}$, the distance function as $d(\boldsymbol{x}_i, \boldsymbol{x}_j)$. The SC measures the cohesion within clusters and the separation among clusters. Suppose $\boldsymbol{x}_i$ belongs to the jth cluster, then the cohesion of $\boldsymbol{x}_i$ is its mean distance from other samples in $\mathcal{S}_j$.

$$a_i = \frac{\sum_{\boldsymbol{x} \in \mathcal{S}_j} d(\boldsymbol{x}_i, \boldsymbol{x})}{|\mathcal{S}_j|} \tag{5.29}$$

In the formula above, $|\cdot|$ returns the size of a set. The separation of $\boldsymbol{x}_i$ is its mean distance from all samples in $\mathcal{S}_p$, where the center $\boldsymbol{c}_p$ is the nearest center to $\boldsymbol{x}_i$ except $\boldsymbol{c}_j$.

$$b_i = \frac{\sum_{\boldsymbol{x} \in \mathcal{S}_p} d(\boldsymbol{x}_i, \boldsymbol{x})}{|\mathcal{S}_p|}; \quad p = \arg\min_j d(\boldsymbol{x}_i, \boldsymbol{c}_j) \tag{5.30}$$

The SC of $\boldsymbol{x}_i$ is

$$SC_i = \frac{b_i - a_i}{\max(a_i, b_i)} \tag{5.31}$$

For the whole dataset, the SC equals to the mean of each point,

$$SC = \frac{\sum_{i=1}^{N} SC_i}{N} \tag{5.32}$$

The range of SC is $[-1, 1]$. A high SC indicates samples of the same category are close, and samples of different categories are distant. Since SC is highly relevant to the data distribution, we care more about its trend with respect to different k rather than its absolute value.

The SSE equals to the sum of squared errors between samples and their clustering centers, which is defined as

$$SSE = \sum_{i=1}^{k} \sum_{x \in \mathscr{S}_i} \|x - c_i\|^2 \tag{5.33}$$

In the formula above, $\|\cdot\|$ returns the l_2 norm of a given vector. When the clustering number k increases, the classifying result would be more meticulous, and the cohesion of each cluster would gradually increase, so the SSE will decrease. It should be noted that when k is smaller than the optimum, the cohesion of each cluster increases fast. Thus the SSE decreases quickly. When k reaches the optimum, the cohesion tends to be stable, so the decrease of the SSE would slow down. That is to say, the relationship pattern between SSE and k is like the shape of an elbow, and the k value corresponding to the elbow is optimum. After the optimal k is determined, we use the clustering result to label load curves.

5.3.2 *Model Training*

The training of GAN models includes three steps: initialization, iteration, and generation.

First, we initialize trainable parameters in the network and set hyperparameters that control the training process. Initialization configurations of different GAN variants are shown in Table 5.5. In the table, epoch is the number of times that the training set being traversed. Batch size is the number of samples trained per iteration. Optimizer is the algorithm of gradient descending during the training. Learning rate and betas are parameters of the optimizer. Noise dim is the length of the noise vector. Latent dim is the length of latent information vector (for InfoGAN only). N_{critic} is the ratio of discriminator training frequency to the generator training frequency. Trainable parameters in convolution and fractional-convolution layers are initialized according to normal distribution with the mean and standard deviation in *Conv Initial*. Trainable parameters in fully-connected layers are initialized according to normal distribution with the mean and standard deviation in *Dense Initial*. The hyper-parameters in Table 5.5 is determined by suggestions in existing literature which have been found to work well across a variety of architectures and datasets.

Second, we iterate trainable parameters in the model over batched samples. The iteration algorithm is determined by the optimizer. Two optimizers are used in this chapter, RMSprop for the WGAN and Adam for others. They are both based on the gradient descent algorithm. Denote the loss function of the network as l, trainable

Table 5.5 Initialization configurations of different GAN variants

Params	BEGAN	BGAN	WGAN-GP	CGAN	InfoGAN	ACGAN
Epochs	15	30	30	30	50	50
Batch size	64	64	64	64	64	64
Optimizer	Adam	Adam	RMSprop	Adam	Adam	Adam
Learning rate	0.0002	0.0002	0.0002	0.0002	0.0002	0.0002
Betas	(0.5, 0.999)	(0.5, 0.999)	–	(0.5, 0.999)	(0.5, 0.999)	(0.5, 0.999)
Noise dim	100	100	100	100	62	100
Latent dim	–	–	–	–	2	–
N_{critic}	1	5	5	5	1	1
Conv initial	(0, 0.02)	(0, 0.02)	(0, 0.02)	(0, 0.02)	(0, 0.02)	(0, 0.02)
Dense initial	(1, 0.02)	(1, 0.02)	(1, 0.02)	(1, 0.02)	(1, 0.02)	(1, 0.02)

parameters in the network as θ. Then the gradient at the tth step is

$$g^t = \nabla_\theta l(\theta) \tag{5.34}$$

The first and second-order momentum determined by historical gradients is

$$m^t = \phi(g^1, g^2, \ldots, g^t) \tag{5.35}$$

$$v^t = \varphi(g^1, g^2, \ldots, g^t) \tag{5.36}$$

θ is updated according to

$$\theta^{t+1} = \theta^t - \frac{m^t}{\sqrt{v^t + \varepsilon}} \tag{5.37}$$

In the formula above, ε is a term added to the denominator to improve numerical stability, its default value is 1e-8. The difference between the Adam optimizer and the RMSprop optimizer is the definition of m^t and v^t.

In the Adam optimizer, m^t is the moving average of m^{t-1} and g^t, v^t is the moving average of v^{t-1} and $diag(g^t g^{t\top})$. The updating formula is

$$m^t = \eta[\beta_1 m^{t-1} + (1 - \beta_1)g^t] \tag{5.38}$$

$$v^t = \beta_2 v^{t-1} + (1 - \beta_2)diag(g^t g^{t\top}) \tag{5.39}$$

In the formula above, β_1 and β_2 are hyper-parameters *betas* in Table 5.5. η is the learning rate.

In the RMSprop optimizer, the updating formula is

$$m^t = \eta g^t \tag{5.40}$$

$$v^t = \gamma v^{t-1} + (1-\gamma) diag(g^t g^{t\top}) \tag{5.41}$$

In the formula above, γ is the smoothing constant. Its default value is 0.99. At the beginning of training, $m^0 = 0$, $v^0 = 0$.

Take Adam optimizer as an example, the training process of a GAN model is shown in Algorithm 1. It should be noted that the presented algorithm is the general case. It needs fine-tuning for different GAN variants if necessary. For example, when it comes to the InfoGAN, the generator and discriminator need to be updated once again by the gradient of the information loss.

Algorithm 1 Algorithm of Training the GAN model

Input: training set $\{load : x, label : y\}$, epochs N_e, batch size bs, learning rate η, betas (β_1, β_2), noise dim N_z, ratio of training frequency N_{critic}, 'Conv Initial' (m_c, s_c), 'Dense Initial' (m_d, s_d)

Output: optimal generator and discriminator parameters θ_G and θ_D

Initialize trainable parameters in networks. For convolutional layers, $\theta \sim \mathcal{N}(m_c, s_c)$; for fully-connected layers, $\theta \sim \mathcal{N}(m_d, s_d)$. Denote initialized parameters in the generator and discriminator as θ_G^0 and θ_D^0.
Initialize the first and second order momentum of the generator and discriminator: $m_G^0 = 0$, $v_G^0 = 0$, $m_D^0 = 0$, $v_D^0 = 0$.
Shuffle the training set and pack it into $N_b = \lfloor N/bs \rfloor$ batches (N is the volume of training set).
for each $i = 1, 2, \cdots, N_e$ **do**
  for each $j = 1, 2, \cdots, N_b$ **do**
    $t = (i-1) * N_b + j$
    Get batched real samples $\boldsymbol{x}_r$ and $\boldsymbol{y}_r$.
    Get random noise vectors and labels, denoted as z and $\boldsymbol{y}_s$.
    Generate synthetic samples $\boldsymbol{x}_s = G(z, \boldsymbol{y}_s; \theta_G)$.
    Get discriminative results of real and synthetic samples $D(\boldsymbol{x}_r, \boldsymbol{y}_r; \theta_D)$ and $D(\boldsymbol{x}_s, \boldsymbol{y}_s; \theta_D)$.
    if $t \% N_{critic} == 0$ **then**
      Calculate $l_G(D(\boldsymbol{x}_s, \boldsymbol{y}_s; \theta_D))$.
      Update g_G^t by Eq. (5.34).
      Update m_G^t v_G^t by Eqs. (5.38) and (5.39).
      Update θ_G^t by Eq. (5.37).
    end if
    Calculate $l_D(D(\boldsymbol{x}_s, \boldsymbol{y}_s; \theta_D), D(\boldsymbol{x}_r, \boldsymbol{y}_r; \theta_D))$.
    Update g_D^t by Eq. (5.34).
    Update m_D^t v_D^t by Eqs. (5.38) and (5.39).
    Update θ_D^t by Eq. (5.37).
  end for
end for
return θ_G, θ_D

5.3.3 *Metrics*

In this part, we will introduce the metrics to evaluate generation performance.

5.3.3.1 Metrics to Evaluate the Distribution

To evaluate the statistical characteristics of generated samples, we compare the distribution of load values and load curves, respectively. The JS divergence and Precision and Recall for Distributions (PRD) are applied in this chapter.

Jensen–Shannon Divergence

The JS divergence is widely used to compute the distance between two distributions. Denote the real load values as $\{x_r^i\}_{i=1}^N$, the synthetic load values as $\{x_s^i\}_{i=1}^N$. First, we regularize load values to the range of [0, 1] as

$$\hat{x}_r^i = \frac{x_r^i}{max(\{x_r^i\}_{i=1}^N)} \quad \hat{x}_s^i = \frac{x_s^i}{max(\{x_s^i\}_{i=1}^N)} \tag{5.42}$$

Set the number of discrete intervals as K and divide [0, 1] into K segments. Then the range of kth interval is $[\frac{k-1}{K}, \frac{k}{K}]$. Compute the number of real load values and synthetic load values within the kth interval, denoted as N_{rk} and N_{sk} respectively. Then the discrete distributions of the real and synthetic samples are

$$P_r = \left[\frac{N_{r1}}{N}, \frac{N_{r2}}{N}, \ldots, \frac{N_{rK}}{N}\right] \tag{5.43}$$

$$P_s = \left[\frac{N_{s1}}{N}, \frac{N_{s2}}{N}, \ldots, \frac{N_{sK}}{N}\right] \tag{5.44}$$

The JS divergence between P_r and P_s is

$$JS(P_r, P_s) = \frac{1}{2}\sum_{k=1}^{K}\left[P_r(k)\log\frac{2P_r(k)}{P_r(k)+P_s(k)} + P_s(k)\log\frac{2P_s(k)}{P_r(k)+P_s(k)}\right] \tag{5.45}$$

In the formula above, $P_r(k)$ and $P_s(k)$ represent the kth element in P_r and P_s.

Precision and Recall for Distributions

The PRD is a novel definition of precision and recall that can disentangle the divergence of image data distributions [23]. It is originated from but superior to recent evaluation metrics that can measure the distribution of images such as Inception Score and FID. The PRD can quantify the degree of mode dropping and mode invention on two separate dimensions, which we called PRD curves.

Denote the real load curves as $\boldsymbol{x}_r$, the synthetic curves as $\boldsymbol{x}_s$, merge them to form a new dataset $\{\boldsymbol{x}_r^1, \boldsymbol{x}_r^2, \ldots, \boldsymbol{x}_r^N, \boldsymbol{x}_s^1, \boldsymbol{x}_s^2, \ldots, \boldsymbol{x}_s^N\}$. Then use k-means to classify the dataset and label the curves. Denote the number of real and synthetic samples in each type denoted as $[N_{r1}, N_{r2}, \ldots, N_{rk}]$ and $[N_{s1}, N_{s2}, \ldots, N_{sk}]$. The discrete distributions of the real and synthetic samples are

$$P_r = \left[\frac{N_{r1}}{N}, \frac{N_{r2}}{N}, \ldots, \frac{N_{rk}}{N}\right] \tag{5.46}$$

$$P_s = \left[\frac{N_{s1}}{N}, \frac{N_{s2}}{N}, \ldots, \frac{N_{sk}}{N}\right] \tag{5.47}$$

Next we compute the PRD curve for P_s with respect to P_r. The PRD will be computed for an equiangular grid of angle θ values between $[0, \pi/2]$. For a given threshold θ, we compute

$$\hat{P}_s(\theta) = \left[\frac{N_{s1}}{N}\tan\theta, \frac{N_{s2}}{N}\tan\theta, \ldots, \frac{N_{sk}}{N}\tan\theta\right] \tag{5.48}$$

Then we compare $\hat{P}_s(\theta)$ with P_r entry by entry and retain the smaller one to form a new vector. The precision at θ equals to the sum of the new vector

$$p(\theta) = \sum_{i=1}^{k} \min(\hat{P}_s(\theta)_i, P_{ri}) \tag{5.49}$$

Mathematically, it measures how much of the synthetic distribution can be generated by a part of the real distribution. The recall at θ is

$$r(\theta) = p(\theta)/\tan\theta \tag{5.50}$$

It measures how much of the real distribution can be generated by a part of the synthetic distribution. When two distributions are highly similar, both the precision and recall are close to 1. It should be noted that different thresholds lead to different trade-offs between precision and recall. If we compute $p(\theta)$ and $r(\theta)$ at every θ from 0 to 2π, we have the precision vector and recall vector. Plot precision on the vertical Y-axis against recall on the horizontal X-axis, then we get the PRD curve. The PRD equals the area under the PRD curve. It is given as follows

$$PRD = \int_{r(0)}^{r(2\pi)} p(\theta)dr(\theta) \tag{5.51}$$

In order to summarize the PRD curves, we also compute the maximum F_1 score, which corresponds to the harmonic mean of the precision and the recall as a single-number summary. It is given as follows

$$F_\beta(\theta) = (1+\beta^2)\frac{p(\theta)r(\theta)}{\beta^2 p(\theta) + r(\theta)} \tag{5.52}$$

Since $\beta \geq 1$ weighs recall higher than precision while $\beta \leq 1$ on the contrast, thus we compute a pair of values for the PRD curve: F_β and $F_{1/\beta}$. Select the maximum $F_\beta(\theta)$ and $F_\beta(\theta)$ when θ ranges from 0 to 2π. In this chapter, we choose $\beta = 8$ as suggested in [23]. As mentioned above, F_8 weighs the recall higher than precision while $F_{1/8}$ on the contrast. If the maximum $F_8 \leq$ the maximum $F_{1/8}$, then the model is with higher precision than recall. On the opposite, if the maximum $F_8 \geq$ the maximum $F_{1/8}$, then the model is with higher recall than precision. Considering the problem of privacy leakage of customers, we believe that a higher precision and a lower recall is better in residential load generation, which indicates that the synthetic distribution is easy to recover from the real while the contrary is difficult.

5.3.3.2 Metrics to Evaluate the Fidelity

Besides comparing the real and synthetic distributions, we also inspect the visual characteristics of generated load curves, which is called the fidelity. For example, the generated weekly load curve should exhibit reasonable periodicity, peak-valley property, and volatility. The root mean squared error (RMSE) and structural similarity (SSIM) are applied in this chapter.

Root Mean Squared Error

The RMSE is used to compute the distance between the vectorized data. It can measure the similarity of shape and value at the same time. Denote the set of synthetic curves as $\{\boldsymbol{x}_s^1, \boldsymbol{x}_s^2, \ldots, \boldsymbol{x}_s^N\}$, set of real curves as $\{\boldsymbol{x}_r^1, \boldsymbol{x}_r^2, \ldots, \boldsymbol{x}_r^N\}$. First, we compute the mean curves of two sets respectively as

$$\bar{\boldsymbol{x}}_s = \frac{\sum_{i=1}^N \boldsymbol{x}_s^i}{N} \quad \bar{\boldsymbol{x}}_r = \frac{\sum_{i=1}^N \boldsymbol{x}_r^i}{N} \tag{5.53}$$

Next, compute the RMSE distance between $\bar{\boldsymbol{x}}_s$ and $\bar{\boldsymbol{x}}_r$ as

$$RMSE(\bar{\boldsymbol{x}}_s, \bar{\boldsymbol{x}}_r) = \sqrt{\frac{\sum_{i=1}^{N_l} (\bar{\boldsymbol{x}}_s(i) - \bar{\boldsymbol{x}}_r(i))^2}{N_l}} \tag{5.54}$$

In the formula above, N_l represents the length of curves; $\bar{\boldsymbol{x}}_s(i)$ and $\bar{\boldsymbol{x}}_r(i)$ represent the load at the ith time slot. The smaller the RMSE, the more similar the synthetic samples and real samples.

Structual Similarity Index

The SSIM index is used to compute the similarity between two images [24]. Denote the 2D images as $\boldsymbol{x}$ and $\boldsymbol{y}$, their width and height as N_w and N_h. First, we compute the mean and variance of the single image and the covariance between two images as follows.

$$\mu_x = \frac{1}{N_w * N_h} \sum_{i=1}^{N_w} \sum_{j=1}^{N_h} \boldsymbol{x}(i, j) \tag{5.55}$$

$$\sigma_x = \sqrt{\frac{1}{N_w * N_h - 1} \sum_{i=1}^{N_w} \sum_{j=1}^{N_h} (\boldsymbol{x}(i, j) - \mu_x)^2} \tag{5.56}$$

$$\sigma_{xy} = \frac{1}{N_w * N_h - 1} \sum_{i=1}^{N_w} \sum_{j=1}^{N_h} (\boldsymbol{x}(i, j) - \mu_x)(\boldsymbol{y}(i, j) - \mu_y) \tag{5.57}$$

Then the luminance, contrast and structure comparison measurement are given as follows

$$l(\boldsymbol{x}, \boldsymbol{y}) = \frac{2\mu_x \mu_y + C_1}{\mu_x^2 + \mu_y^2 + C_1} \tag{5.58}$$

$$c(\boldsymbol{x}, \boldsymbol{y}) = \frac{2\sigma_x \sigma_y + C_2}{\sigma_x^2 + \sigma_y^2 + C_2} \tag{5.59}$$

$$s(\boldsymbol{x}, \boldsymbol{y}) = \frac{\sigma_{xy} + C_3}{\sigma_x \sigma_y + C3} \tag{5.60}$$

where C_1, C_2 and C_3 are small constants given by

$$C_1 = (K_1 L)^2, C_2 = (K_2 L)^2, C_3 = C_2/2 \tag{5.61}$$

respectively. In the formula above, L is the maximum load values, and $K_1 \ll 1$, $K_2 \ll 1$ are two scalar constants. In this chapter, we set $K_1 = 0.01$ and $K_2 = 0.03$, which is found to be suitable for a variety of datasets. Then the SSIM index is defined as:

$$SSIM(\boldsymbol{x}, \boldsymbol{y}) = l(\boldsymbol{x}, \boldsymbol{y}) \times c(\boldsymbol{x}, \boldsymbol{y}) \times s(\boldsymbol{x}, \boldsymbol{y}) \tag{5.62}$$

Value of the SSIM is between 0 and 1. When two images are similar, their SSIM is close to 1.

5.4 Case Studies

In this section, we present the generation and evaluation results of the proposed GAN models trained on the real world residential load data. All numerical experiments are conducted on a PC equipped with 12 Intel(R) Core(TM) i7-8700K CPU @ 3.70GHz and an NVIDIA GeForce RTX 2060 GPU. All programs for GAN variants are written in Python using torch v1.1.0.

5.4.1 Data Description

The training data are from the Smart Metering Electricity Customer Behavior Trials [25]. Electricity consumptions of over 5000 Irish homes and businesses from 14/07/2009/ to 31/12/2010 were collected. The load data are recorded every 30 min. We randomly select 20000 weekly load curves from 1000 residential consumers as our training set after data cleaning.

First, we use k-means to cluster the curves. For each weekly load curve, we convert it into a daily load curve by computing the mean of load curves every day. Then we cluster the averaged daily load curves which can be viewed as vectors with the size of 1×48. The SC and SSE on the vertical Y-axis against k ranging from 2 to 14 on the horizontal X-axis are plotted in Fig. 5.5.

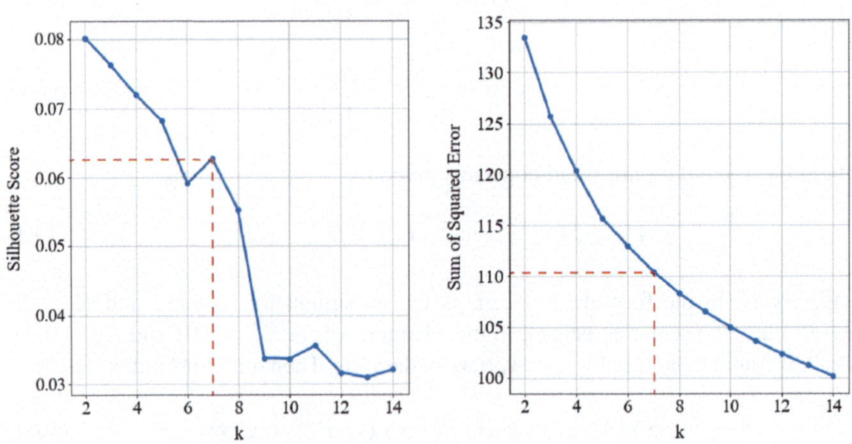

Fig. 5.5 The SC and SSE of different k

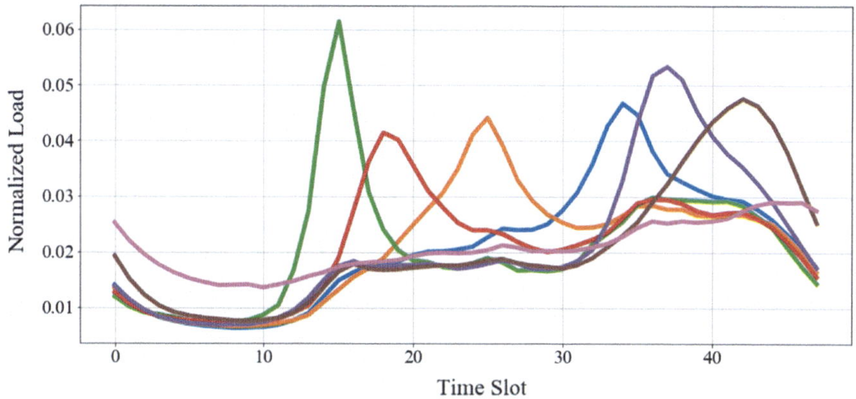

Fig. 5.6 Typical mean daily load of weekly load curves

Table 5.6 Number of load curves in each cluster

Cluster	1	2	3	4	5	6	7
Number	3032	4139	2210	3731	2873	1107	2908

It can be observed that the elbow of the SSE curve appears when k is in [5, 7]. The SC decreases as k increases except for $k = 7$ and $k = 11$. Considering the trade-off between SSE and SC, we set $k = 7$. Centers of 7 clusters are shown in Fig. 5.6. The number of curves in each cluster is given in Table 5.6.

5.4.2 *Unclassified Generation*

In this part, we evaluate unclassified GAN variants presented in Sect. 5.2.3. For each variant, we generate 20000 synthetic weekly load curves, respectively. It should be noted that since we care more about the shape of load curves rather than their absolute values, all real curves are regularized before sending to the discriminator. Thus, the generated curves are also regularized. To recover them to the same value range as the real loads, we multiply the synthetic curves by a constant scalar which is the ratio of averaged real load values to averaged synthetic load values.

First, we inspect the visual characteristics of synthetic load curves. We plot mean curves of real samples and generated samples for BEGAN, BGAN, and WGAN-GP, respectively in Fig. 5.7. It can be observed that all synthetic curves exhibit periodicity which is corresponding to the daily living pattern. Also, their peak-valley positions and values are similar to real ones. In terms of the fidelity, the BEGAN outperforms the other two obviously. We find that there are many spikes in curves generated by the WGAN-GP, which reflects the instability of the training process.

Fig. 5.7 Synthetic and real mean weekly load curves

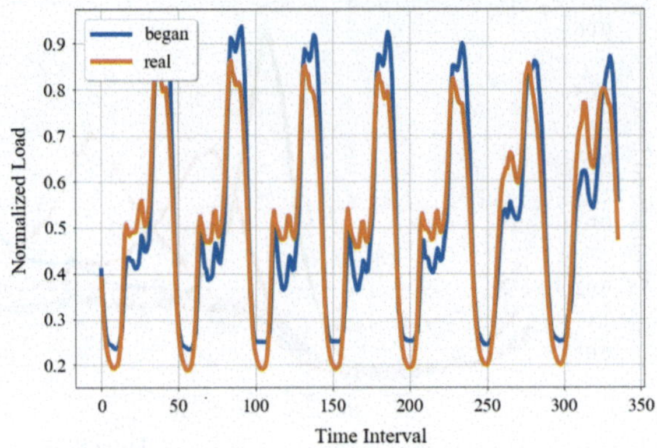

(a) BEGAN

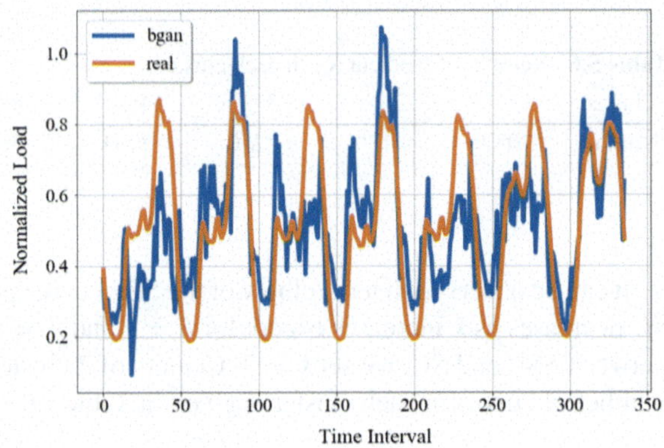

(b) BGAN

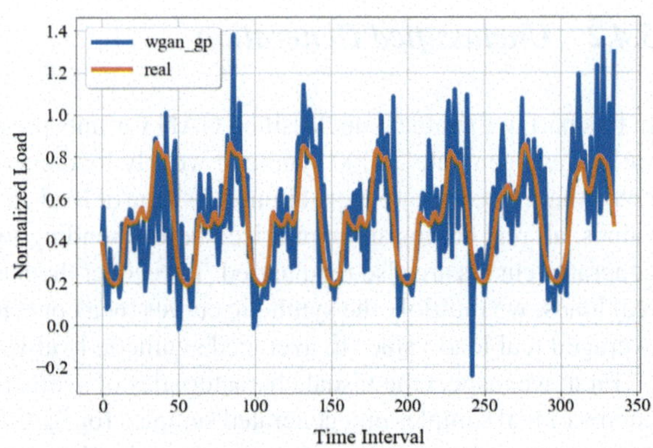

(c) WGAN-GP

Table 5.7 Evaluation metrics of unclassified GAN models

GAN	Jensen–Shannon divergence	RMSE	PRD	F_8	$F_{1/8}$	SSIM
BEGAN	0.2470	0.0767	0.1621	0.1372	0.5645	0.7054
BGAN	0.0743	0.1533	0.3253	0.3446	0.6871	0.5375
WGAN-GP	0.0913	0.1892	0.2506	0.5075	0.4684	0.5747

Second, we inspect the statistical characteristics of load values and load curves. We compute the discrete distribution of real and synthetic load values according to Eqs. 5.43 and 5.44. Plot the probability distribution functions for three GANs respectively in Fig. 5.8. It can be observed that loads generated by the BEGAN deviate from real loads obviously, which indicates that the BEGAN would generate higher loads comparing with real values. Distributions of loads from BGAN and WGAN-GP have similar features as the real. The scatter plot of load curve means on the horizontal X-axis against standard variances on the vertical Y-axis is shown in Fig. 5.9. It actually reflects the diversity of load curves. We find that although the curves from the BEGAN fit the real curves best on the shape, the diversity of synthetic curves is quite poor compared with the BGAN and WGAN-GP. In other words, the BEGAN is easy to run into the modes collapse. We plot the PRD curves in Fig. 5.10. In terms of the PRD, BEGAN and BGAN perform better than the WGAN-GP. The precision is higher compared with the recall in the BEGAN and BGAN, which indicates the synthetic curves are mainly originated from the real curve distribution and the real curves are difficult to be recovered from the synthetic curve distribution.

Quantitative metrics presented in Sect. 5.3.3 are listed in Table 5.7. When it comes to the similarity of visual characteristics, the BEGAN outperforms the other two obviously. On the other hand, in terms of the similarity of statistical characteristics, the BGAN and WGAN-GP perform better. To conclude, unclassified GANs have to make a trade-off between the diversity and fidelity of generated curves.

5.4.3 Classified Generation

In this part, we evaluate classified GANs presented in Sect. 5.2.4. For each category, we generate the same number of synthetic curves as that of real ones. Same as above, each curve is multiplied by a constant scalar.

First, we inspect the visual characteristics of synthetic load curves. Take the 2th category as an example, we plot mean curves of real samples and generated samples for CGAN, InfoGAN, and ACGAN respectively in Fig. 5.11. We can find large ripples in curves from the CGAN. Although they are periodic, the volatility of curves is quite unreasonable. Synthetic curves from the InfoGAN exhibit rational peak-valley positions and values. However, there exist negative load values in generated curves.

Fig. 5.8 Probability distribution functions of real and synthetic load values

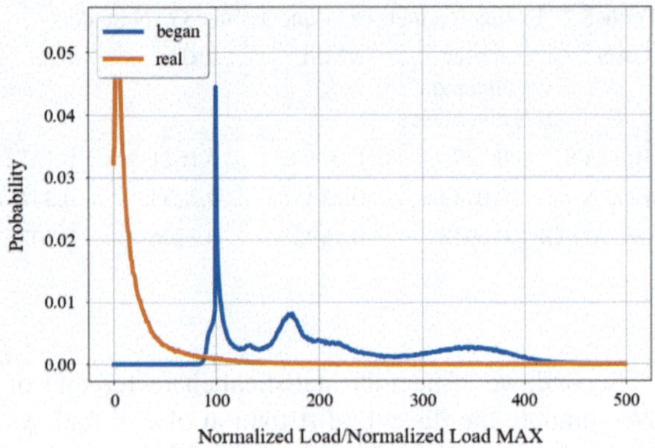

(a) BEGAN

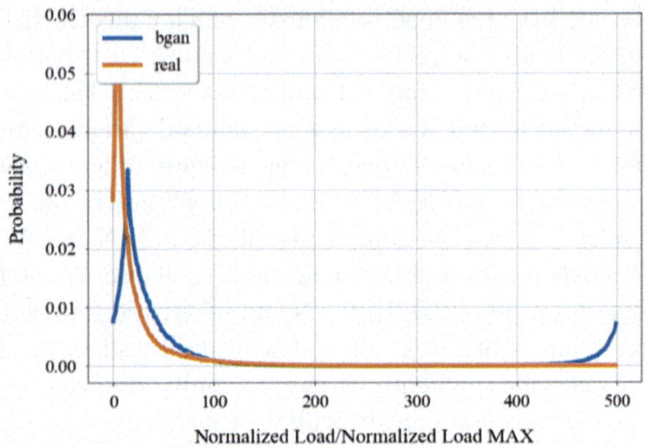

(b) BGAN

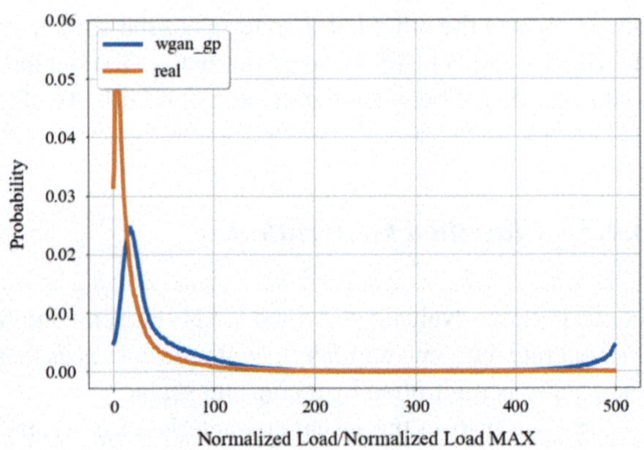

(c) WGAN-GP

Fig. 5.9 Scatter plot of real and synthetic load curve means and standard variances

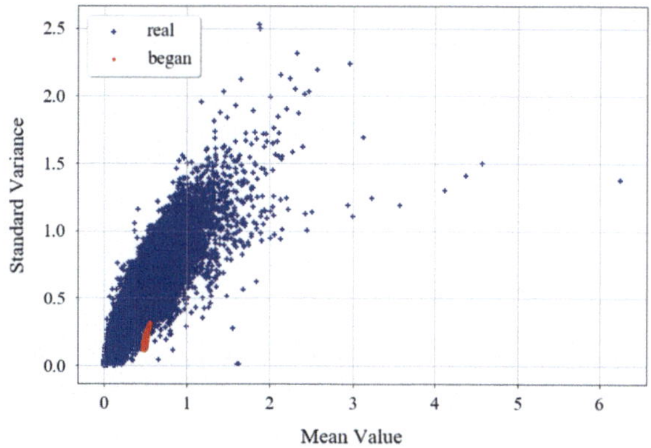

(a) BEGAN

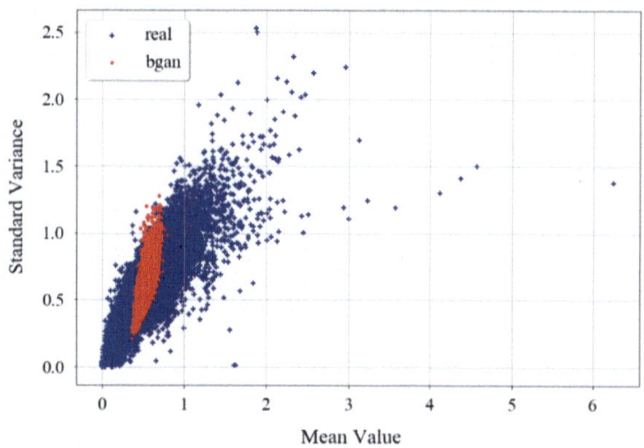

(b) BGAN

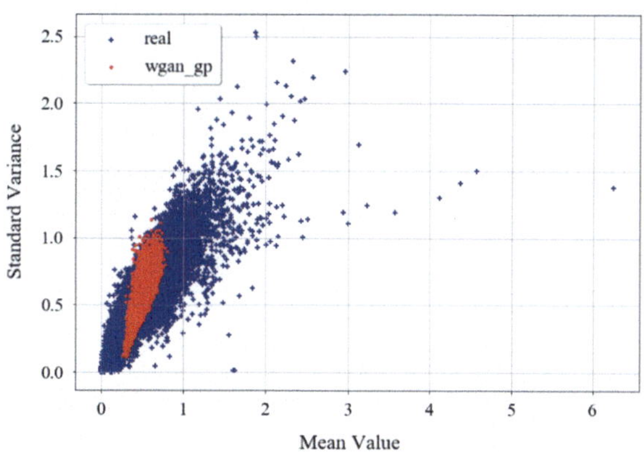

(c) WGAN-GP

Fig. 5.10 Precision and recall for distributions of real and synthetic load curves

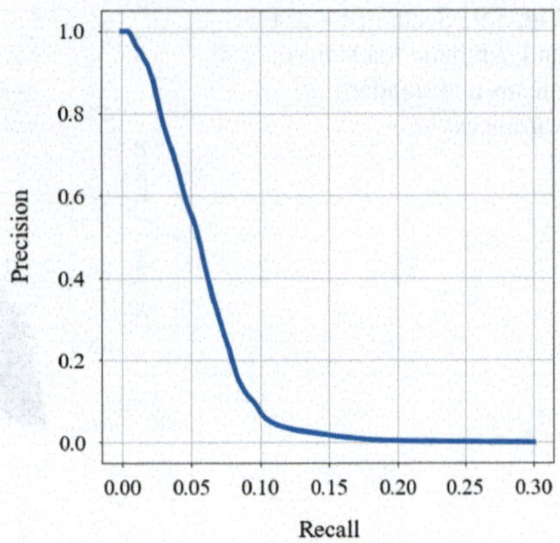

(a) BEGAN

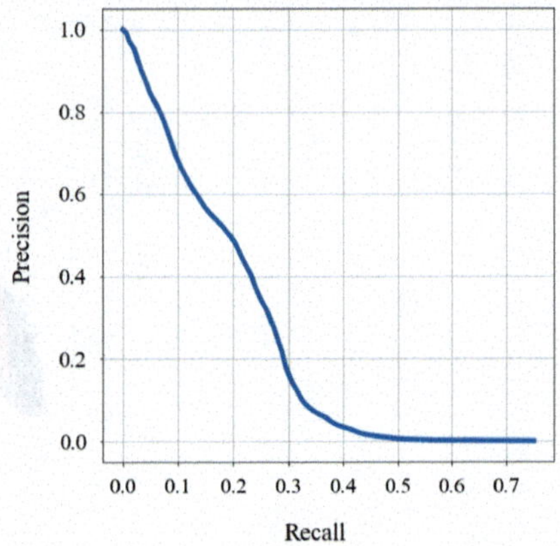

(b) BGAN

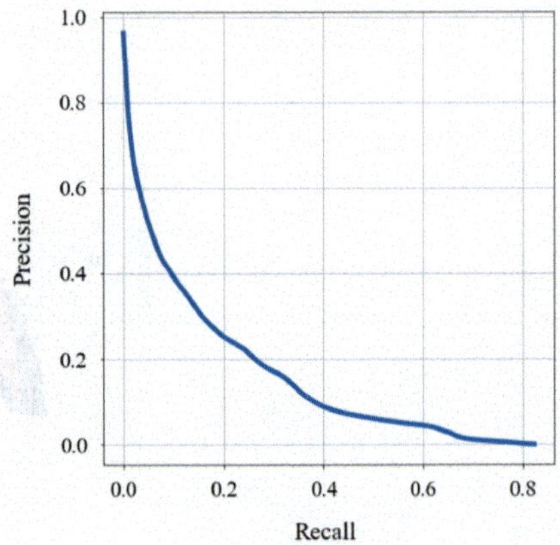

(c) WGAN-GP

Fig. 5.11 Synthetic and real mean weekly load curves of the 2th category

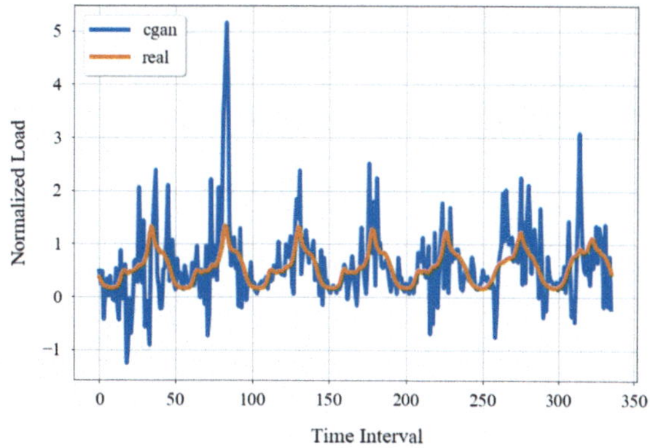

(a) BEGAN

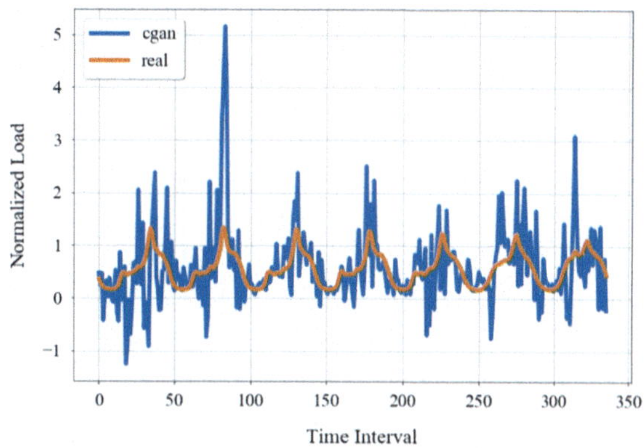

(b) BGAN

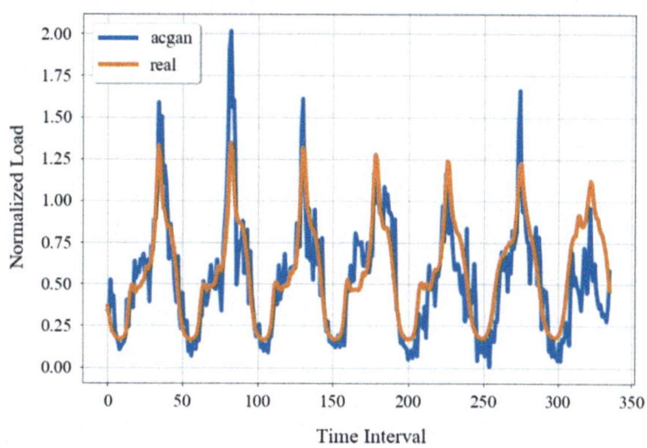

(c) WGAN-GP

Fig. 5.12 Probability distribution functions of real and synthetic load values of the 7th category

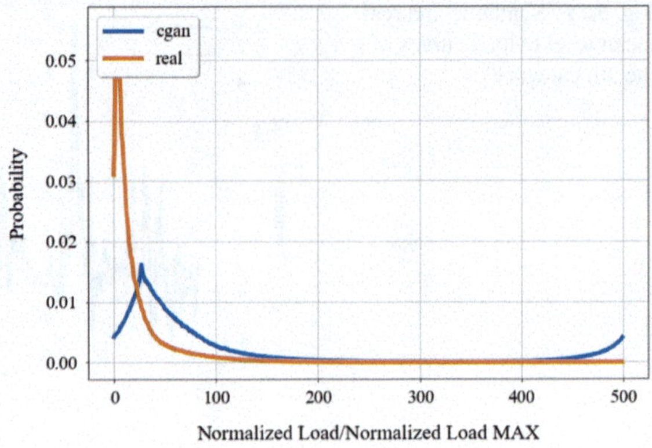

(a) CGAN

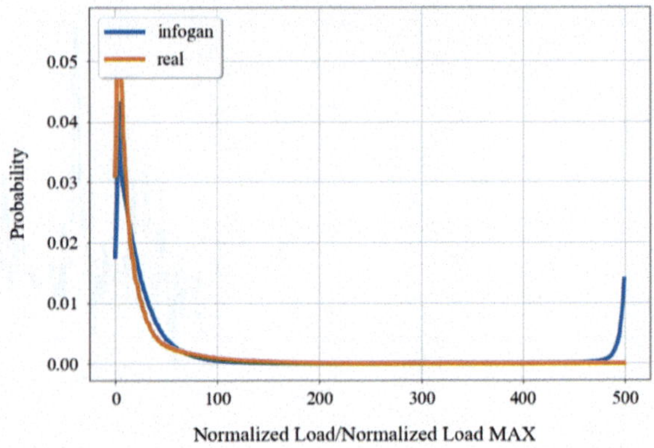

(b) InfoGAN

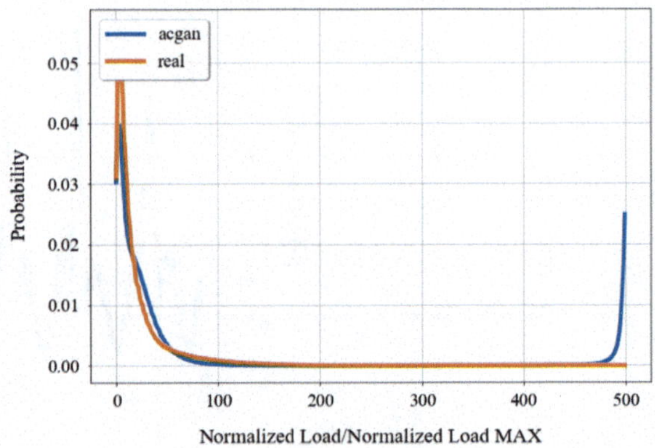

(c) ACGAN

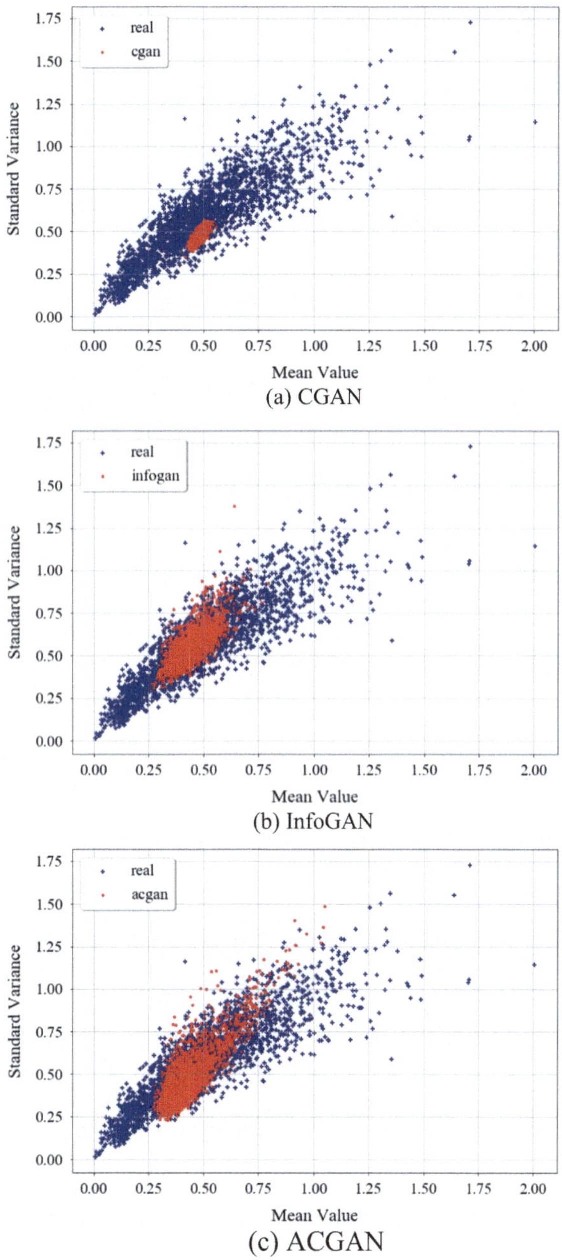

Fig. 5.13 Scatter plot of real and synthetic load curve means and standard variances of the 3th category

Fig. 5.14 Precision and recall for distributions of real and synthetic load curves of the 1th category

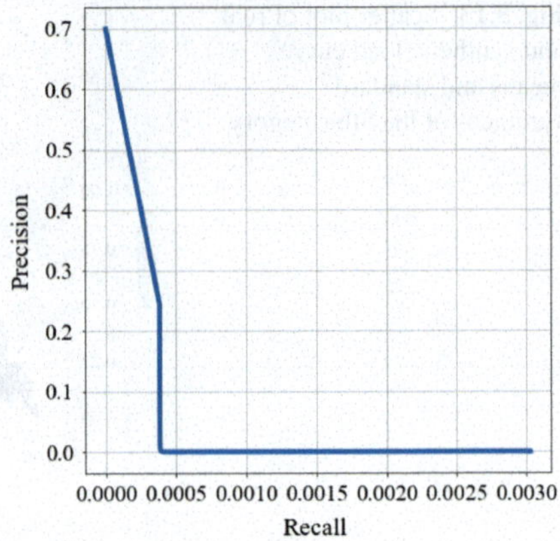

(a) CGAN

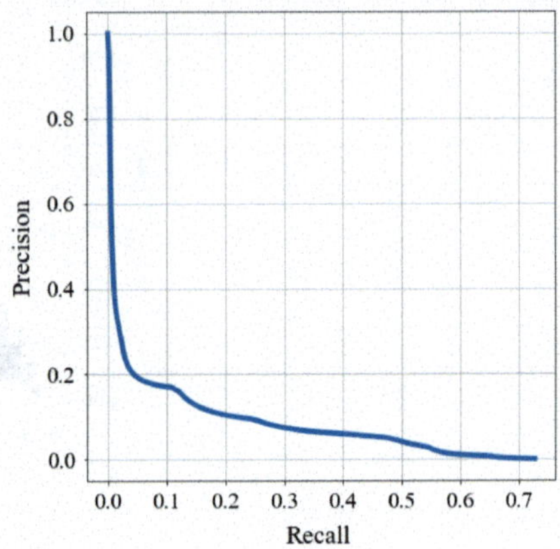

(b) InfoGAN

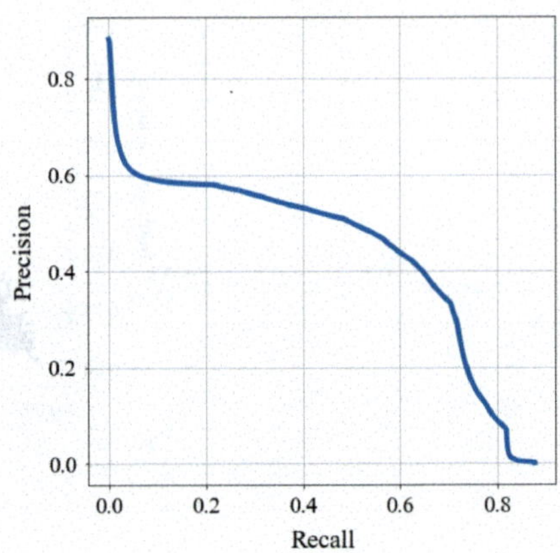

(c) ACGAN

The ACGAN outperforms the other two obviously, mean of synthetic curves is almost the same as that of real curves.

Second, we inspect the statistical characteristics of load values and load curves. Take the 7th category as an example, we plot the probability distribution functions for three GANs respectively in Fig. 5.12. It can be observed that distributions of generated loads from the InfoGAN and ACGAN are similar to that of real loads. However, we can observe an upward tail at the end of the probability distribution function. This might be caused by the supersaturation of the generator neurons. Some parameters in the network are trapped in local optimum and cause relevant neurons output maximum for any input. After the $Tanh$ activation, the output load value is always the maximum. The scatter plot of load curve means and standard variances of the 3th category is shown in Fig. 5.13. The ACGAN is shown to generate load curves with appropriate diversity. However, synthetic curves have not been able to cover all possible real scenarios since the model weighs more on the fidelity than the

Table 5.8 Evaluation metrics of classified GAN models

RMSE	1	2	3	4	5	6	7	Mean
CGAN	0.5031	0.6310	0.3792	0.4625	0.3802	0.4231	0.4002	0.4542
InfoGAN	0.3285	0.3509	0.3044	0.4301	0.3455	0.3626	0.3307	0.3504
ACGAN	0.2006	0.1948	0.1854	0.1963	0.2077	0.2658	0.1895	0.2057
JS divergence	1	2	3	4	5	6	7	Mean
CGAN	0.1287	0.0851	0.2113	0.1645	0.2068	0.1441	0.1660	0.1581
InfoGAN	0.0137	0.0223	0.0256	0.0620	0.0262	0.0546	0.0218	0.0323
ACGAN	0.0299	0.0342	0.0395	0.1151	0.0284	0.0493	0.0277	0.0463
PRD	1	2	3	4	5	6	7	Mean
CGAN	0.0000	0.0000	0.0059	0.0000	0.0021	0.0048	0.0013	0.0020
InfoGAN	0.1462	0.0355	0.0606	0.0054	0.0246	0.0746	0.0344	0.0545
ACGAN	0.4264	0.3867	0.4678	0.3860	0.3847	0.3202	0.4474	0.4027
F_8	1	2	3	4	5	6	7	Mean
CGAN	0.0000	0.0000	0.0050	0.0000	0.0004	0.0024	0.0000	0.0011
InfoGAN	0.0513	0.0136	0.0243	0.0027	0.0085	0.0297	0.0113	0.0202
ACGAN	0.2567	0.2025	0.3064	0.2660	0.2275	0.1547	0.3057	0.2456
$F_{1/8}$	1	2	3	4	5	6	7	Mean
CGAN	0.0000	0.0000	0.0013	0.0000	0.0014	0.0010	0.0003	0.0006
InfoGAN	0.4273	0.3031	0.3646	0.1723	0.3610	0.2351	0.2646	0.3040
ACGAN	0.7180	0.6630	0.7550	0.4607	0.7104	0.4455	0.7490	0.6431
SSIM	1	2	3	4	5	6	7	Mean
CGAN	0.5637	0.5067	0.5630	0.5022	0.5548	0.5231	0.5641	0.5397
InfoGAN	0.5867	0.5935	0.5522	0.5243	0.5565	0.6419	0.5727	0.5754
ACGAN	0.5808	0.5734	0.5745	0.5552	0.5743	0.5541	0.5916	0.5720

diversity when making the trade-off. The PRD curves of the 1th category are plotted in Fig. 5.14. In terms of the PRD, the ACGAN outperforms the other two obviously. The area under the PRD curve is far greater than the former GANs. It reflects that the synthetic curve distribution and real curve distribution overlap mainly.

Quantitative metrics for all categories are listed in Table 5.8. It can be found that the ACGAN wins on most indexes in terms of fidelity and diversity. For different categories, the ACGAN has stable performance. It should also be noted that the maximum $F_{1/8}$ is far greater than the maximum F_8 for the ACGAN, which indicates the model is with high precision and low recall. Thus the synthetic distribution is easy to recover from the real while the contrary is difficult, which prevents the privacy leakage of customers.

In summary, the ACGAN balances well between the diversity and fidelity of generated load curves. Comprehensive comparisons on different metrics between the ACGAN and other 5 widely used GANs reveal the superiority of the ACGAN. With the ACGAN, we are able to generate residential load curves of different categories.

5.5 Conclusion

Due to technical barriers and rising privacy concerns, acquire abundant available residential load data becomes a big challenge both for the academia and industry. To solve the problem, various generative models are used to produce synthetic residential loads for use. However, as one of the most popular generative models, GANs are rarely used in this area. In this chapter, we conduct a comprehensive investigation of 6 widely used GAN models with regard to their performance on load generation. For every GAN variant, we design the proper network architecture and loss functions. The standard process of data preprocessing, model training, and evaluation is also presented. Case study results demonstrate that the ACGAN outperforms others significantly. It can balance well between the fidelity and diversity of generated loads. With the ACGAN, we are able to generate residential load under the specific consumption type, which might be helpful in the generation, delivery, and distribution of the electrical power.

References

1. McDaniel, P., & McLaughlin, S. (2009). Security and privacy challenges in the smart grid. *IEEE Security & Privacy*, *7*(3), 75–77.
2. Swan, L. G., & Ugursal, V. I. (2009). Modeling of end-use energy consumption in the residential sector: A review of modeling techniques. *Renewable and Sustainable Energy Reviews*, *13*(8), 1819–1835.
3. Capasso, A., Grattieri, W., Lamedica, R., & Prudenzi, A. (1994). A bottom-up approach to residential load modeling. *IEEE Transactions on Power Systems*, *9*(2), 957–964.

4. McKenna, K., & Keane, A. (2016). Open and closed-loop residential load models for assessment of conservation voltage reduction. *IEEE Transactions on Power Systems*, *32*(4), 2995–3005.
5. Tsagarakis, G., Collin, A. J., & Kiprakis, A. E. (2012). Modelling the electrical loads of UK residential energy users. In *2012 47th International Universities Power Engineering Conference (UPEC)* (pp. 1–6). Uxbridge: IEEE.
6. Dickert, J., & Schegner, P. (2011). A time series probabilistic synthetic load curve model for residential customers. In *2011 IEEE Trondheim PowerTech* (pp. 1–6). Stockholm: IEEE.
7. Collin, A. J., Tsagarakis, G., Kiprakis, A. E., & McLaughlin, S. (2014). Development of low-voltage load models for the residential load sector. *IEEE Transactions on Power Systems*, *29*(5), 2180–2188.
8. Stephen, B., Tang, X., Harvey, P. R., Galloway, S., & Jennett, K. I. (2015). Incorporating practice theory in sub-profile models for short term aggregated residential load forecasting. *IEEE Transactions on Smart Grid*, *8*(4), 1591–1598.
9. Labeeuw, W., & Deconinck, G. (2013). Residential electrical load model based on mixture model clustering and markov models. *IEEE Transactions on Industrial Informatics*, *9*(3), 1561–1569.
10. Xu, F. Y., Wang, X., Lai, L. L., & Lai, C. S. (2013). Agent-based modeling and neural network for residential customer demand response. In *2013 IEEE International Conference on Systems, Man, and Cybernetics* (pp. 1312–1316). Manchester: IEEE.
11. Uhrig, M., Mueller, R., & Leibfried, T. (2014). Statistical consumer modelling based on smart meter measurement data. In *2014 International Conference on Probabilistic Methods Applied to Power Systems (PMAPS)* (pp 1–6). Durham: IEEE.
12. Gu, Y., Chen, Q., Liu, K., Xie, L., & Kang, C. (2019). Gan-based model for residential load generation considering typical consumption patterns. In *2019 IEEE Power & Energy Society Innovative Smart Grid Technologies Conference (ISGT)* (pp. 1–5). (Washington, D.C.: IEEE).
13. Goodfellow, I. J., Pouget-Abadie, J., Mirza, M., Xu, B., Warde-Farley, D., Ozair, S., Courville, A., & Bengio, Y. (2014). Generative Adversarial Networks. arXiv:1406.2661.
14. Radford, A., Metz, L., & Chintala, S. (2015). Unsupervised Representation Learning with Deep Convolutional Generative Adversarial Networks. arXiv:1511.06434.
15. Tompson, J., Goroshin, R., Jain, A., LeCun, Y., & Bregler, C. (2014). Efficient Object Localization Using Convolutional Networks. arXiv:1411.4280.
16. Berthelot, D., Schumm, T., & Metz, L. (2017). BEGAN: Boundary Equilibrium Generative Adversarial Networks. arXiv:1703.10717.
17. Hjelm, R. D., Jacob, A. P., Che, T., Trischler, A., Cho, K., & Bengio, Y. (2017). Boundary-Seeking Generative Adversarial Networks. arXiv:1702.08431.
18. Arjovsky, M., Chintala, S., & Bottou, L. (2017). Wasserstein GAN. arXiv:1701.07875.
19. Gulrajani, I., Ahmed, F., Arjovsky, M., Dumoulin, V., & Courville, A. (2017). Improved Training of Wasserstein GANs. arXiv:1704.00028.
20. Yi, W., Chen, Q., Kang, C., & Xia, Q. (2017). Clustering of electricity consumption behavior dynamics toward big data applications. *IEEE Transactions on Smart Grid*, *7*(5), 2437–2447.
21. Mirza, M. & Osindero, S. (2014). Conditional Generative Adversarial Nets. arXiv:1411.1784.
22. Chen, X., Duan, Y., Houthooft, R., Schulman, J., Sutskever, I., & Abbeel, P. (2016). InfoGAN: Interpretable Representation Learning by Information Maximizing Generative Adversarial Nets. arXiv:1606.03657.
23. Sajjadi, M. S. M., Bachem, O., Lucic, M., Bousquet, O., & Gelly, S. (2018). Assessing Generative Models via Precision and Recall. arXiv:1806.00035.
24. ZHOU, W. (2004). Image quality assessment: From error measurement to structural similarity. *IEEE Transactions on Image Processing*, *13*, 600–613.
25. Commission for Energy Regulation (CER). (2012). CER Smart Metering Project - Electricity Customer Behaviour Trial 2009–2010.

Chapter 6
Partial Usage Pattern Extraction

Abstract Massive amounts of data are being collected owing to the popularity of smart meters. Two main issues should be addressed in this context. One is the communication and storage of big data from smart meters at a reduced cost which has been discussed in Chap. 3. The other one is the effective extraction of useful information from this massive dataset. In this chapter, the K-SVD sparse representation technique, which includes two phases (dictionary learning and sparse coding), is used to decompose load profiles into linear combinations of several partial usage patterns (PUPs), which allows the smart meter data to be compressed and hidden electricity consumption patterns to be extracted at the same time. Then, a linear support vector machine (SVM)-based method is used to classify the load profiles into two groups, residential customers and small and medium-sized enterprises (SMEs), based on the extracted patterns. Comprehensive comparisons with the results of k-means clustering, the discrete wavelet transform (DWT), principal component analysis (PCA), and piecewise aggregate approximation (PAA) are conducted on real datasets in Ireland. The results show that our proposed technique outperforms these methods in both compression ratio and classification accuracy. Further analysis is also conducted on the PUPs.

6.1 Introduction

Data compression techniques should be carefully selected for specific applications according to the characteristics of the dataset. For example, the high-frequency PMU dataset is low-dimensional. The SVD and low-rank techniques perform well in this situation [1]. Regarding the smart meter data, two major characteristics can be observed. One is sparsity, i.e, a daily load profile essentially consists of several partial usage patterns (PUPs). For example, relatively high electricity consumption occurs only a small fraction of the time, while the rest of the data are approximately zero for residential customers. The other characteristic is diversity, i.e, even though there is a set of PUPs, they are combined in a variety of ways in the load profiles of different customers and on different days. If we can effectively identify the PUPs with definite physical significance, we can realize high-performance data compres-

Y. Wang et al., *Smart Meter Data Analytics*,
https://doi.org/10.1007/978-981-15-2624-4_6

sion. In addition, a load profile's pattern can be recognized automatically. In fact, sparse coding is a data compression and feature extraction technique that has been used in many fields, including image processing and language recognition, in recent years [2]. It codes and reconstructs a signal efficiently by exploiting its sparsity and therefore, is quite applicable to compressing the data from individual smart meters.

The proposed sparse coding-based data compression technique can not only reduce the size of a dataset but also effectively extract PUPs from massive load profiles for different applications. In the field of electricity consumption pattern extraction, many studies have been conducted. Clustering is the most commonly used technique for identifying the typical pattern of each customer to enable that customer's consumption behavior to be described in terms of several typical load patterns. However, it should be noted that the load patterns learned by clustering techniques are quite different from the PUPs learned by sparse coding-based techniques. Clustering-based techniques generally consider a daily load profile as a whole, whereas the proposed sparse coding-based technique decomposes the daily load profile into different PUPs. In this case, diverse load profiles can be described as linear combinations of the PUPs. Because the electricity consumption of an individual customer is more random and fluctuating than that of an aggregation of customers, depicting the variety of individual customers in an effective way and exploiting the PUPs of each load profile can be easily performed in this way [3].

The K-SVD algorithm is an efficient sparse coding algorithm for creating a redundant dictionary and representing a signal in a sparse way [4]. In this chapter, the K-SVD algorithm is adopted to compress data from an individual smart meter and extract the diversified PUPs from the load profile. It attempts to identify redundant PUPs using a certain number of data-training and optimization processes and decomposes each load profile into a linear combination of only a few PUPs to guarantee its sparsity. Furthermore, to ensure that the extracted PUPs have clear physical meanings, the coefficient of each PUP is constrained to be non-negative; therefore, this technique is called non-negative K-SVD here [3]. Thus, the sparsity and diversity of individual load profiles are simultaneously exploited.

Customer classification is much necessary and useful in practical work [5]. Even though the electric company knows beforehand what the use of the electricity will be, and the customer will be put into a predefined group at the beginning. It is difficult for the service provider to detect the changes of customer groups only based on the billing information [3, 6]. To verify that the extracted PUPs have definite physical significance, a simple customer classification is conducted based on the PUPs which is much significant in practice. Specifically, a linear SVM-based method is used to classify the load profiles into two groups, residents and small and medium-sized enterprises (SMEs), based on the PUPs with the assumption that electricity consumption behaviors are closely related to the socio-economic backgrounds of the corresponding customers. Then, both data compression-based and classification-based indices are defined and quantified to verify the effectiveness of the proposed technique.

6.2 Non-negative K-SVD-Based Sparse Coding

Sparse coding is firstly applied to load profiles to extract PUPs, which is implemented by K-SVD. In this section, the idea of sparse coding is firstly introduced and the non-negative K-SVD algorithm is also given.

6.2.1 The Idea of Sparse Representation

Research on sparse representations was inspired by the mechanism underlying networks of neurons in the brain [7]. The basic assumption of sparse coding is that a signal $x_i = [x_{i,1}, x_{i,2}, \ldots, x_{i,N}]^T$ with N dimensions which refers to one load profile in this chapter, can be represented in terms of a linear combination of K basic vectors. The basic vectors are called PUPs in this chapter. The representation of x_i may either be exacted,

$$x_i = \sum_{k=1}^{K} a_{i,k} d_k \tag{6.1}$$

or approximated,

$$x_i \approx \sum_{k=1}^{K} a_{i,k} d_k \tag{6.2}$$

where $d_k = [d_{k,1}, d_{k,2}, \ldots, d_{k,N}]^T$ denotes the kth PUP, which has N dimensions, $a_i = [a_{i,1}, a_{i,2}, \ldots, a_{i,K}]^T$ denotes the coefficient vector of K PUPs, which has K dimensions. These K PUPs form a redundant dictionary, $D \in R^{N \times K}$, where K is greater than N.

Generally, the lossy data compression algorithms include two parts, coding and reconstruction. The encoder transforms the original load profile into another format that requires less storage space, and the decoder recovers the load profile with minimal reconstruction loss. From a data compression perspective on sparse coding, given a certain dictionary, D, searching the coefficient vector, a_i, is load profile encoding, while the linear combination of basis vectors is essentially load profile reconstruction. Then, the original load profile, x_i, is transformed into a_i. Sparse coding attempts to obtain a sparse and redundant dictionary set for using in characterizing the original load profile. Sparsity means that only a few elements of a_i are nonzero; redundancy means that $K > N$. Figure 6.1 presents a visualization of sparse coding. The known redundant dictionary can be used to obtain the coefficient of each PUP by K-SVD which is introduced in the next part. It shows that only the first, third, fifth, and Kth coefficients are non-zero. That is to say, among these K PUPs, the presented load profile is a linear combination of only four (the 1st, 3rd, 5th, and Kth) PUPs. Therefore, in the encoding stage, the 48-dimensional load profile is transformed into four coefficients. Then, in the reconstruction stage, the load profile is restored according to Eq. (6.1).

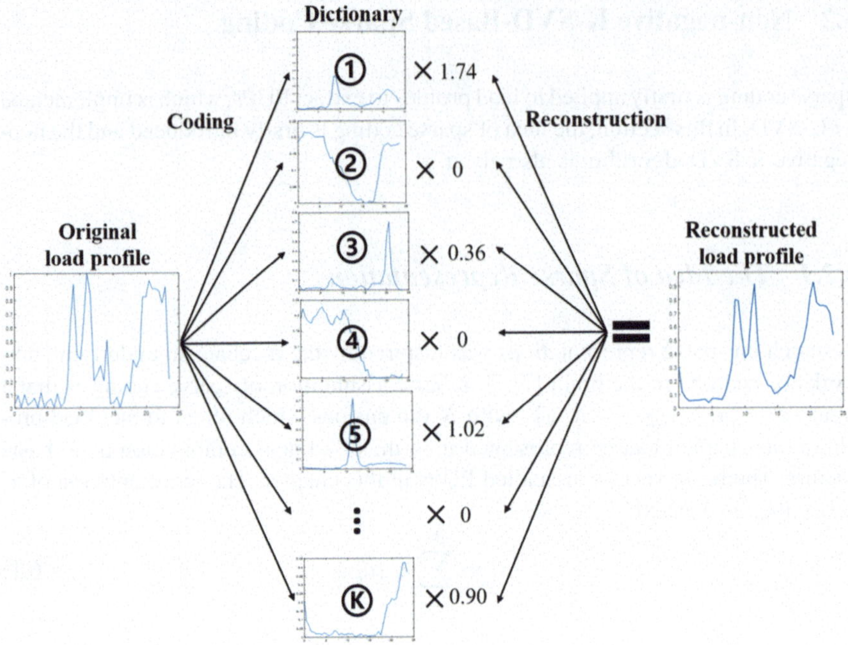

Fig. 6.1 A visualization of sparse coding

6.2.2 *The Non-negative K-SVD Algorithm*

For the sparse coding of M load profiles, Eq. (6.1) can be rewritten in matrix form as follows:

$$X = DA \tag{6.3}$$

where $X = [x_1, x_2, \cdots, x_M]$ denotes the set of M load profiles and $A = [a_1, a_2, \cdots, a_M]$ denotes the corresponding coefficient vectors. The electricity consumption behavior is influenced by various factors with large uncertainty and variation which can be viewed as noise. This chapter tries to compress these data by extract typical partial usage patterns. Thus, Eq. (6.3) is only approximately valid due to noise; that is to say, there is reconstruction loss, which should be minimized. Therefore, given a set of load profiles, X, sparse coding can be formulated as the following optimization problem:

$$\begin{aligned} &\min \|X - DA\|_F^2 \\ &s.t. \ \|a_i\|_0 \le s_0, \quad 1 \le i \le M \\ &\quad a_{i,k} \ge 0, \quad 1 \le i \le M, \quad 1 \le k \le K \\ &\quad d_{k,n} \ge 0, \quad 1 \le k \le K, \quad 1 \le n \le N \end{aligned} \tag{6.4}$$

where s_0 denotes the maximum number of nonzero elements in each coefficient vector, a_i; $\|\cdot\|_0$ denotes the l_0 norm.

The first constraint ensures that each load profile is represented with the target sparsity s_0, which is predetermined [3]; the second and third non-negative constraints on the coefficient vectors and the dictionary should be guaranteed because each customer's actual electricity consumption is non-negative. The root mean squared error (RMSE) is used to evaluate the representation performance of the algorithms. The Frobenius norm, $\|\cdot\|_F$, in Eq. (6.4) is defined as follows:

$$\|E\|_F = \sqrt{\sum_i \sum_j |e_{ij}|^2} \tag{6.5}$$

where, e_{ij} is the elements of E.

The optimization problem has two tasks: (1) search a redundant dictionary that captures the features or PUPs of load profiles as well as possible and (2) optimize the coefficient vector of each load profile to guarantee its sparsity and an acceptable reconstruction loss. The non-negative K-SVD algorithm proposed by Michal Aharon et al. [3] is an effective algorithm for solving Eq. (6.4) using a SVD based approach. The non-negative K-SVD algorithm can be considered a generalization of the k-means clustering algorithm, and it works by iteratively alternating between sparse coding and updating the dictionary [3]. During the sparse coding stage, the dictionary, D, is frozen and the set of load profiles, X, is coded by A. During the dictionary update stage, each basis vector is updated sequentially by defining the non-zero coefficients set ω_k and further minimizing the reconstruction error. Thus, the relevant coefficients are changed. There are three parameters that must be carefully determined for the non-negative K-SVD algorithm: the size of the dictionary, K, the sparsity constraint, s_0, and the number of iterations, J. When $s_0 = 1$, K-SVD is reduced to the traditional k-means clustering.

6.3 Load Profile Classification

Based on the extracted PUPs, load profile classification can be conducted. The effectiveness of sparse coding based feature extraction can be verified by the boost of classification accuracy. In addition, linear SVM can be used for feature selection and ranking. In this way, the most relevant PUPs can be selected.

6.3.1 *The Linear SVM*

SVMs, which are popular in classification techniques, attempt to find separating hyperplanes that maximize the distance between two classes of data, as shown in Fig. 6.2. The coefficient vector, a_i, can be regarded as the set of features extracted by sparse coding. Each load profile has a label, $y_i \in \{-1, 1\}$, that corresponds to "SME" or "resident" respectively. Thus, features-label pairs, (a_i, y_i), are obtained.

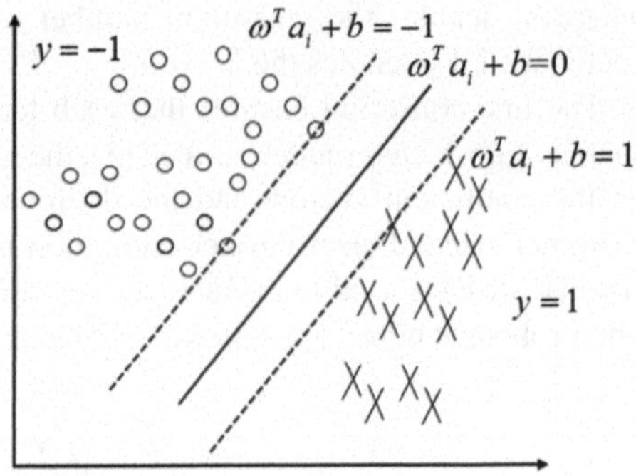

Fig. 6.2 A sketch map of an SVM

Then, the linear SVM is formulated as an optimization problem as follows [8]:

$$\min_{\gamma,\omega,b} \frac{1}{2}\|\omega\|^2 + C\sum_{i=1}^{m}\xi_i$$
$$s.t.\ y_i(\omega^T a_i + b) \geq 1 - \xi_i,\quad \xi_i \geq 0 \tag{6.6}$$

where, ω denotes the weights of the features; $C > 0$ is a penalty parameter for the training error; ξ_i denotes the loss function; b is the bias term in SVM. When the optimal value of C and weights of the features ω for any testing instance a_i have been found, the decision function is defined as

$$f(a_i) = \mathrm{sgn}(\omega^T a_i + b) \tag{6.7}$$

where sgn(·) is the sign function; its value is 1 or -1 when the input is positive or negative, respectively. There are four reasons to use a linear SVM for load profiles classification: (i) a linear SVM does not need to compute a kernel value for each pair of load profiles, which makes it run faster than other kernel-based SVMs. This means that a linear SVM is able to address large datasets; (ii) a linear SVM has only one parameter, C, that must be determined. The optimal value of C can be found quickly, in contrast to other types of SVMs that have two or more parameters that must be determined; (iii) the cross-validation accuracy of a linear SVM is as good as that of some kernel-based SVMs when the number of load profiles is large enough; and (iv) the weights of the features, ω, can be used to determine the relevance and importance of each feature in the linear SVM-based model.

6.3.2 Parameter Selection

The weights of the features, ω_i, represent the importance of the features in the decision function. A larger absolute value of ω_j means that the jth feature is more important and relevant in the classification model [9]. Note that only ω in linear SVM is meaningful. Thus, the features can be ranked according to the absolute value of ω. The most relevant features will be analyzed and presented in the section that describes the numerical experiments.

6.4 Evaluation Criteria and Comparisons

In this section, five criteria are proposed to evaluate the performance of the proposed method from the perspective of data compression and load profile classification. Besides, the theory of four commonly used data compression and feature extraction methods including k-means, DWT, PCA, and PAA, are briefly introduced.

6.4.1 Data Compression-Based Criteria

Lossy compression is essentially a compromise between the compression ratio (CR) and the loss of information.

(1) The CR is defined as the ratio between the sizes of the uncompressed data and compressed data. It is the ratio between the number of nonzero coefficients, s_0, and the dimension of original load profile, N, or

$$CR = s_0/N \tag{6.8}$$

(2) The RMSE (root mean squared error) is a frequently used measure of the reconstruction error,

$$RMSE = \sqrt{\frac{1}{NM}\sum_{i=1}^{M}\left(x_i - \sum_{j=1}^{K} a_j d_j\right)^2} \tag{6.9}$$

(3) The MAE (mean absolute error), another index that quantifies the reconstruction error, is defined as follows:

$$MAE = \frac{1}{NM}\sum_{i=1}^{M}\left|x_i - \sum_{j=1}^{K} a_j d_j\right| \tag{6.10}$$

It is worth noting that the relative error is not suitable for evaluating the loss of information because when the original data are close to zero, little absolute error will result in a great deal of relative error. Usually, the smaller the CR, the RMSE, and the MAE are, the better the compression algorithm is.

Table 6.1 Confusion matrix of the binary classifier

		Actual	
		SME	Resident
Predicted	SME	*TP*	*FP*
	Resident	*FN*	*TN*

6.4.2 Classification-Based Criteria

The electricity consumption of each customer is affected by various factors, which leads to greater uncertainty. Therefore, to a certain extent, there is a great deal of noise in each load profile. The proposed method should guarantee low reconstruction error and extract useful information. This chapter designs a binary SVM-based classifier (for residents and SMEs) to test and quantify whether and, if so, how much useful information is extracted. A confusion matrix can be obtained, as shown in Table 6.1. TP, FN, FP, and TN are defined as the numbers of positives correctly predicted as positives, positives incorrectly predicted as negatives, negatives incorrectly predicted as positives, and negatives correctly predicted as negatives, respectively.

(4) The accuracy of the classifier is the proportion of the data that are correctly labeled,

$$Accuracy = \frac{TP + TN}{TP + TN + FP + FN} \tag{6.11}$$

(5) The F1 score is essentially the harmonic mean of the recall and the precision. It is used to evaluate the classifier that corresponds to's performance on a dataset with an imbalance of labeled data.

$$recall = \frac{TP}{TP + FN} \tag{6.12}$$

$$precision = \frac{TP}{TP + FP} \tag{6.13}$$

$$F1 = \frac{2 \times precision \times recall}{precision + recall} \tag{6.14}$$

Both the accuracy and F1 score are values between 0 and 1. The higher the accuracy and F1 score are, the better the classifier performs. So far, we have proposed five indexes used in evaluating the performance of the proposed data compression and feature extraction method.

6.4.3 Comparisons

To verify the superiority of the proposed technique, we compare K-SVD with some other common data compression and pattern extraction algorithms, including k-means clustering, the DWT, PCA, and PAA, which are briefly introduced in this part.

(1) k-means clustering
k-means clustering is a clustering method that is commonly used to obtain typical electricity consumption patterns. Each load profile can be approximated as the center of the cluster it belongs to. From this point of view, k-means clustering is an approach to data compression. As stated above, k-means clustering is a particular case of K-SVD when the sparsity constrained is 1.

(2) The DWT
The DWT is an efficient signal processing technique for data compression and characterization [10]. In a Haar basis u_i, the coefficients, c_i, are computed as shown in Eq. (6.15) and sorted in order of decreasing magnitude. Then, the first s_0 coefficients are retained and the others are set to zero. The value of is s_0 also predetermined according to the required compression ratio. This reduction in the number of nonzero coefficients provides the compression. Finally, the inverse DWT (IDWT) is applied to the compressed coefficients to reconstruct the load profile.

$$x_i = \sum_{i=1}^{s_0} c_i u_i \tag{6.15}$$

(3) PCA
PCA is another technique that is commonly used for data compression and time series analysis [11]. PCA is a linear transformation technique that attempts to identify a new set of orthogonal coordinates for its original dataset. A new set of uncorrelated variables are derived from the actual interrelated variables in the data. These new variables, or principal components (PCs), are also sorted in decreasing order so that the front few capture more of the variations present in the original variables.

(4) PAA
PAA is an intuitive data compression technique that is often used with time series [12]. It first segments the time horizon into several equal parts and then, approximates the load profile by replacing the real values that fall in each time interval with their average values. By piecewise averaging, the "spikes" are filtered out, and the outline is retained.

(5) Lossless compression methods
A-XDR coding, DEGA (Differential Exponential Golomb and Arithmetic) coding, and LZMH (Lempel Ziv Markov Chain Huffman) coding are three state-of-the-art lossless compression algorithms for smart meter data proposed in [13]. For the datasets with a granularity of 15 min and one hour, the excepted compression ratios of these methods vary from 0.14 to 1 for the REDD load data set [13]. These methods have also been tested in the numerical experiments on the same dataset.

6.5 Numerical Experiments

We implement the numerical experiments using Matlab R2015a on a standard PC with an Intel® Core^{TM} i7-4770MQ CPU running at 2.40 GHz and 8.0 GB RAM. For data compression and pattern extraction, we employ the KSVD toolbox; for our classification method, we use LIBLINEAR toolbox.

6.5.1 Description of the Dataset

The dataset used in our study was provided by Electric Ireland and SEAI (Sustainable Energy Authority of Ireland). We select the load profiles of 500 customers (300 residents, 200 SMEs) over 100 days at a granularity of 30 min. After cleaning and normalization, the entire dataset, X, consists of 49,232 daily load profiles.

Figure 6.3 shows the average daily load profiles of residential customers and SMEs. The electricity consumption of residential customers increases gradually from 6:00 to 8:00, reaches a steady-state until 9:00, and remains approximately constant between 8:00 and 16:00. Then, the consumption continues increasing and peaks at approximately 20:00. The electricity consumption of SMEs remains high during working hours, from 9:00 to 17:00. The consumption the rest of the time is relatively low in comparison to that of residential customers.

Figure 6.4 shows the daily load profiles of a resident and an SME for one week. The electricity-consuming behavior of resident #1002 is significantly different from that of SME #1021. Resident #1002's consumption reaches its peak at noon and is higher at 20:00 and 24:00. In contrast, there are only two short-duration peaks at approximately 8:00 and 21:00 in SME #1021's consumption. The rest of the

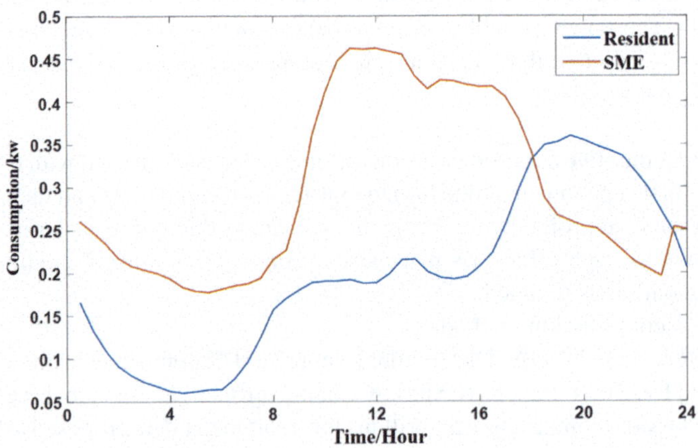

Fig. 6.3 Averaged daily load profiles of residential customers and SMEs

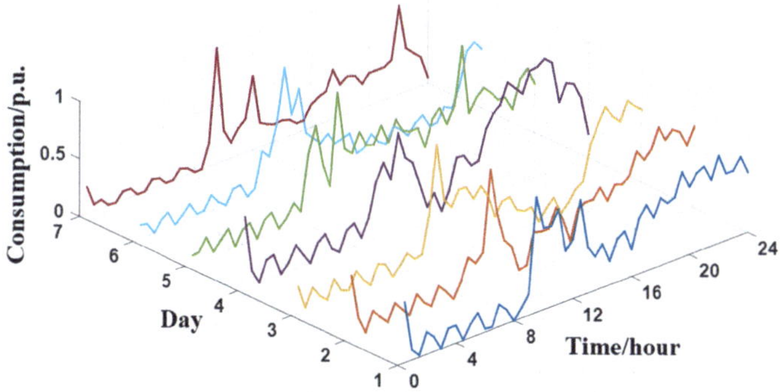

(a) Resident#1002

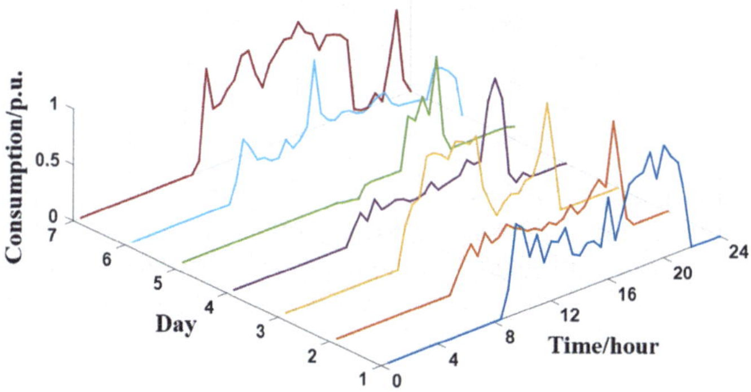

(b) SME#1021

Fig. 6.4 Daily load profiles of a resident and an SME for one week

time, electricity consumption is much lower due to some constantly running electric appliances, such as refrigerators. Each customer has different usage patterns on different days in terms of peak hours and peak durations. The peaks in morning and at night can be decomposed, which shows the sparsity of these load profiles.

6.5.2 Experimental Results

As explained above, the RMSE and the MAE are determined by the size of the dictionary K and the CR. The value of s_0 depends on the requirements of compression ratio. While the value of K is indeed determined by several times of trials by considering its influence on recovery error (RMSE) and classification accuracy. Larger K will result smaller RMSE. However, when the value of K is much smaller, it will have larger influence on the RMSE; when the value of K gets larger, the influence

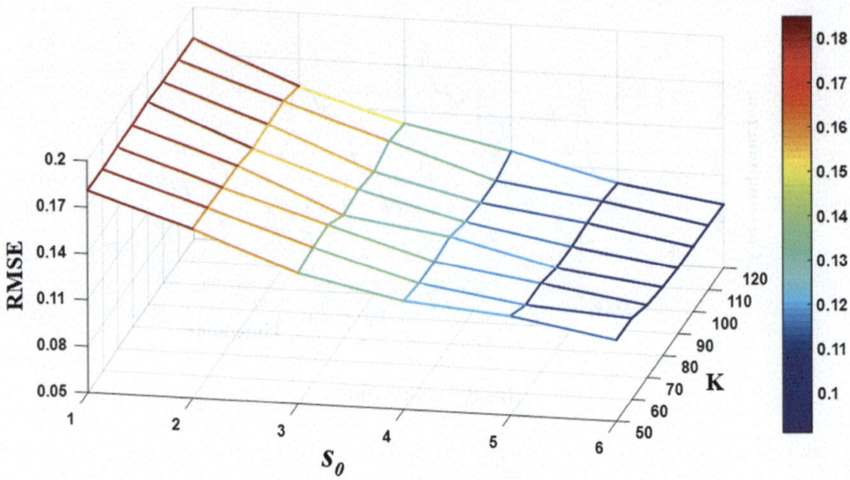

Fig. 6.5 The RMSE of the K-SVD algorithm as the parameters vary

of it will be much weaker. We vary K from 60 to 120 in intervals of 10, s_0 from 1 to 6 in steps of one unit and run the K-SVD algorithm for 42 iterations to determine how these two parameters influence the compression quality and classification accuracy. As shown in Fig. 6.5, the RMSE decreases as s_0 and K increase. s_0 is a better indicator of the compression quality than K is because the dictionary becomes more descriptive as K increases.

Then, using the linear SVM-based classification model described above, we classify the load profiles based on the extracted patterns by varying K from 20 to 120 in steps of 10 units and s_0 from 1 to 6 in steps of one unit. The accuracy of each case is shown in Fig. 6.5. When K is too large, many non-typical or meaningfulness PUPs may be extracted; while when K is too small, enough typical PUPs cannot be captured due to the limitation of the size of dictionary. The relationship between accuracy and value is much complex and is naturally non-linear and non-monotonous. However, when K varies from 50 to 90, the accuracy fluctuates up and down with little difference. Among the trials on the values of K and s_0 in this experiment, when $s_0 = 5$ and $\mathrm{K} = 80$, which is a "knee point" in Fig. 6.5, the classification accuracy is the highest as shown in Fig. 6.6. This means that five PUPs are enough to describe the customers' consumption patterns and that a better trade-off between s_0 and the RMSE can be achieved. Therefore, K should not be too large for three reasons: (i) the impact of K on the RMSE is much smaller than that of s_0; (ii) a large dictionary will increase the time complexity of the sparse coding-based method; and (iii) typical PUPs cannot be captured effectively when K is too large as stated before.

If the smart meter data are stored in float type. 49,232 daily load profiles will take up a storage space of 9.015 MB (49232*48*4 Byte = 9.015 MB). When $s_0 = 4$ and $K = 80$, the size of the dictionary and compressed data set will be 15 KB (80*48*4 Byte = 15 KB) and 961.56 KB (49232*5*4 Byte = 961.56 KB).

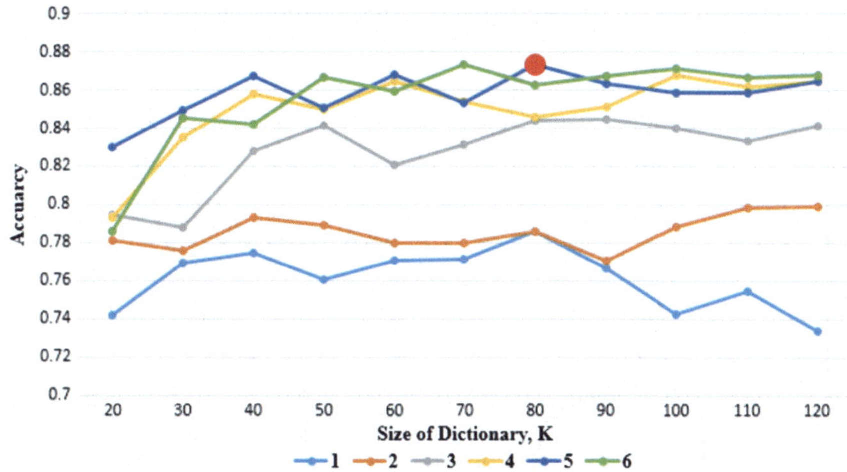

Fig. 6.6 The accuracy of the K-SVD algorithm as the parameters vary

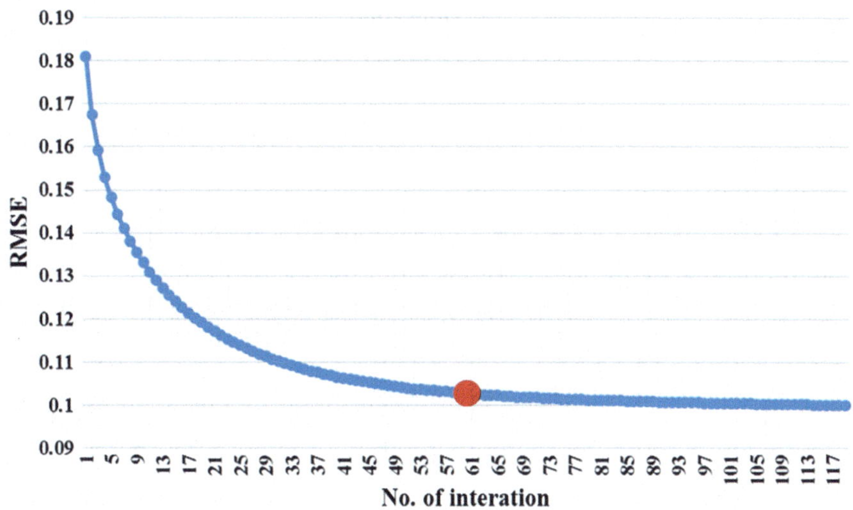

Fig. 6.7 The RMSE of the K-SVD algorithm for different numbers of iterations

We also record the RMSE at each iteration of the K-SVD algorithm, as shown in Fig. 6.7. The RMSE decreases slightly when the number of iterations is greater than 60. Therefore, we choose 60 for the number of iterations, J, in our case studies. Figure 6.8 shows the reconstructions of four typical loads using the K-SVD algorithm. The solid and dotted lines are the original and reconstructed load profiles, respectively. The overall trend of each load profile is identified, and most of the peaks are reproduced.

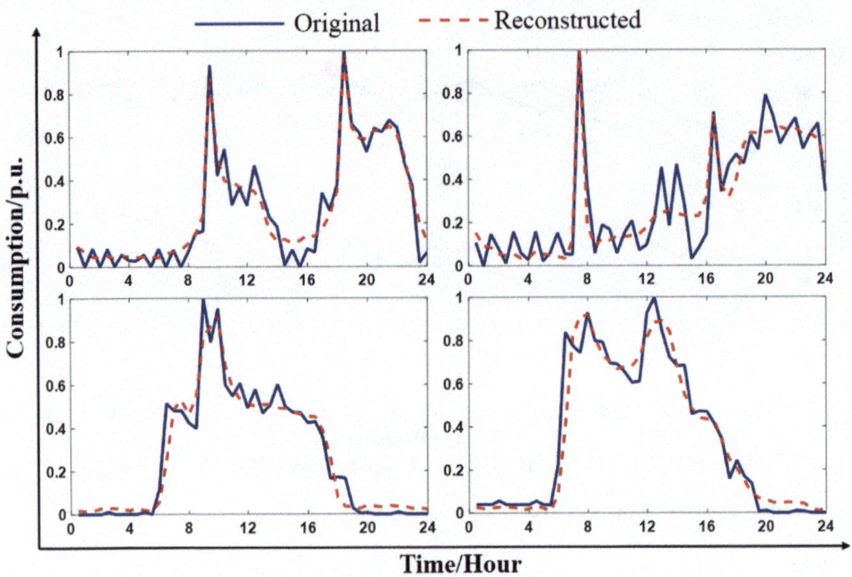

Fig. 6.8 Load profiles reconstructed using the K-SVD algorithm

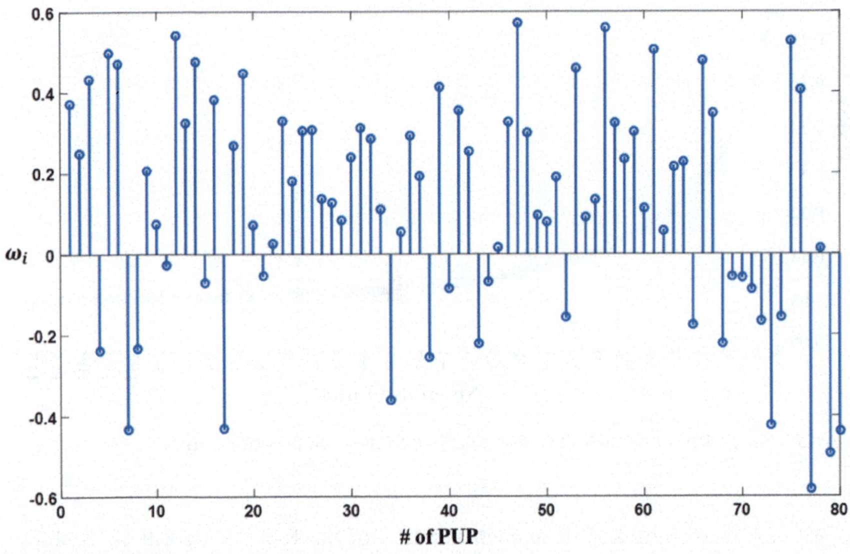

Fig. 6.9 The ω_i of different PUPs identified by the K-SVD algorithm

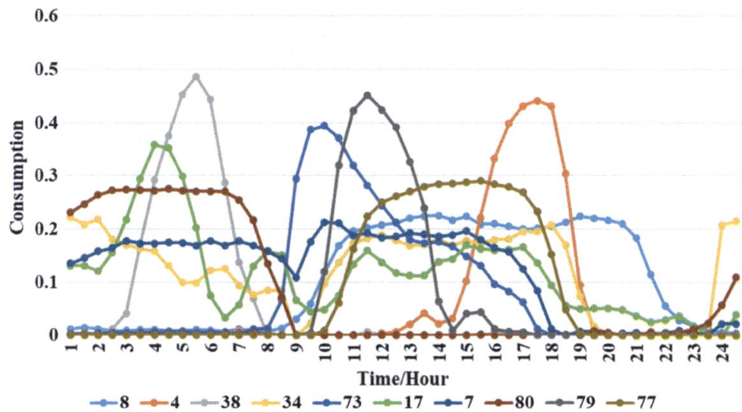

Fig. 6.10 The ten most relevant and important PUPs for SMEs

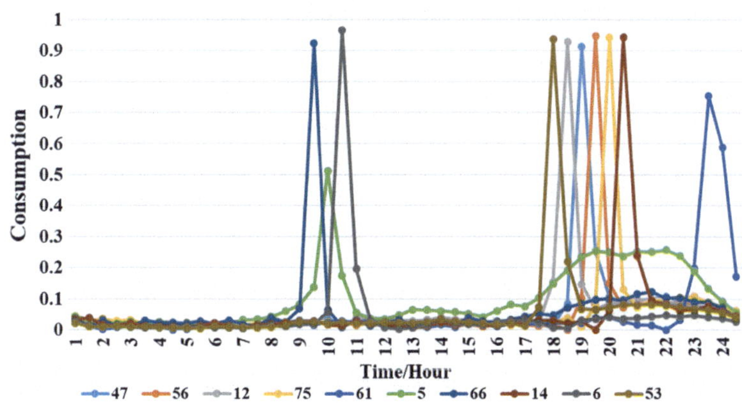

Fig. 6.11 The ten most relevant and important PUPs for residential customers

Table 6.2 Comparison of the PUPs of SMEs and residential customers

	Shape	Duration	Peak times
SME	Vaulted	Long	Dawn, working hours
Resident	Sharp peak	Short	Morning, night

Based on the extracted PUPs, the classification is performed with $s_0 = 5$, $K = 80$, and $J = 60$. The ω_i for the 80 PUPs are shown in Fig. 6.9. The number of negative ω_i is much smaller than the number of positive ω_i, which means that the consumption patterns of SMEs are less diverse than those of residential customers.

Figures 6.10 and 6.11 show the ten most relevant and important PUPs for SMEs and residential customers, respectively, according to the absolute value of each ω_i. We summarize the differences between the PUPs for SMEs and residents in Table 6.2.

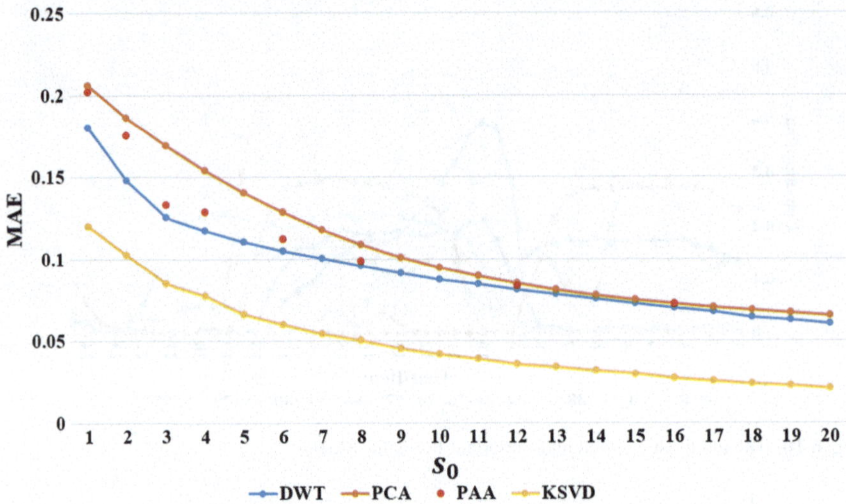

Fig. 6.12 The compression quality of the K-SVD algorithm, the DWT, PCA, and PAA

Table 6.3 Compression rations (CRs) of typical lossless compression methods

	DEGA coding	LZMH coding	A-XDR coding
CR	0.257	2.75	0.693

6.5.3 *Comparative Analysis*

We retain the largest s_0 coefficients of the DWT and the PCs of the PCA, and then, calculate the MAE of each case by varying s_0 from 1 to 20 in steps of one unit. We also perform PAA by dividing the 48 time periods into 1, 2, 3, 4, 6, 8, 12, and 16 parts. As shown in Fig. 6.12, the K-SVD algorithm provides the best compression quality for all values of s_0. We also have tested the performance of three state-of-the-art lossless compression methods on the dataset from Electric Ireland and SEAI. These methods include DEGA coding, LZMH coding, and A-XDR coding. Their CRs are summarized in Table 6.3. The results show that DEGA coding performs best among these three state-of-the-art lossless compression methods. The compression ratios of DEGA coding is 0.257. Compared with DEGA coding, the proposed sparse coding-based method can achieve a compression ratio of 0.083 when s0 is set to 4 with very little reconstruction error (only 0.066 measured by MAE).

Figure 6.13 shows reconstructions of the load profile shown in Fig. 6.8 performed using the DWT, PCA and PAA when $s_0 = 6$. The performance is worse than that of the K-SVD algorithm when $s_0 = 5$. PAA and PCA can identify trends in the load profiles; the DWT can retain the peak value of each load profile. However, the K-SVD algorithm can capture the trend and the peak of each load profile simultaneously, as shown in Fig. 6.8. The K-SVD algorithm performs better because individual load

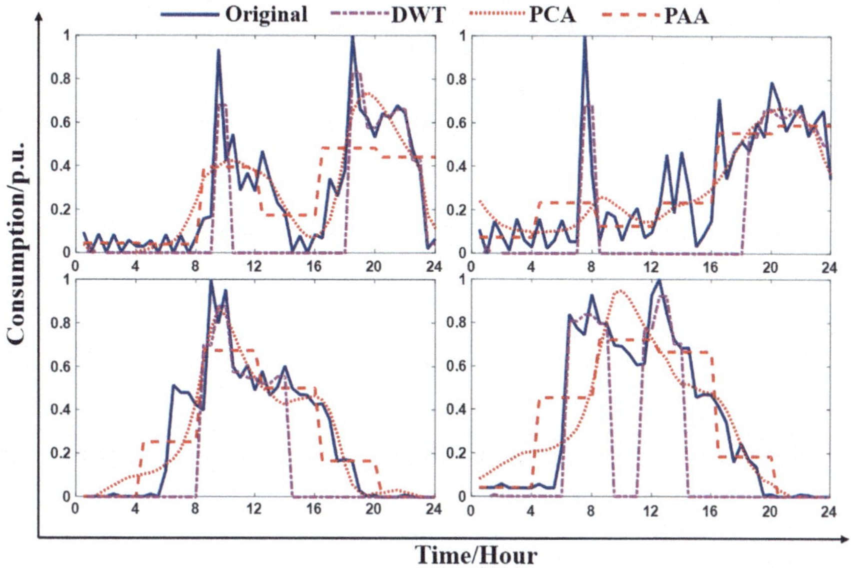

Fig. 6.13 A load profile reconstructed using the DWT, PCA, and PAA

profiles vary significantly and have fixed consumption patterns, which makes them suitable for sparse coding.

As compression algorithms, the K-SVD algorithm and the DWT are very similar because they can be viewed as using a linear combination of several basis vectors. The basis vectors of the DWT are predefined and orthogonal. Those of the K-SVD algorithm are non-orthogonal and can be adapted to the characteristics of the set of load profile. These are all lossy compression techniques. The DWT and PCA can recover a load profile without information loss when all 48 elements are used; however, information is still lost by the K-SVD algorithm when $s_0 = 48$.

Despite its time complexity, the coding performed in a PCA is explicit, while those of the DWT and the K-SVD algorithm are implicit, which means that optimization or another certain operation is necessary. PCA and the DWT involve only linear operations. The time required for coding in the K-SVD algorithm is about 6 hours which is much higher than that required by PCA and the DWT, but it is still acceptable in practice. A compressed load profile does not exactly require real-time acquisition but does require that the data be transferred daily.

Table 6.4 compares the performance of the K-SVD algorithm with those of k-means clustering ($k = 80$), PCA ($s_0 = 5$), the DWT ($s_0 = 5$), and PAA ($s_0 = 6$) in terms of the five proposed criteria. Except for k-means clustering and the original load profiles, the K-SVD algorithm has lower reconstruction error and higher classification accuracy. In particular, accuracy is significantly improved. k-means clustering, as a special case of the K-SVD algorithm, performs better than the other techniques and has larger reconstruction errors at the same time. The original load profiles are also

Table 6.4 Comparisons with different techniques

	Parameter	RMSE	MAE	Accuracy	F1 Score
K-SVD	5, 80	0.099	0.060	0.874	0.793
k-means	80	0.120	0.180	0.786	0.752
PCA	5	0.111	0.167	0.771	0.764
DWT	5	0.141	0.327	0.667	0.688
PAA	6	0.112	0.181	0.706	0.725
Original	48	/	/	0.735	0.724

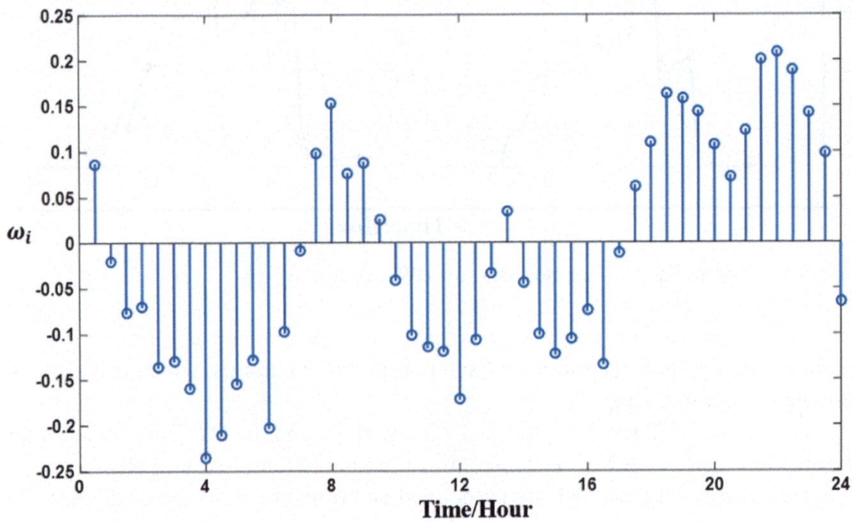

Fig. 6.14 ω_i at different times of the day

classified and Fig. 6.14 shows the ω_i for different times of day. The negative values of ω_i are mainly concentrated in the morning and at night, and the positive value of ω_i are mainly concentrated during working hours and at dawn, which is consistent with the results of the K-SVD algorithm.

6.6 Further Multi-dimensional Analysis

6.6.1 Characteristics of Residential & SME Users

From the dataset, we can clearly see that residential users and SME users have significant consumption preference. Figure 6.15 shows the four load profiles for the two kinds of users, which are drawn by simply applying *k*-means clustering to the dataset.

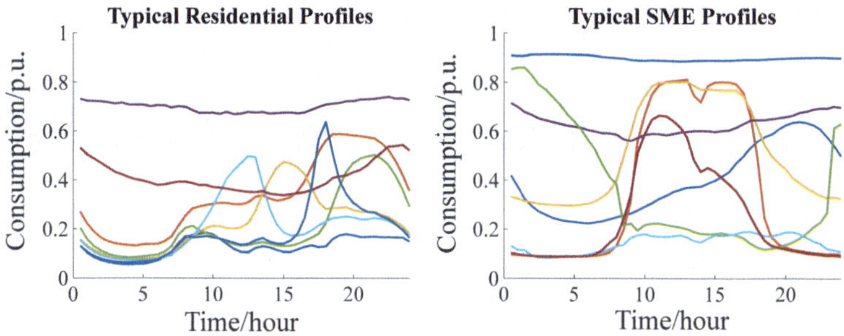

Fig. 6.15 Residential & SME load profiles generated by k-means

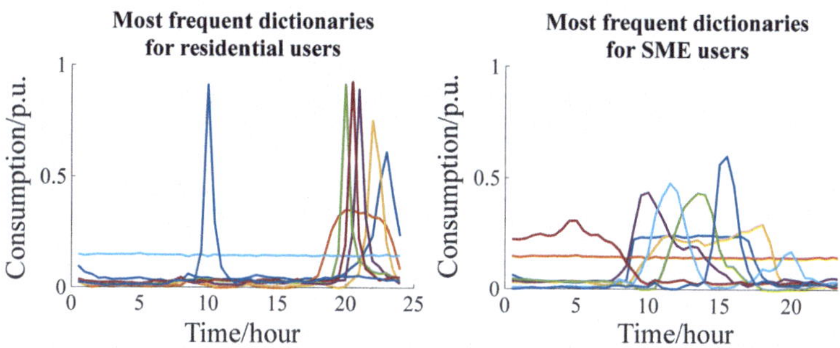

Fig. 6.16 The 8 most frequently used dictionaries or PUPs

Actually, residential profiles usually have short and strong peaks at certain periods of time in a day, while SME users have less variability and their consumptions last longer. This is not properly presented in Fig. 6.15, because the means of profiles reduce the fluctuation. Most traditional clustering methods usually require a step of calculating the centroid of a cluster. However, a centroid is sometimes not representative enough, and we can see that PUPs perform well in capturing the features and keeping the original information of the load profiles from Figs. 6.15 and 6.16.

Figure 6.16 shows the 8 most frequently used PUPs for residential and SME users. For residential users, some persistent PUPs at night like the orange line can be considered as the usage of television or personal computers before sleep. Some peak shape PUPs might correspond to using microwave oven or dryer at a certain time. For SME users, most PUPs have persistent load during office hour but the peak time is usually different. There is a PUP that lasts for a whole day for both kinds of users, which corresponds to appliances like refrigerators or fresh air systems.

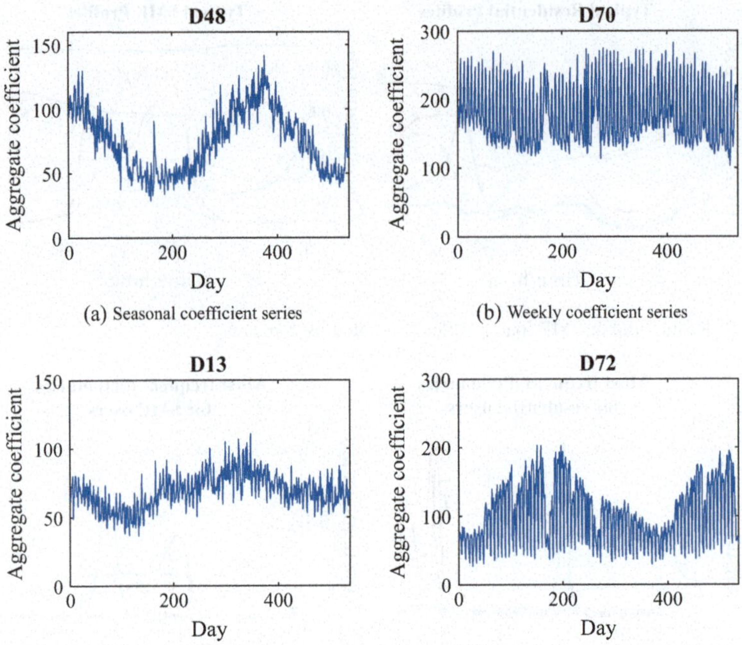

(a) Seasonal coefficient series (b) Weekly coefficient series (c) Constant coefficient series (d) Mixed type coefficient series

Fig. 6.17 Coefficient series of four typical PUPs

6.6.2 Seasonal and Weekly Behaviors Analysis

Periodic patterns can be extracted from users' sparse coding, and seasonal-related PUPs can be defined according to their coefficient. Residential users are considered to have stronger seasonal patterns, so we use them as an example. We aggregate the whole residential dataset with 3411 users and add up their coefficient for each of the 536 days. Our analysis can be extended to user aggregates of the other size, and the results are similar in most cases. Typically we can see three kinds of PUPs: seasonal PUPs, weekly PUPs, and constant PUPs. The coefficient of seasonal PUPs has an approximate period of 365 days. As for the weekly PUPs the period is 7 days, and for constant PUPs the variation of the coefficient is relatively small. Figure 6.17 shows the coefficient series of four typical PUPs. Figure 6.17a shows the residential aggregate uses less PUP#48 during winter. Figure 6.17b illustrates a weekly pattern of PUP#70. Figure 6.17c is a constant PUP and Fig. 6.17d is a seasonal-weekly mixed PUP.

To quantify the periodic characteristics of the PUPs, time series decomposition methods can be applied to the coefficient series. Since the length of the series is 536 which is not long enough for most decomposition methods to deal with a period of 356, here we use Discrete Fourier Transform (DFT) to extract the spectra of the coefficient series. Figure 6.18 shows the spectrum of PUP# 70 using DFT. Due to

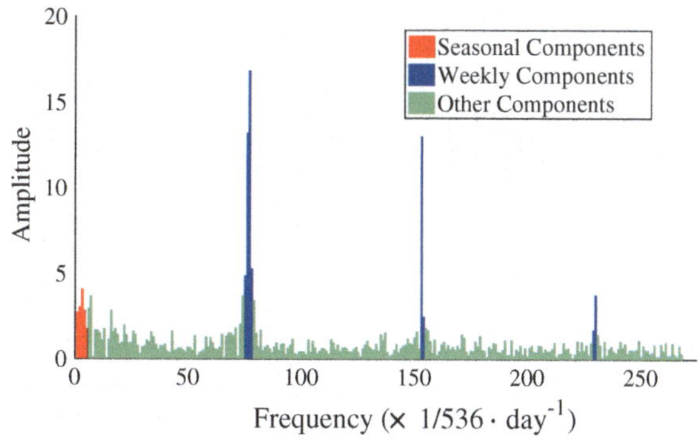

Fig. 6.18 The spectrum of PUP#70

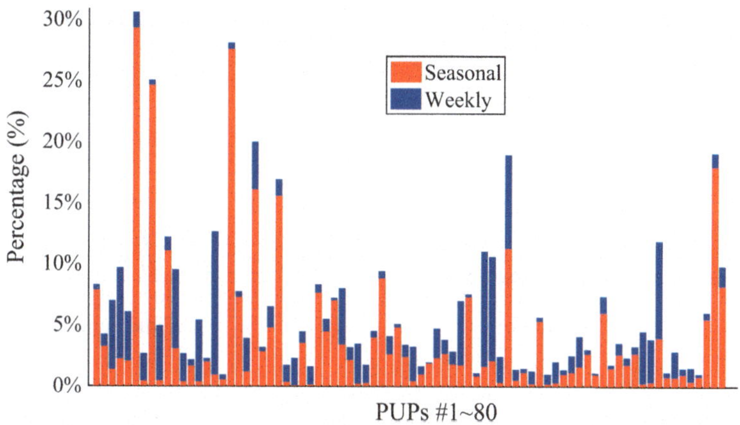

Fig. 6.19 Energy percentage of periodic components for PUPs

the non-sinusoidality of the periodic components and leakage error, seasonal and weekly components correspond to several lines marked with different colors in the spectrum. Note that we only show the half spectrum because of symmetry and that the DC component is not plotted.

The energy of a component in DFT is defined as its squared amplitude in the spectrum. We calculate the proportion of energy for the 80 PUPs and Fig. 6.19 shows the results. Based on the amplitudes and phases of the DFT results, we can determine which PUPs residential use more in winter or summer, as well as in weekdays or weekends. Figure 6.20 shows typical winter PUPs and summer PUPs for residential users. Residential users tend to consume more electricity in the evening during winter. In one of the summer PUPs, we can see people start consuming electricity at midnight. To some extent, this PUP marks air-conditional usage during bedtime. Also, if we look carefully at its coefficient series, we can see there is a sudden increase around Christmas every year. This is likely to correspond to events such as night parties

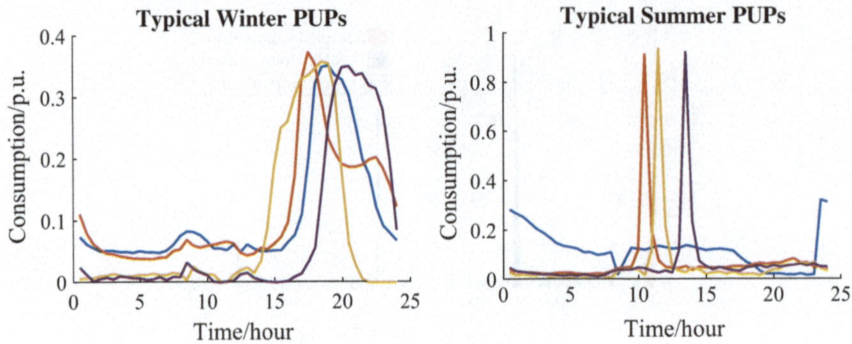

Fig. 6.20 Typical seasonal PUPs

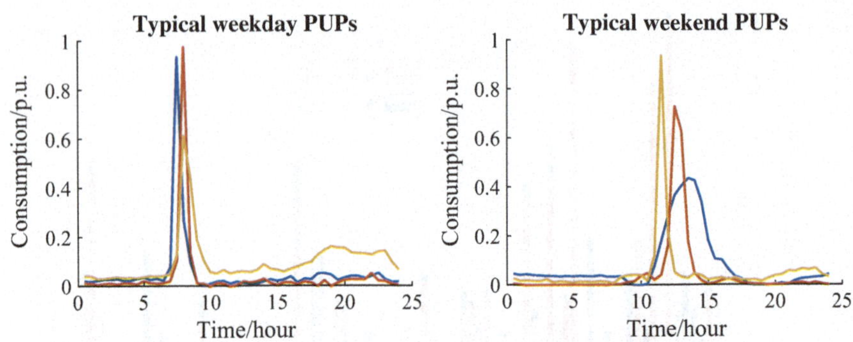

Fig. 6.21 Typical weekly PUPs

which happens more during summer or winter holidays. Figure 6.21 shows the typical weekly PUPs. Residential users clearly consume electricity earlier on weekdays when they have to go to work.

6.6.3 Working Day and Off Day Patterns Analysis

For every load profile, the dominant PUP is defined with the biggest coefficient. In the SME pre-trail survey, Question 61022 gave us the information about whether a user works during weekends. However, the survey only covers 290 of the 347 SME users. We counted their dominant PUPs on their working day and off day respectively. The average frequency a PUP_i appears on a working day and on an off day is defined as w_i and o_i. And p_i defined as $w_i/(w_i + o_i)$ measures the possibility of a working day when PUP_i appears. A working day pattern of SME Users has a greater p_i and an off day pattern has a smaller one. Figure 6.22 shows the typical SME PUPs for

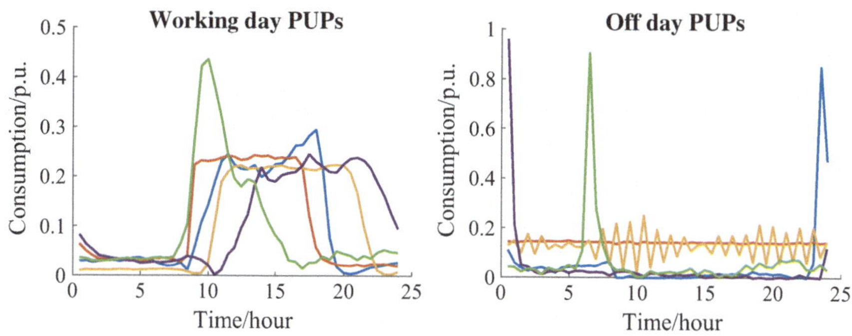

Fig. 6.22 Working day and off day patterns of SME users

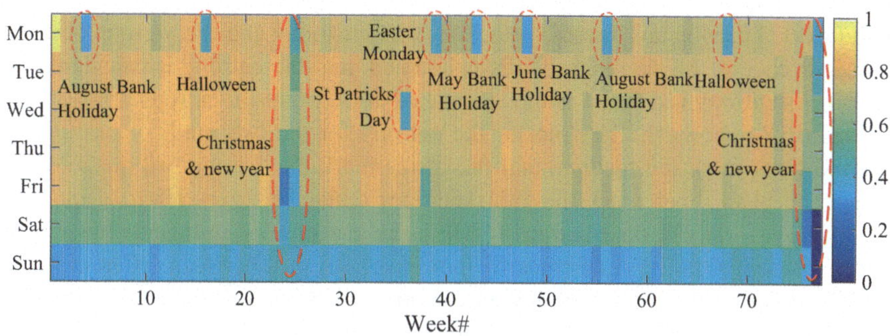

Fig. 6.23 Prediction of working day or off day

working day and off day. The working/off patterns can be applied in designing price packages and load forecasting.

To demonstrate the effectiveness of the two kinds of PUPs, we do a simple test on the remaining 57 users. The color bar in Fig. 6.23 marks the proportion of working day PUPs on a specific day for the 57 users. A yellow one indicates a working day, and a blue one indicates an off day. The results in Fig. 6.23 is consistent with weekends and all the public holidays in Ireland without any prior knowledge. As we can see, some SMEs work on Saturdays but fewer on Sundays. The prediction of working/off day is not only useful in energy services but also a good reference for economy and labor market.

6.6.4 *Entropy Analysis*

While periodic pattern analysis is useful in load forecasting, entropy analysis measures the variability of a customer, which can help find potential targets for demand response programs. The 536 days are classified into 7 groups according to the day

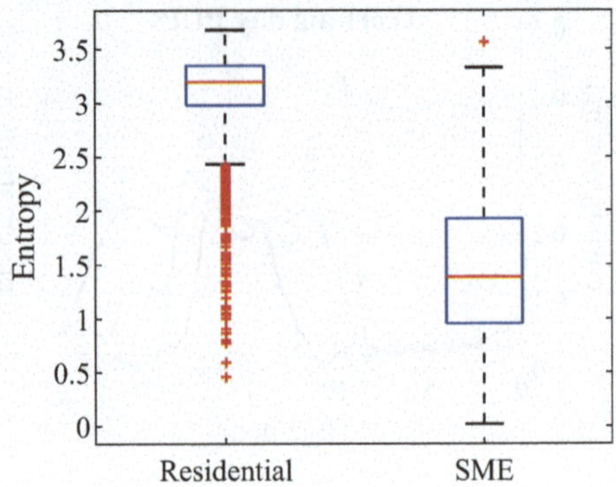

Fig. 6.24 Box-plot of entropy for residential and SME users

of the week, i.e. Monday, Tuesday, etc. For every customer and every group of days, the occurrence of his dominant PUPs is counted, and the entropy is calculated. A customer's entropy is defined as the average of the 7 groups of entropy.

Figure 6.24 shows the box-plot of the entropy for the two kinds of customers. The red lines mark the median, and the boxes mark the 1st and 3rd quantiles q_1 & q_3. The black lines mark the Whisker lines defined as $q_3 + 1.5(q_3 - q_1)$ and $q_1 - 1.5(q_3 - q_1)$. The distribution of SME users' entropy is significantly lower than that of residential users. A customer is more likely to shift between different PUPs on a fixed day of the week with higher entropy, indicating that his consumption is more flexible. Also, a lot of residential entropy is below the lower Whisker lines and marked as outliers. In some cases, this is due to bad data with zero measurements. In other cases, it indicates that there is usually nobody at home so that the consumptions are consistently low.

6.6.5 Distribution Analysis

The representation coefficient is very important in the characterization of a user's behavior. During a period of time, some PUPs are preferred by a customer, and this information can be utilized by electricity retailers to offer some personalized electricity price packages to the customer. The distribution of the coefficient can show the preference of the customer.

Figure 6.25 shows the coefficient distribution for user #6665. Summer and winter distribution is calculated by adding up the coefficient in summer and winter days, respectively. PUP#29, #78, #20 are preferred in summer and PUP#79, #24, #20 are preferred in winter. Consumption preferences can be used to design more personalized price packages or energy services.

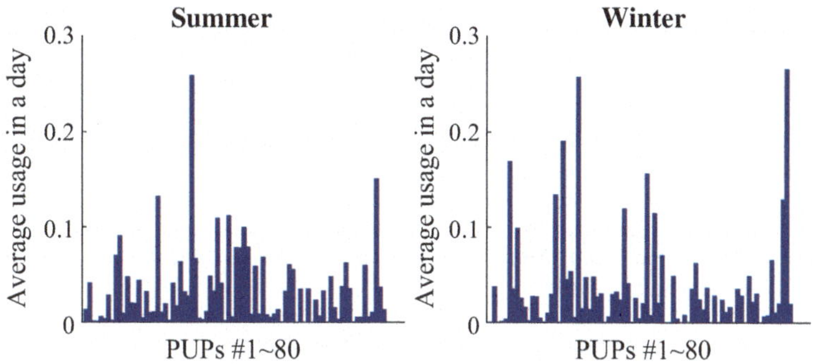

Fig. 6.25 User#6665's coefficient distribution in summer/winter

6.7 Conclusions

This chapter proposes a non-negative K-SVD-based sparse coding technique for electricity consumption data compression and pattern extraction. The load profiles are decomposed into typical PUPs, which reveal the usage patterns of customers. To demonstrate the effectiveness of the technique, comprehensive comparisons with the results of *k*-means clustering, the DWT, PCA, and PAA are conducted. The results show that the proposed non-negative K-SVD-based technique can achieve higher compression ratios and lower information losses. In addition, the atoms of the dictionary can be interpreted well based on their shapes.

Multi-dimensional analyses show that PUPs are able to capture important features of a load profile including periodic behaviors, working and off day patterns as well as consumption entropy and distribution. The PUPs help in modeling consumption behaviors for different user aggregates, which is very meaningful in load forecasting. The entropy and distribution measure the consumption variation and preferences of an individual user, helping in finding potential targets for demand response programs and offering personalized electricity services.

References

1. Gao, P., Meng, W., Ghiocel, S. G., Chow, J. H., Fardanesh, B., & Stefopoulos, G. (2016). Missing data recovery by exploiting low-dimensionality in power system synchrophasor measurements. *IEEE Transactions on Power Systems*, *31*(2), 1–8.
2. Yoshua, B., Aaron, C., & Pascal, V. (2013). Representation learning: a review and new perspectives. *IEEE Transactions on Pattern Analysis & Machine Intelligence*, *35*(8), 1798–1828.
3. Aharon, M., Elad, M., & Bruckstein, A. M. (2005) K-SVD and its non-negative variant for dictionary design. In *Wavelets XI*
4. Piao, M., & Ryu, K. H. (2016) Subspace frequency analysis-based field indices extraction for electricity customer classification. *ACM Transactions on Information Systems*, *34*(2), 1–18.

5. Wang, Y., Chen, Q., Kang, C., Zhang, M., Wang, K., & Zhao, Y. (2015). Load profiling and its application to demand response: A review. *Tsinghua Science and Technology*, *20*(2), 117–129.
6. Piao, M., Shon, H. S., Lee, J. Y., & Ryu, K. H. (2014) Subspace projection method based clustering analysis in load profiling. *IEEE Transactions on Power Systems*, *29*(6), 2628–2635.
7. Olshausen, B. A., & Field, D. J. (1996) Emergence of simple-cell receptive field properties by learning a sparse code for natural images. *Nature*, *381*(6583), 607.
8. Chang, K. W., Hsieh, C. J., & Lin, C. J. (2008) Coordinate descent method for large-scale L2-loss linear support vector machines. *Journal of Machine Learning Research*, *9*(3), 1369–1398.
9. Chang, Y. W., & Lin, C. J. (2008) Feature ranking using linear SVM. In *Causation and prediction challenge* (pp. 53–64)
10. Ning, J., Wang, J., Gao, W., & Liu, C. (2011). A wavelet-based data compression technique for smart grid. *IEEE Transactions on Smart Grid*, *2*(1), 212–218.
11. Mehra, R., Bhatt, N., Kazi, F., & Singh, N. M. (2013) Analysis of pca based compression and denoising of smart grid data under normal and fault conditions. In *2013 IEEE International Conference on Electronics, Computing and Communication Technologies* (pp. 1–6). IEEE.
12. Lin, J., Keogh, E., Lonardi, S., & Chiu, B. A. (2003) Symbolic representation of time series, with implications for streaming algorithms. In *Proceedings of the 8th ACM SIGMOD Workshop on Research Issues in Data Mining and Knowledge Discovery* (pp. 2–11). ACM.
13. Unterweger, A., Engel, D., & Ringwelski, M. (2015) The effect of data granularity on load data compression. In *DA-CH Conference on Energy Informatics* (pp. 69–80). Springer.

Chapter 7
Personalized Retail Price Design

Abstract Designing customizing prices is an effective way to promote consumer interactions and increase customer stickiness for retailers. Fueled by the increased availability of high-quality smart meter data, this chapter proposes a novel data-driven approach for incentive-compatible customizing time-of-use (ToU) price design based on massive historical smart meter data. Consumers' ability to choose freely and consumers' willingness are fully respected in this framework. The Stackelberg relationship between the profit-maximizing retailer (leader) and the strategic consumers (followers) in an incentive-compatible market is modeled as a bilevel optimization problem. Smart meter data are used to estimate consumer satisfaction and predict consumer behaviors and preferences. Load profile clustering is also implemented to cluster consumers with similar preferences. The bilevel problem is integrated and reformulated as a single mixed-integer nonlinear programming (MINLP) problem and then simplified to a mixed-integer linear programming (MILP) problem. To validate the proposed model, the smart meter dataset from the Commission for Energy Regulation (CER) in Ireland is adopted to illustrate the whole process better.

7.1 Introduction

How to make full use of the smart meter data to promote better demand-side management has been a major focus area for utility companies with the increasing installation of smart meters [1, 2]. Smart meters can provide the retailer with more detailed high-quality information about the electricity consumption activities, and the retailer can use the data to extract the electricity consumption patterns of consumers and develop innovative customizing retailing strategies. It is appealing to the retailer that it could increase profits and market penetration while maintaining consumers' willingness through personalized pricing or customizing pricing [3]. There has been a surge need in researches on how to effectively and practically implement customizing pricing for retailers in the power market [4].

The study of customizing pricing originates from the researches of demand-side energy management. To promote better energy management, different types of time-varying tariffing approaches have been proposed by giving consumers incentives,

Y. Wang et al., *Smart Meter Data Analytics*,
https://doi.org/10.1007/978-981-15-2624-4_7

among which time-of-use (ToU) pricing is more widely adopted for its less volatility and risk for consumers [5]. The energy management problems faced by the retailer focuses on the bidding strategy [6–9] and retailing price design [10–12], both of which are highly correlated to each other. These two kinds of energy management problems are commonly designed as a Stackelberg game [6–12] which is a hierarchical control problem in which sequential decision making of the retailer and the consumers occurs. Smart meter data enables us to know the detailed differences among consumers, and further improvement could be made in the situation where every different individual strategic consumer may behave differently.

Some interesting works related to customizing pricing design have been developed in recent years. [13, 14] focus on how to identify the differences among consumers through different clustering methods and statistical analysis. [15] uses appliance identification to find a fine-grained method to simulate consumer behaviors. [16] proposes a game between the retailer and different types of consumers (residential,industrial,commercial). In a market consisting of a single type of users, as one of the vital considerations of energy management, the market mechanism must be *incentive compatible* [17, 18]. Consumers should be allowed to choose freely, and consumers' willingness should be fully respected. Incentive compatibility states in economics and game theory that the incentives should motivate the actions of individual participants (consumers) to behave in a manner consistent with the rules established by the agent (retailer) [19]. In a voluntary optional market, each consumer evaluates the benefits of each tariff scheme provided by the retailer and selects the one that offers the greatest benefits, so the consumer is supposed to prefer his/her designated scheme than any other one [17]. Problems may arise in these models which design pricing schemes for different individuals separately because the retailer still needs to check each consumers' satisfaction for all the schemes to guarantee that each consumer prefers the pricing scheme just tailored for him. Otherwise, it will result in ill incompatible incentive design which also suggests a huge deviation between consumer's will and the retailer's expectation.

Fueled by the increased availability of high-quality smart meter data, this chapter proposes a novel data-driven approach for incentive-compatible customizing time-of-use (ToU) price design based on massive historical smart meter data. The Stackelberg relationship between the profit-maximizing retailer (leader) and the strategic consumers (followers) along with the considerations for the incentive-compatible market is modeled as a bilevel optimization problem. Smart meter data is used to estimate the satisfaction of the consumer, predict consumer behaviors and preferences inspired by the work of [20]. Load profile clustering is also implemented to gather consumers of similar preferences. The bilevel problem is integrated and reformulated as a single mixed-integer nonlinear programming (MINLP) problem and then simplified to a mixed-integer linear programming (MILP) problem. To validate the proposed model, The smart meter dataset from Commission for Energy Regulation (CER) in Ireland is adopted to better illustrate the whole process.

7.2 Problem Formulation

7.2.1 Problem Statement

We consider the customizing price design problem in a one-to-many situation: one retailer and many consumers. The retailer determines the price schemes to schedule its optimal profits (upper-level problem), while the consumer adjusts its flexible consumption (lower-level problem) according to his choice of these pricing schemes.

The structure of an electricity retailing market is shown in Fig. 7.1. The retailer is considered as a price-maker in the market of consumer-level, who can change the price without losing consumers as long as the retailer's pricing schemes don't make consumers discontent. Meanwhile the retailer is also a price-taker in the market of distribution-level because distributional market-clearing involves behaviors of different stakeholders and relies on complex schedules and the price to buy electricity from the distributional-level market mainly depends on the contract price and the distribution locational marginal prices (DLMPs). For consumers, they are assumed to operate on the principle of utility maximization. A consumer will only choose the optimal pricing scheme to maximize his utility. The consumer is also a habitual decision-maker whose preference will not change drastically.

Each day is divided into T periods which depends on the smart meter sampling interval. The subscript $t = 1, 2, \ldots, T$ denotes the corresponding time slot. $\boldsymbol{q} = (q_1, q_2, \ldots, q_T)$ is consumers' electricity consumption in the corresponding slot, and $\boldsymbol{p} = (p_1, p_2, \ldots, p_T)$ is price in the corresponding slot. $\boldsymbol{p}_{(0)} = \left(p_{1(0)}, p_{2(0)}, \ldots, p_{T(0)}\right)$, $\boldsymbol{q}_{(0)} = \left(q_{1(0)}, q_{2(0)}, \ldots, q_{N(0)}\right)$ denotes price and power consumption before

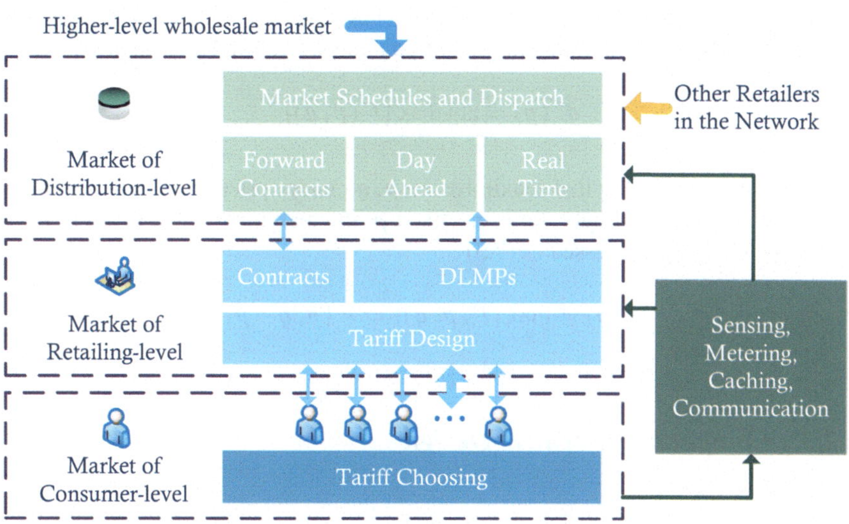

Fig. 7.1 The structure of a electricity retailing market

new pricing schemes take effect respectively. The retailer designs $r = 1, 2, \ldots, R$ different pricing schemes for all the consumers $k = 1, 2, \ldots, K$ and each consumer chooses one of them.

In addition, we also suppose smart meter data are shared by the retailer and consumers. Some common knowledge about initial selling price and consumers' demand elasticity is also pre-known. The ToU prices developed in this chapter can be used on a monthly or weekly basis within a rolling window framework, and the ToU price repeats every day during each window. Considering volatile customer behavior, we use the average load profiling during each window for every consumer.

7.2.2 Consumer Problem

Consumer behaviors are formulated mainly in terms of consumer preferences, and utility function is a way to describe preferences [19]. For every individual, this chapter adopts the consumers' utility function widely accepted in other researches [16, 18, 20–22] defined as below

$$F(\boldsymbol{p}, \boldsymbol{q}) = u(\boldsymbol{q}) - \sum_{t=1}^{T} p_t q_t \tag{7.1}$$

where $u(\boldsymbol{q})$ corresponds to the consumer' satisfaction gained from using certain amount of power. $u(\boldsymbol{q})$ often takes the form of a concave function to simulate the diminishing marginal utility [19, 21].

The consumers follow the principle of utility maximization, for any given $\boldsymbol{p}$, a consumer would adapt his consumption to the best response under $\boldsymbol{p}$. That means $\boldsymbol{q}$ will be set to the value that maximize $F(\boldsymbol{p}, \boldsymbol{q})$ under any fixed $\boldsymbol{p}$. Namely,

$$\boldsymbol{q}^*(\boldsymbol{p}) = \arg\max_{\boldsymbol{q}} \{F(\boldsymbol{p}, \boldsymbol{q})\} \tag{7.2}$$

where $\boldsymbol{q}^*(\boldsymbol{p})$ is the value that maximize $F(\boldsymbol{p}, \boldsymbol{q})$ under price $\boldsymbol{p}$. Clearly, $\boldsymbol{q}^*(\boldsymbol{p})$ is a function of $\boldsymbol{p}$. Accordingly, for every fixed $\boldsymbol{p}$, the maximum value of $F(\boldsymbol{p}, \boldsymbol{q})$ is also a function of $\boldsymbol{p}$ expressed as (7.3)

$$U(\boldsymbol{p}) = \max_{\boldsymbol{q}} \{F(\boldsymbol{p}, \boldsymbol{q})\} = F(\boldsymbol{p}, \boldsymbol{q}^*(\boldsymbol{p})) \tag{7.3}$$

7.2.3 Compatible Incentive Design

If a pricing mechanism is an incentive-compatible, truthfully declaring the preference is the dominant strategy. If the retailer considers scheme r as the true preference of consumer k through his load profiles, but he or she does not choose the designated scheme, the consumer k can be viewed as declaring a false preference. The retailer should ensure that the satisfaction consumers receive is the highest when the designed desired outcome is achieved so that the pricing scheme a consumer like is exactly the one the retailer designs for him. In this way, truthfully and faithful choosing the corresponding pricing scheme is the consumer's dominant strategy. Thus for every consumer k, if the retailer designs the pricing scheme r for him, Eq. (7.4) should be satisfied

$$U_k\left(\boldsymbol{p}_r\right) \geq U_k(\boldsymbol{p}') \ \forall k \tag{7.4}$$

where $\boldsymbol{p}'$ denotes any other pricing scheme. Thus, choosing a pricing scheme r is the dominant strategy. Besides, if the retailer wants consumers to adopt the new pricing schemes, the utility gained from the new pricing should exceed the old one, expressed as follows

$$U_k\left(\boldsymbol{p}_r\right) \geq U_k(\boldsymbol{p}_0) \ \forall k \tag{7.5}$$

where Eq. (7.5) is a key to associate the price and power consumption in different time slots. If the retailer desires to raise the price during some hours, then it must bring down the price during other periods to hold Eq. (7.5). Equation (7.5) makes the power consumption of different time act as substitutes and also makes load shifting during different periods possible.

7.2.4 Retailer Problem

The retailer wants to maximize its profit by providing diverse types of price schemes. The retailer's daily profit function is given by

$$\begin{aligned} R = \sum_{k=1}^{K}\sum_{t=1}^{T} p_{k,t} \times q_{k,t} - \sum_{t=1}^{T}\sum_{n=1}^{N_F} p_n^F \times L_n^F \times o_{n,t}^F \times o_n \\ - \sum_{t=1}^{T} p_t^{D,est} \times L_t^D - \xi \times CVaR \end{aligned} \tag{7.6}$$

The exact meanings of the terms are as follows:

(1) The first term is the daily monetary revenue payed by the consumers $k = 1, 2, \ldots, K$.

(2) The second and third terms are the cost for forward contracts and day-ahead markets. $n = 1, 2, \ldots, N_F$ is the contract number. p_n^F is the price of forward contracts in contract n. L_n^F is the quantity of power for contract n. $o_{n,t}^F$ is a binary parameter which equals to 1 when contract n covers period t. o_n is a binary variable which equals to 1 when contract n is chosen. $p_t^{D,est}$ is the estimation for day-ahead price at period t throughout the duration of the time window. L_t^D is the quantity bought to

supply load at period t from day-ahead markets. The retailer will buy all predictable consumers' load through forward contracts and day-ahead markets to avoid price fluctuation in real-time markets. The real-time market cost is due to the need to balance unpredictable random load. The purchasing strategy to balance predictable load is as follows

$$\sum_{k=1}^{K} q_{k,t} = \sum_{n=1}^{N_F} L_n^F \times o_{n,t}^F \times o_n + L_t^{D,est}, \forall t \tag{7.7}$$

(3) The third term is the risk of loss calculated by using conditional value at risk (CVaR). CVaR is a risk measure frequently used in current risk management for the retailer [11, 23] As discussed above, Eq. (7.7) does not include any purchasing from real-time market. But cost for balancing random load must be included. Besides, we use estimation price $p_t^{D,est}$ for day-ahead cost. However, the real day-ahead price contains a certain deviation and fluctuation. These two cost cannot be predicted before new price schemes takes effective and we considers as pure risk. ξ, the risk weighting factor, measures the degree of importance the retailer attaches to risk. This chapter uses CVaR to represent these two risk loss as follows

$$\begin{aligned} CVaR = \inf_{u \in R} \{u + \frac{1}{\left(1 - \alpha^{CVaR}\right) \cdot N_S} \\ \sum_{n_S=1}^{N_S} [(-\Delta R^D - \Delta R^{RT}) - u]^+\} \end{aligned} \tag{7.8}$$

where $n_S = 1, 2, \ldots, N_S$ is the ordinal number of historical loss, α^{CVaR} is the given confidence level, ΔR^D and ΔR^{RT} are the n_S recorded loss in day-ahead markets and real-time markets respectively.

7.2.5 Data-Driven Clustering and Preference Discovering

The number of pricing schemes could reach as high as the number of consumers but it will lead to serious price discrimination which is too hard to implement in reality. A more reasonable approach is to offer a relatively small number of choices compared to the number of consumers out of considerations for two aspects: (1) it reduces the complexity brought by the number of consumers for the retailer; (2) it is more practical for consumers to choose among relatively small number of price schemes. In order to achieve this goal, consumers with similar preferences will be designated the same price scheme and we should cluster consumers of similar preferences. Figure 7.2 shows the basic idea how smart meter data is used.

Before clustering, concrete expressions of utility should be specified because it indicates the level of preference. Different choices of utility function itself is a

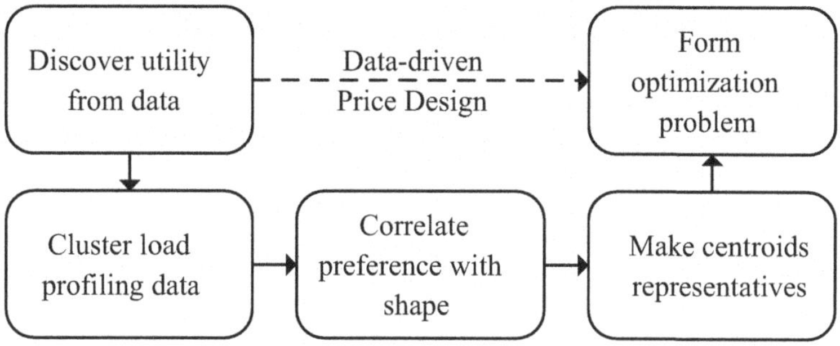

Fig. 7.2 The flowchart to illustrate how smart meter data is used

widely-discussed complicated problem which is beyond the scope of this chapter. The specific choice of form of utility function doesn't influence the model discussed above and the solution methods to be discussed in the next chapter but only affects the consumer preference and reaction. This chapter uses power function ($f(q) = \beta q^{\alpha} + \gamma$, if $\alpha > 0, \alpha \in (0, 1)\&\beta \in (0, \infty)$; if $\alpha < 0, \alpha \in (-\infty, 0)\&\beta \in (-\infty, 0)$) as the basic form of $F(\boldsymbol{p}, \boldsymbol{q})$ similar to [16]. It is a simple form of the pure numerical utility in [20] and has a closed form to be easily dealt with.

Because what matters is the relative values of $F(\boldsymbol{p}, \boldsymbol{q})$ instead of the absolute value [19], $F(\boldsymbol{p}_{(0)}, \boldsymbol{q}_{(0)})$ is set to 0 without loss of generality. To fully use smart meter load profiling data, a consumer is assumed to have adapted his original behaviors as the best response to the original price, even when the original price is flat. One one hand, different levels of price signal give consumer a marginal price signal how much they will pay if they use one more kWh electricity. Consumers are very likely to reduce their electricity consumption during high-price hours and increase consumption during low-price hours. On the other hand, consumers' preferences for different hours incentivize load shifting, which largely depends on the consumer's daily routine. The both reactions to price are optimization processes. So when $\boldsymbol{p} = \boldsymbol{p}_{(0)}$, the maximum of $F(\boldsymbol{p_0}, \boldsymbol{q})$ is at $\boldsymbol{q}_{(0)}$. The two conditions are expressed as follows:

$$F(\boldsymbol{p}_{(0)}, \boldsymbol{q}_{(0)}) = 0 \qquad \frac{\partial F(\boldsymbol{p}_{(0)}, \boldsymbol{q}_{(0)})}{\partial q_t} = 0, \forall t \tag{7.9}$$

Combining all above, one of the easy feasible solutions is as follows

$$u(\boldsymbol{q}) = \sum_{t=1}^{T} \left[\frac{p_{t(0)} q_{t(0)}}{\alpha} \left[\left(\frac{q_t}{q_{t(0)}} \right)^{\alpha} - 1 \right] + p_{t(0)} q_{t(0)} \right] \tag{7.10a}$$

$$q_t^* = \left(\frac{p_t}{p_{t(0)}} \right)^{\frac{1}{\alpha - 1}} \times q_{t(0)} \tag{7.10b}$$

$$U(\boldsymbol{p}) = \sum_{t=1}^{T} \left(\frac{1}{\alpha} - 1 \right) \left[\left(\frac{p_t}{p_{t(0)}} \right)^{\frac{\alpha}{\alpha - 1}} - 1 \right] \times q_{t(0)} p_{t(0)} \tag{7.10c}$$

where $\alpha \in (-\infty, 0) \cup (0,1)$ is a parameter pre-known through elasticity ε by $\varepsilon = 1/(\alpha - 1)$. Please see Appendix I for the proof. The personality or habits reside behind the historical smart meter data and does not need to be expressed explicitly in numbers. Smart meter data helps us deal with the concrete form of the utility function when it is not about cost minimization, but utility maximization.

It is not necessary that the absolute values of utility are the same to indicate similar preferences but only need to preserve the inequality relation in Eq. (7.4) since $U(\boldsymbol{p})$ is always in the comparison form to indicate preferences. For each load profile, we process data as follows to cluster consumers of similar preferences

$$\tilde{q}_{k,t(0)} = \frac{q_{k,t(0)}}{\max\limits_{t} \left\{ q_{k,t(0)} \right\}}, \forall k \tag{7.11}$$

and cluster the processed load profiling into $r = 1, 2, \ldots, R$ clusters and each cluster contains $k_r = 1, 2, \ldots, K_r$ load profiling.

Theorem 7.1 *Load profiling of consumers who have the same preferences are of the same shape after being processed by Eq. (7.11).*

Proof See Appendix II.

From Theorem 7.1, we know clustering can cluster consumers of similar preferences. Notice there may be some deviations between different load profiling in real clustering so that Theorem 7.1 may only approximately holds true. Furthermore, the mean value of the original load profile in each cluster (centroid) can represent all the members in the corresponding cluster in terms of both preferences and quantity of load. Since the centroid is in similar shape of the cluster members, so they have similar preferences. In terms of the quantity of the load in the cluster, we have

$$\begin{aligned}
\sum_{k=1}^{K_r} q_{k,t} &= \sum_{k=1}^{K_r} \left(\frac{p_{r,t}}{p_{r,t(0)}} \right)^{\frac{1}{\alpha-1}} \times q_{k,t(0)} \\
&= \left(\frac{p_{r,t}}{p_{r,t(0)}} \right)^{\frac{1}{\alpha-1}} \times \sum_{k=1}^{K_r} q_{k,t(0)} \\
&= \left(\frac{p_{r,t}}{p_{r,t(0)}} \right)^{\frac{1}{\alpha-1}} \times K_r \times q_{k,t(0)} = K_r \times q_{r,t}
\end{aligned} \tag{7.12}$$

So the centroid can represent the members in terms of electricity quantity. It simplifies the problem of the retailer by equivalently reducing the number of consumers. Subscript r is used to represent the cluster centroid. Equations (7.6) and (7.7) are conversed to equations below

$$R = \sum_{r=1}^{R} \sum_{t=1}^{T} K_r \times p_{r,t} \times q_{r,t} - \sum_{t=1}^{T} \sum_{n=1}^{N_F} p_n^F \times L_n^F \times o_{n,t}^F$$

$$\times o_n - \sum_{t=1}^{T} \widehat{p_t^D} \times L_t^D - \xi \times CVaR \tag{7.13a}$$

$$\sum_{r=1}^{R} q_{r,t} \times K_r = \sum_{n=1}^{N_F} L_n^F \times o_{n,t}^F \times o_n + L_t^D, \forall t \tag{7.13b}$$

For detailed clustering methods, in this chapter, different clustering methods are adopted to find the best clustering results [24]. These methods include: hierarchical clustering, k-means clustering, fuzzy C-means clustering, Gaussian mixture model. Both within-cluster compactness and between-cluster separation of different clustering results contribute to the different results of the model. Clustering results are evaluated by the Davies–Bouldin index, which represents the worst-case within-to-between cluster ratio for all clusters. For a detailed discussion about how clustering affects the whole model sees Sensitivity Analysis.

7.2.6 Integrated Model

Before formulating the integrated model, some other constraints are given below to fix the ToU structure. We assume each ToU has $m = 1, 2, \ldots, M$ blocks and p_r^m is the price of the block m for pricing scheme r

$$\sum_{m=1}^{M} e_{r,t}^m = 1, \quad \sum_{t=1}^{T} e_{r,t}^m \geq D_{\min}, \forall m, r \tag{7.14a}$$

$$\left|e_{r,T}^m - e_{r,1}^m\right| + \sum_{t=2}^{T} \left|e_{r,t-1}^m - e_{r,t}^m\right| = 2, \forall m, r \tag{7.14b}$$

$$p_{r,t} = \sum_{m=1}^{M} e_{r,t}^m \times p_r^m, \forall t, r \tag{7.14c}$$

where $e_{t,r}^m$ is a binary variable. For pricing scheme r, if period t belongs to its ToU block m, $e_{t,r}^m = 1$. $D_{\min}$ is the minimum duration periods of each block. Generally, a ToU price contains three blocks but m can be other values rather than just 3 in our framework. Equation (7.14a) restricts each time slot belongs to one block. Equation (7.14b) restricts each block only changes two times.

Here, we can give the integrated model of designing customizing ToU to maximize the retailer's profit while ensuring consumers' willingness

$$\max R \tag{7.15a}$$

$$\text{s.t. } (7:4)(7:5)(7:10b)(7:10c)(7:13a)(7:13b)(7:14a)(7:14b)(7:14c) \tag{7.15b}$$

Notice subscript r will be added for Eqs. (7.4), (7.5), (7.10b) and (7.10c). Clearly, this model is a MINLP model.

7.3 Solution Methods

7.3.1 Framework

The integrated optimization problem is nonlinear and may be difficult to find the optimal solution. This model is nonlinear mainly due to the power function in Eqs. (7.10b) and (7.10c), the product of two decision variables in Eqs. (7.14c), (7.13a), the expression of CVaR Eq. (7.8) and the absolute values in Eq. (7.14b). Piece-wise linear approximation is used to deal with power function. Equivalent linear transformation is used to eliminate binary variable product, simplify CVaR and eliminate absolute values.

7.3.2 Piece-Wise Linear Approximation

One of the reasons for being a nonlinear model is the power function term in constraints and the term $p_{r,t} \times q_{r,t}$ of two decision variables' product in objective function Eq. (7.13a). If the term $p_{r,t} \times q_{r,t}$ is treated as a whole and $q_{r,t}$ is substituted by Eq. (7.10b), the whole term can be expressed as

$$p_{r,t} \times q_{r,t} = p_{r,t}^{\frac{\alpha}{\alpha-1}} \times \left(\frac{1}{p_{t(0)}}\right)^{\frac{1}{\alpha-1}} \times q_{r,t(0)}, \tag{7.16}$$

which is a function of $p_{r,t}$ and relates to the term $p_{r,t}^{\alpha/(\alpha-1)}$. Meanwhile, the term $p_{r,t}^{\alpha/(\alpha-1)}$ also appears in consumers' utility function Eq. (7.10c).

It is not a coincidence or a special case that just fits this model. Taking $p_{r,t} \times q_{r,t}$ as two decision variables' product in the retailer's profit function is a common practice and $q_{r,t}$ should be affected by $p_{r,t}$ somehow to simulate demand response in related works [7, 13]. Considering things above, this chapter treats the term $p_{r,t} \times q_{r,t}$ as a whole and uses the piece-wise linear approximation of $q_{r,t}$ and $p_{r,t} \times q_{r,t}$ respectively for linearizing the model. In this chapter, we assign

$$\Psi\left(p_{r,t}\right) = p_{r,t}^{\frac{\alpha}{\alpha-1}}, \quad \Theta\left(p_{r,t}\right) = p_{r,t}^{\frac{1}{\alpha-1}} \tag{7.17}$$

The first term appears in profit and utility function and indicates how profit and utility will change as price changes. The second term appears in the consumer's reaction to price and indicates how behaviors change along as price changes.

Piece-wise linear approximation of $\Psi\left(p_{r,t}\right)$ and $\Theta\left(p_{r,t}\right)$ are as follows

$$p_{r,t}=\sum_{j=1}^{J+1} w_{j,r,t}a_{j,r,t} \tag{7.18a}$$

$$\phi_{r,t}=\sum_{j=1}^{J+1} w_{j,r,t}\Psi\left(a_{j,r,t}\right) \tag{7.18b}$$

$$\theta_{r,t}=\sum_{j=1}^{J+1} w_{j,r,t}\Theta\left(a_{j,r,t}\right) \tag{7.18c}$$

$$w_{1,r,t}\le z_{1,r,t},\quad w_{J+1,r,t}\le z_{J,r,t} \tag{7.18d}$$

$$w_{j,r,t}\le z_{j-1,r,t}+z_{j,r,t}\ ,\ j=2\ldots J \tag{7.18e}$$

$$\sum_{j=1}^{J} z_{j,r,t}=1,\quad \sum_{j=1}^{J+1} w_{j,r,t}=1 \tag{7.18f}$$

where $j=1,2,\ldots,J$ is the piece-wise segment number. $a_{1,r,t}<a_{2,r,t}<\cdots<a_{J+1,r,t}$ are segment connection endpoints. Positive continuous variables $w_{j,r,t}$ and binary variables $z_{j,r,t}$ are intermediate variables. $\phi_{r,t}$ and $\theta_{r,t}$ are the piece-wise linear approximation of $\Psi\left(p_{r,t}\right)$ and $\Theta\left(p_{r,t}\right)$ respectively. The specific method to find segment connection endpoints is referred to Ref. [25].

7.3.3 Eliminating Binary Variable Product

The product of a binary variable and a continuous variable in Eq. (7.14c) is conversed to linear constraints below:

$$\sigma_{r,t}^{m}\le M\times e_{t,r}^{m},\quad \sigma_{r,t}^{m}\le p_{r}^{m} \tag{7.19a}$$

$$\sigma_{r,t}^{m}\ge p_{r}^{m}-M\times\left(1-e_{t,r}^{m}\right),\quad \sigma_{r,t}^{m}\ge 0 \tag{7.19b}$$

where $\sigma_{r,t}$ is the result of the product operation, M is a sufficiently large number compared to p_{r}^{m}.

7.3.4 CVaR

Equation (7.8) is conversed to linear constraints below:

$$CVaR \geq u + \frac{1}{\left(1 - \alpha^{CVaR}\right) \cdot N_S} \sum_{ns=1}^{N_S} W_{ns} \tag{7.20a}$$

$$W_{ns} \geq 0, \quad W_{ns} \geq \left[\left(-\Delta R_{ns}^{D} - \Delta R_{ns}^{RT}\right) - u\right] \tag{7.20b}$$

where W_{ns} is an intermediate variable.

7.3.5 *Eliminating Absolute Values*

The absolute values in (7.14) is conversed to linear constraints below:

$$e_1 - e_2 \leq A \leq e_1 - e_2 + 2 \times B \tag{7.21a}$$

$$e_2 - e_1 \leq A \leq e_2 - e_1 + 2 \times (1 - B) \tag{7.21b}$$

For simplicity, the subscripts are omitted and e_1,e_2 represent any two terms that take absolute values in Eq. (7.14). A is the result of modulus value operations. B is an intermediate variable. A and B are both binary variables.

To sum up the arguments above, the objective function and all constraints are conversed to linear form and a MILP model is finally reformulated. The problem is coded into General Algebraic Modelling System (GAMS) model solved with the MILP solver Cplex. To compare the performance of the linear and nonlinear model, the nonlinear model is also coded into GAMS model solved with the MINLP solver BARON. The programs are run on a personal computer with Intel Core i5 2.80 GHz CPU and 8 GB RAM.

7.4 Case Study

7.4.1 *Data Description and Experiment Setup*

The smart meter electricity trial data of 6435 consumers from Commission for Energy Regulation (CER) based in Ireland are used for case study. The data were collected every 30 min and $T = 48$ is set in this case. Before new pricing schemes take effect, flat rate $p_{t(0)} = 0.2$ \$/kWh for all t is adopted by the retailer. National Institute of Economic and Industry Research estimated the mean long run elasticity of demand as -0.37 for residential consumers and could rise to -0.4 [26]. We set demand elasticity as $\varepsilon = -0.4$ and get $\alpha = -1.5$. Each different ToU scheme has 3 segments so $M = 3$. D_{min} is set to 4 so that the minimum duration time of each ToU block is 2 h. $\Psi(p)$ and $\Theta(p)$ is approximated by $J = 15$ lines in the range of $p \in [0.04, 0.8]$ just as shown in Fig. 7.3. ξ is set to 1.

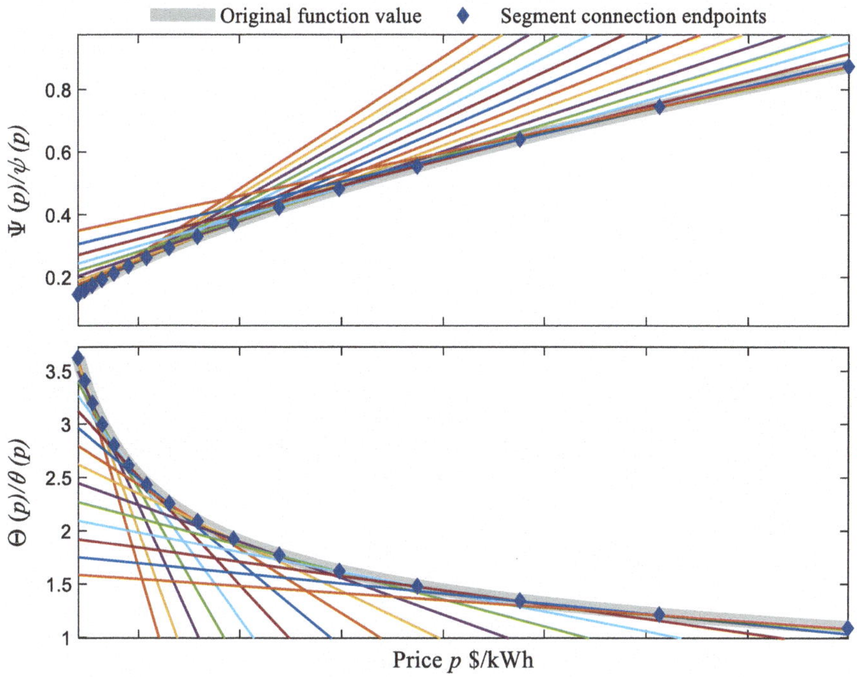

Fig. 7.3 Piece-wise linear approximation of $\Psi(p)$ and $\Theta(p)$

7.4.2 *Basic Results*

7.4.2.1 Clustering

The data are clustered into 5−10 clusters by various methods and Davies–Bouldin index is used to choose the best result among them. The evaluation result is shown in Fig. 7.4. For hierarchical clustering methods, we use complete-linkage (HIA-COMPLETE) and ward method (HIA-WARD) to perform agglomerative clustering. For k-means clustering, we use sample method (KM-SAMPLE), uniform scattering (KM-UNIFORM), k-means++ (KM-PLUS) to initialize centroids. For fuzzy C-means clustering (FCM), we set the hyper-parameter m that controls how fuzzy the cluster will be equal to 1.1, 1.2, 1.3 respectively. For Gaussian mixture model, expectation maximization algorithm (GMEM) is used to perform iterations with initial points set by k-means++ (PLUS) or random scattering (RAND).

The optimum value of the varying cluster numbers is chosen such that the Davies–Bouldin index is minimized. So the optimum value is $R = 6$ clusters in total and the detailed method is HIA-COMPLETE. The correspondent load profiling of the six clusters are shown in Fig. 7.5.

Figure 7.5 provides a glimpse at the detailed load patterns of each cluster: The load profiling in cluster 1 is scheduled to a nine-to-five peak while cluster 2 peaks only

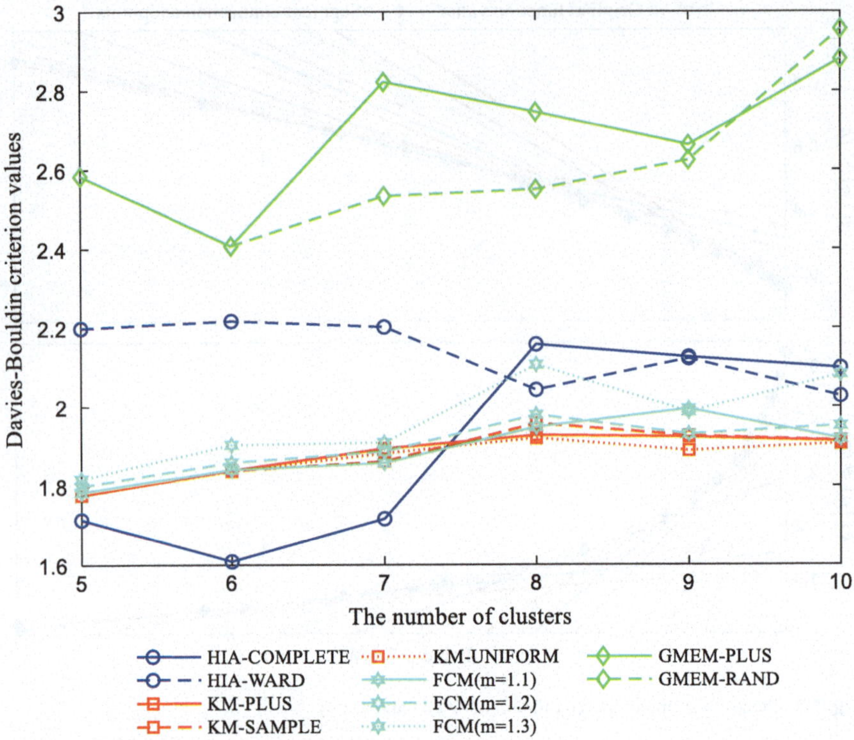

Fig. 7.4 The Davies–Bouldin criterion values of different clustering methods and cluster numbers. The minimum is chosen as the final result

in the morning. Load in cluster 3 is evenly distributed on the whole day. Consumers in cluster 4 and 5 both prefer to consume in the evening, but cluster 4 consume as much in the afternoon as in the evening. Consumers of cluster 6 regularly stay up late at night.

7.4.2.2 Pricing Design and Load Response

Figure 7.6 shows the designed ToU schemes for different clusters by solving the proposed MILP model and Fig. 7.7 shows a comparison between loads under flat pricing and ToU pricing schemes. The retailer encourages consumers to reschedule their electricity consumption to use more during low-pool-price hours by lowering the ToU price during these hours for all clusters. Generally, low-pool-price periods are usually off-peak of the total load as well. In Fig. 7.7, consumers indeed respond to the off-peak retailing price fall but for different clusters of consumers, how much the retailing price will fall and how long this block will last may not be the same. The retailer needs to balance consumers' utility decline when retailing price rises

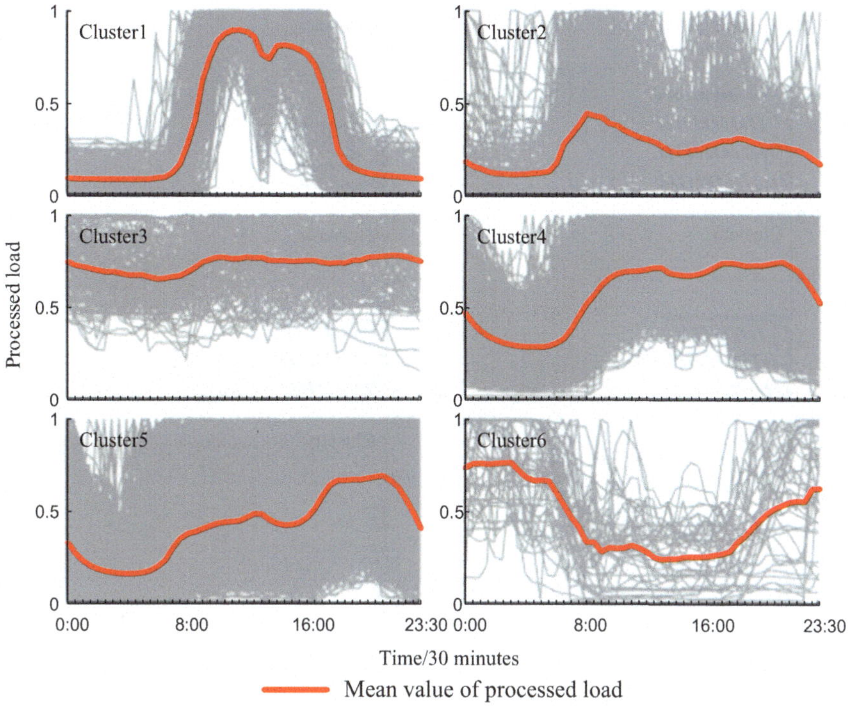

Fig. 7.5 The optimal clustering result under the method HIA-COMPLETE

and utility increase when retailing price falls. The retailer will not unduly raise the retailing price for fear of losing consumers since consumers' utility may drop sharply with a high price, so the price of all clusters is below 0.3 \$/kWh. Some details which ensure each consumer will be more satisfied with the pricing scheme designed for him than any other one can be directly seen in Fig. 7.6. Take cluster 6 for an example. The lowest retailing price is designed for cluster 6 whose original peak is just during low-pool-price periods. To shave more price during peak time is inviting for cluster 6 and the retailer can benefit from increase in consumption of cluster 6 though retailing price falls in these hours.

7.4.2.3 Linearization

For the MILP model, it takes 125.6 s to find the optimal solution 1186.01\$. For the MINLP model, it converges to a solution 1008.7\$ after it runs out of time resources of 2 h. Linearization enhances the speed of solving the problem, and the MILP model does not fall into a local optimum in a two-hour time limit. Linearization may bring relaxing errors, but increasing linear segments can help to fix the gap.

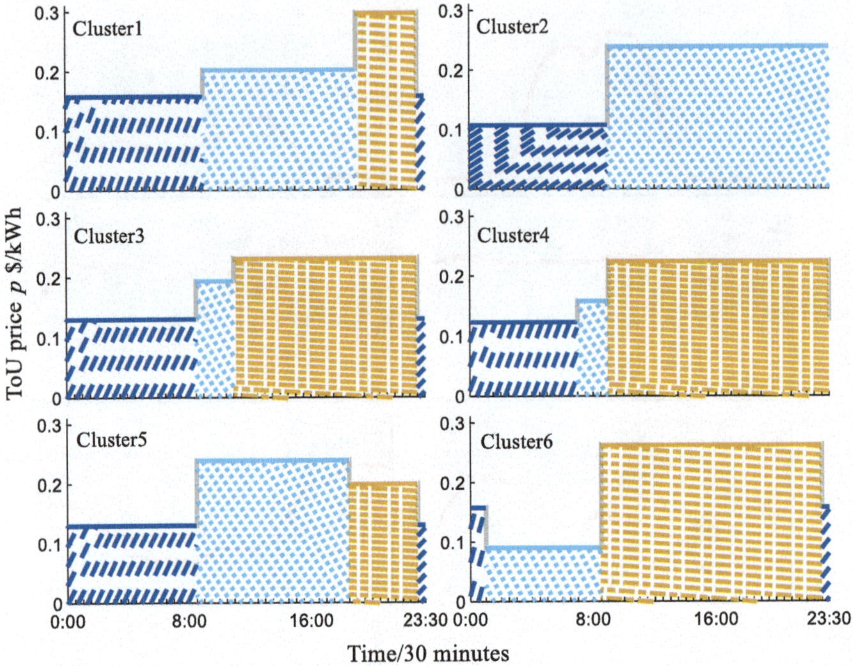

Fig. 7.6 The optimal ToU pricing schemes of the six clusters

7.4.3 *Sensitivity Analysis*

7.4.3.1 Elasticity ε

We set ε equal to -0.3, -0.4, -0.5 respectively to perform sensitivity analysis of α. Conventionally, elasticity is compared by its absolute value but not the original value and this chapter follows the convention in the discussions below.

Figure 7.8 shows the comparison between ToU schemes under different elasticity ε. The general trend of ToU price of the lowest pricing block is increasing when elasticity decreases. For off-peak periods, when elasticity is high, the retailer can bring down the price to encourage consumers to consume more. If the increase in revenue brought by the increase in consumption (which is called yield effect) is bigger than the loss along with the price down (which is called price effect), lowering price is profitable. However, as elasticity becomes smaller, consumers are less sensitive to price changes so they are not so willing to adjust their consumption, which may result in the yield effect being offset by the price effect very soon. So the off-peak price is higher when elasticity decreases. The ToU price in the highest pricing block doesn't change so much to keep consumers utility values high.

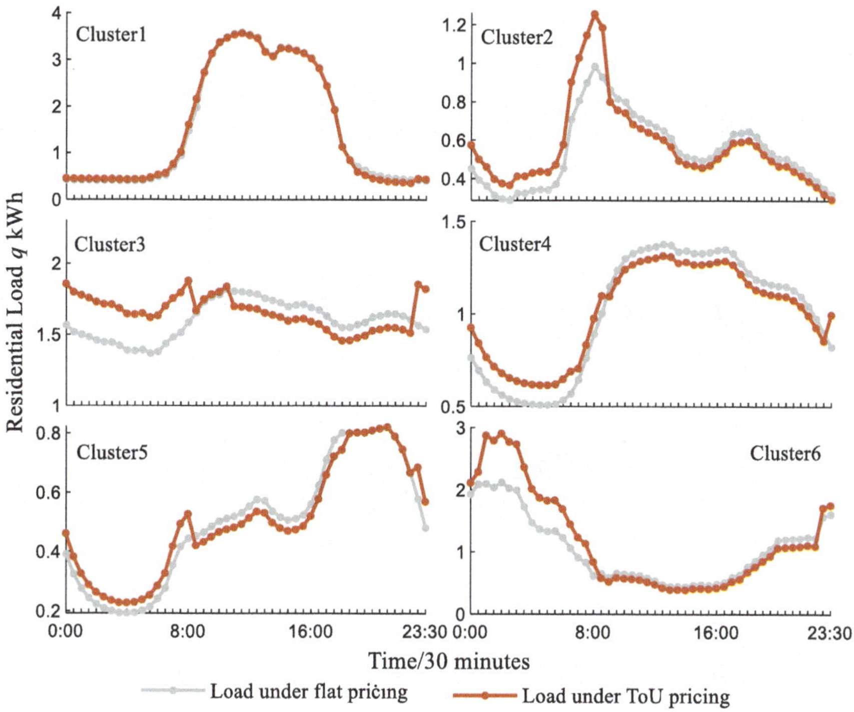

Fig. 7.7 Load response of consumers of the six clusters

Figure 7.9 shows the comparison between total loads of these 6435 consumers under different elasticity ε. Peak shaving and valley filling are more notable when elasticity is higher. Table 7.1 shows if consumers have higher elasticity, they can be motivated to use more energy. Table 7.2 shows if consumers have higher elasticity, the retailer also can make more profit.

7.4.3.2 Risk Weighting Factor

The risk-weighting factor is set as different values to study the effect of CVaR, forward contracts, and day-ahead market and the result is shown in Fig. 7.10. As the retailer attaches more importance to risk, it tends to purchase more electricity through forwarding contracts rather than from day-ahead market because the retailer faces risk in day-ahead market. When $\xi = 100$, the retailer nearly considers all its cost as risk. The trend of CVaR is decreasing as risk-weighting factor rises because the CVaR becomes the decisive factor of the value of the revenue function as ξ rises.

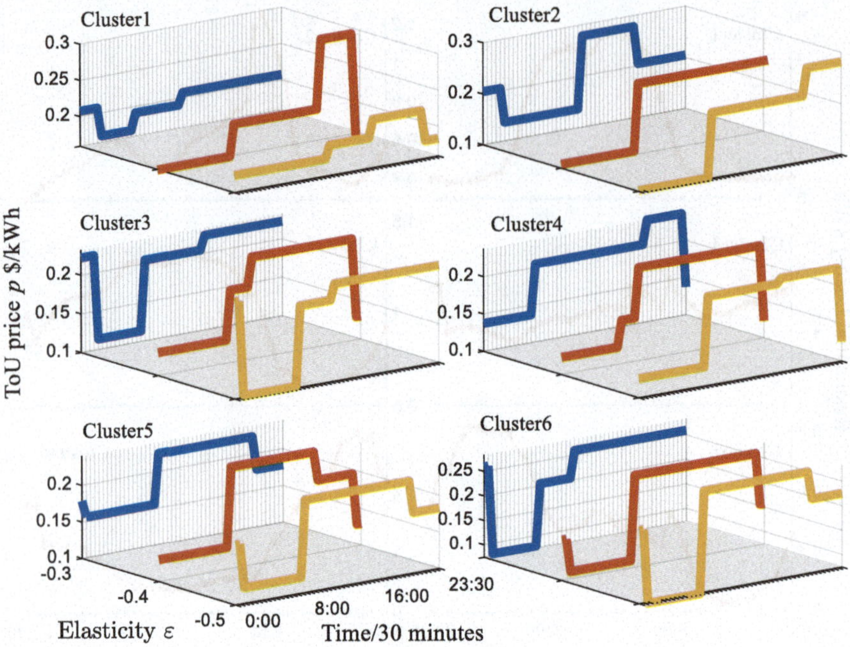

Fig. 7.8 Different ToU schemes under different elasticity ε

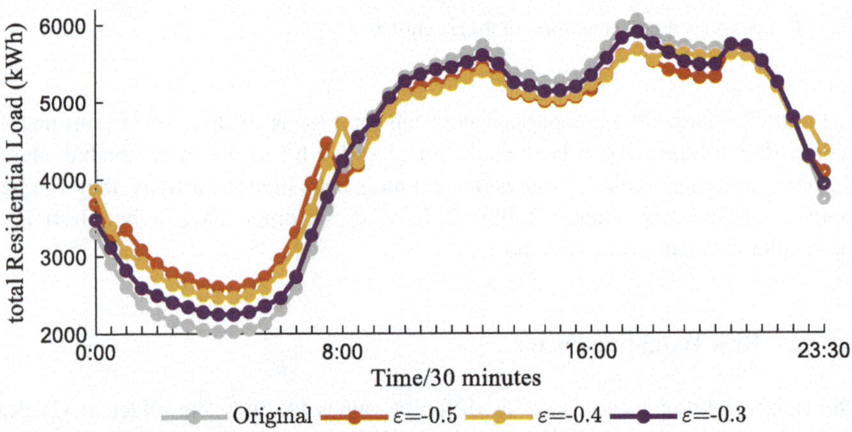

Fig. 7.9 Load under different elasticity ε

Table 7.1 Total energy use under different elasticity

Elasticity ε	Original	−0.2	−0.3	−0.4	−0.5
Total energy use (kWh)	210150.8	211139.2	212144.5	215050.2	215177.2

Table 7.2 Retailing profit under different elasticity

Elasticity ε	Original	−0.2	−0.3	−0.4	−0.5
Retailing profit($)	752.03	833.49	977.00	1186.01	1385.42

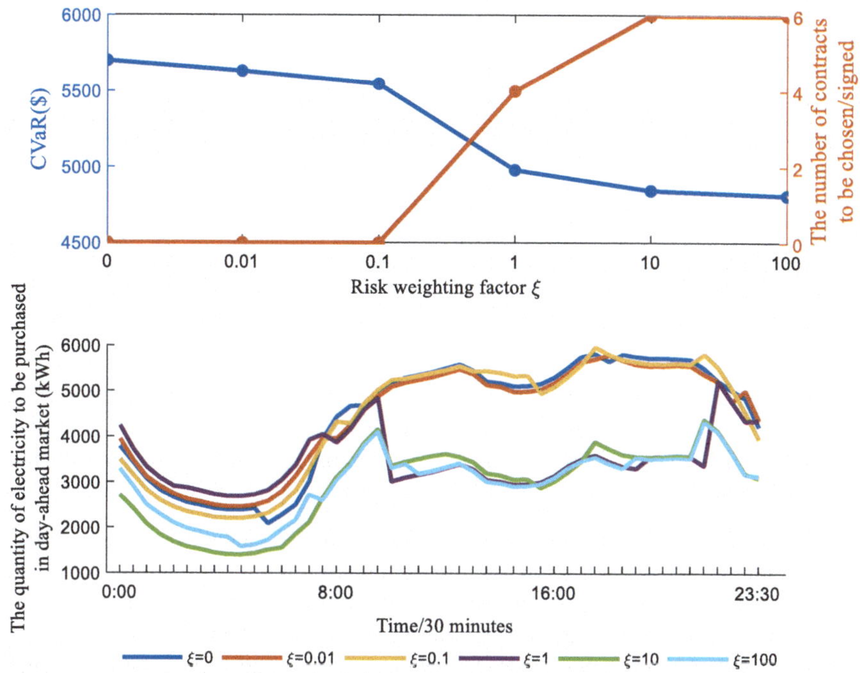

Fig. 7.10 CVaR, the number of forward contracts to be signed/chosen and the quantity of power to be purchased in day-ahead market under different risk-weighting factors

7.4.3.3 Clustering Methods

Different clustering methods are adopted to group load profiles, and consumers in the same cluster have similar preferences. A statistical index, Davies–Bouldin, is used to choose the best cluster result. But how different cluster methods influence the performance of the whole model is worthy of discussion. Table 7.3 shows the performance of different clustering methods when maintaining the number of clusters $R = 6$.

In Table 7.3, the first column is the retailer's retailing profit. The second column is the total consumers' welfare calculated by the sum of the individual utility functions, sometimes referred to as a classical utilitarian [19]. The third column is the average retailing price. We represent all the consumers' preferences as the corresponding cluster centroids' preferences, but there may be some deviations between individuals

Table 7.3 Performance evaluation of clustering methods

	Retailing profit ($)	Social welfare	Average price ($/kWh)	First/Second choice
ORIGINAL	752.03	0	0.2000	–/–
HIA-COMP	**1186.01**	**339.72**	**0.1947**	**65/89**
HIA-WARD	1188.70	10.01	0.1971	33%/59%
KM-PLUS	1145.68	7.01	0.1973	9%/20%
KM-SAMPLE	1137.61	4.50	0.1975	22%/48%
KM-UNIFORM	1142.61	15.76	0.1973	11%/31%
FCM(m = 1.1)	1150.43	9.43	0.1970	30%/47%
FCM(m = 1.2)	1176.08	18.64	0.1968	19%/35%
FCM(m = 1.3)	1208.06	0.64	0.1970	8%/20%
GMEM-PLUS	1145.82	36.01	0.1965	13%/28%
GMEM-RAND	1144.85	46.60	0.1967	10%/24%

and the cluster centroid. If 6435 consumers choose among these 6 pricing schemes by themselves, some members may not select the same pricing scheme as the cluster centroid does because of the deviations. We simulate the real situation with the following steps:

1. First, the utility gained from the six pricing schemes is calculated by Eq. (7.10c) and sorted in descending order for every consumer.
2. Second, since consumers act in the principle of utility-maximization, the top in order for every consumer is the consumer's *first choice* in the real market. The proportion of consumers whose *first choices* are just the same as their corresponding centroids' choices is the index First Choice.
3. Third, the second-highest in order for every consumer is the consumer's *second choice* in real market. The proportion of consumers, one of whose *first choices* or *second choices* are just the same as their corresponding centroids' choices, is the index Second Choice. Second Choice is calculated to extend difference tolerance between individuals and centroids.

According to Table 7.3, all the clustering methods increase both retailing profit and social welfare and decrease average retailing price. By using this model, the retailer will at least get the same profit as it does under flat pricing since flat pricing is a feasible solution where the price of all ToU segments is the same as flat price. HIA-COMP may not lead in the retailing profit, but it is indeed far ahead in satisfying consumers. Thus, a Pareto improvement is achieved compared with the original flat pricing scheme and the result gained by using HIA-COMP is also a Pareto optimum among all the methods. The Pareto optimum coincides more with consumers' interests, and this implies if the retailer wants to increase profit further, consumers are bound to be hurt. The high social welfare of HIA-COMP is due to its wider between-cluster separation. If load profiles in different clusters are very similar, different profiles in different clusters have little difference, so the retailer just arbitrarily keeps consumer utility at a small marginal value near zero to maximize its profit. On the contrary, if wide between-cluster separation is achieved, the retailer must keep consumers' utility large to avoid that consumers will not choose the corresponding pricing scheme. Wider between-cluster separation brings bigger social welfare.

First Choice and Second Choice focus more on within-cluster compactness. A less dispersed within-cluster compactness implies low variance within each cluster so the centroid can be a qualified representative in preferences. It can be inferred from Table 7.3 that HIA-COMP has denser within-cluster compactness. The retailer needs to condense within-cluster compactness because it wants to predict its profit as accurate as possible, so HIA-COMP is the optimal choice.

7.5 Conclusions and Future Works

This chapter proposes a data-driven optimization-based approach to design ToU tariffs explicitly dealing with compatible incentive. The Stackelberg game between the retailer and the strategic consumers, the considerations for the incentive-compatible market, the retailer's cost, risk, and purchasing strategy are considered in this model. Smart meter data is used to dig into consumers' preferences, and clustering method is used to gather consumers of similar preferences. Then through linear conversions, a mixed-integer linear programming (MILP) problem is finally formulated to design optimal personalized pricing schemes. Case study results confirm the ToU tariff can achieve the effect of peak shaving and valley filling, increasing the retailer's profitability and ensuring consumers' willingness and preferences at the same time.

Appendix I

Proof **Proof for** $\varepsilon = 1/(\alpha - 1)$

Equation (7.10b) can be changed to:

$$
\begin{aligned}
q_t^* &= \left(\frac{p_t}{p_{t(0)}}\right)^{\frac{1}{\alpha-1}} \times q_{t(0)} \Rightarrow \\
\left(\frac{q_{t(0)} + \Delta q_t}{q_{t(0)}}\right)^{\alpha-1} &= \frac{p_{t(0)} + \Delta p_t}{p_{t(0)}} = 1 + \frac{\Delta p_t}{p_{t(0)}}
\end{aligned}
\tag{7.22}
$$

where Δp_t and Δq_t is the incremental change of p_t and q_t.

If Δp_t and Δq_t are small enough compared with p_t and q_t. The left side of (7.22) is expanded by using first-order Taylor-series as

$$\left(\frac{q_{n(0)} + \Delta q_n}{q_{n(0)}}\right)^{\alpha-1} = \left(1 + \frac{\Delta q_n}{q_{n(0)}}\right)^{\alpha-1} \approx 1 + (\alpha - 1)\frac{\Delta q_n}{q_{n(0)}} \tag{7.23}$$

Combining (7.22) and (7.23), we can get the following equations

$$1 + (\alpha - 1)\frac{\Delta q_n}{q_{n(0)}} \approx 1 + \frac{\Delta p_n}{p_{n(0)}} \tag{7.24}$$

and (7.24) can be also expressed as

$$\frac{\Delta q_n / q_{n(0)}}{\Delta p_n / p_{n(0)}} \approx \frac{1}{(\alpha - 1)} \tag{7.25}$$

The left side is just the definition of elasticity ε.

Appendix II

Proof For any consumer k, if he prefers $\boldsymbol{p}_r$ to $\boldsymbol{p}'$, namely (7.4) should be satisfied. Take the concrete expression of $U(\boldsymbol{p})$ of (7.10c) into Eq. (7.4) and we get

$$\begin{aligned}&\sum_{t=1}^{T} \mu\left(p_{r,t}, p_{k,t(0)}\right) \cdot q_{k,t(0)} \geq \sum_{t=1}^{T} \mu\left(p'_t, p_{k,t(0)}\right) \cdot q_{k,t(0)} \Rightarrow \\ &\boldsymbol{\mu}(\boldsymbol{p}_r, \boldsymbol{p}_{k(0)}) \cdot \boldsymbol{q}_{k(0)}^{\mathrm{T}} \geq \boldsymbol{\mu}(\boldsymbol{p}', \boldsymbol{p}_{k(0)}) \cdot \boldsymbol{q}_{k(0)}^{\mathrm{T}}\end{aligned} \tag{7.26}$$

where $\boldsymbol{\mu}(\boldsymbol{p}, \boldsymbol{p}_{k(0)}) = \left(\mu(p_1, p_{k,1(0)}), \ldots, \mu(p_T, p_{k,T(0)})\right)$ represents the terms unrelated to $q_{t(0)}$ in the expression of $U(\boldsymbol{p})$, namely

$$\mu\left(p_t, p_{k,t(0)}\right) = \left(\frac{1}{\alpha} - 1\right)\left[\left(\frac{p_t}{p_{k,t(0)}}\right)^{\frac{\alpha}{\alpha-1}} - 1\right] \times p_{k,t(0)} \tag{7.27}$$

$\boldsymbol{\mu}(\boldsymbol{p}, \boldsymbol{p}_{k(0)})$ is a function of the old price scheme $\boldsymbol{p}_{k(0)}$ and the new price scheme $\boldsymbol{p}$. Equation (7.26) for consumer k_1 and k_2 are displayed as follows respectively

$$\boldsymbol{\mu}(\boldsymbol{p}_r, \boldsymbol{p}_{k_1(0)}) \cdot \boldsymbol{q}_{k_1(0)}^{\mathrm{T}} \geq \boldsymbol{\mu}(\boldsymbol{p}', \boldsymbol{p}_{k_1(0)}) \cdot \boldsymbol{q}_{k_1(0)}^{\mathrm{T}} \tag{7.28a}$$

$$\boldsymbol{\mu}(\boldsymbol{p}_r, \boldsymbol{p}_{k_2(0)}) \cdot \boldsymbol{q}_{k_2(0)}^{\mathrm{T}} \geq \boldsymbol{\mu}(\boldsymbol{p}', \boldsymbol{p}_{k_2(0)}) \cdot \boldsymbol{q}_{k_2(0)}^{\mathrm{T}} \tag{7.28b}$$

If k_1 and k_2 have similar preferences, they will choose the same pricing scheme most of the time including the last time when they chose among various pricing schemes. So $\boldsymbol{q}_{k_1(0)} = \boldsymbol{q}_{k_2(0)}$ is satisfied and the following should be true

$$\mu(\boldsymbol{p}_r, \boldsymbol{p}_{k_1(0)}) = \mu(\boldsymbol{p}_r, \boldsymbol{p}_{k_2(0)}) \;\; \mu(\boldsymbol{p}', \boldsymbol{p}_{k_1(0)}) = \mu(\boldsymbol{p}', \boldsymbol{p}_{k_2(0)}) \tag{7.29}$$

Aiming at finding consumers who have similar preferences through load profiling clustering, it is significant to find the relationship between two load profiling so that it is guaranteed when Eq. (7.28a) is satisfied, Eq. (7.28b) is also satisfied. Thus, considering Eq. (7.29), $\boldsymbol{q}_{k_1(0)}$ needs to vary proportionally with $\boldsymbol{q}_{k_2(0)}$

$$\boldsymbol{q}_{k_2(0)} = \eta_{k_1,k_2}\boldsymbol{q}_{k_1(0)}, \forall t \tag{7.30}$$

Equation (7.30) is processed by Eq. (7.11), and we get

$$\tilde{\boldsymbol{q}}_{k_1(0)} = \tilde{\boldsymbol{q}}_{k_2(0)}, \forall t \tag{7.31}$$

Equation (7.31) means the shape of the load profiling after being processed is the same.

References

1. Grid 2030 (2013). A national vision for electricity's second 100 years. Technical report, United States Department of Energy Office of Electric Transmission and Distribution.
2. Akhavan-Hejazi, H., & Mohsenian-Rad, H. (2018). Power systems big data analytics: An assessment of paradigm shift barriers and prospects 4, 91–100, 11.
3. Adam Elmachtoub, Vishal mname Gupta, and Michael mname Hamilton. The value of personalized pricing. *SSRN Electronic Journal*, 1–46, 1.
4. Yang, J., Zhao, J., Luo, F., Wen, F., & Yang Dong, Z. (2017). Decision-making for electricity retailers: A brief survey. *IEEE Transactions on Smart Grid*, *9*(5), 4140–4153.
5. Celebi, E., & David Fuller, J. (2007). A model for efficient consumer pricing schemes in electricity markets. *IEEE Transactions on Power Systems*, *22*(1), 60–67.
6. Zugno, M., Miguel Morales, J., Pinson, P., & Madsen, H. (2013). A bilevel model for electricity retailers' participation in a demand response market environment. *Energy Economics*, *36*, 182–197.
7. Wei, W., Liu, F., & Mei, S. (2014). Energy pricing and dispatch for smart grid retailers under demand response and market price uncertainty. *IEEE Transactions on Smart Grid*, *6*(3), 1364–1374.
8. Song, M., & Amelin, M. (2016). Purchase bidding strategy for a retailer with flexible demands in day-ahead electricity market. *IEEE Transactions on Power Systems*, *32*(3), 1839–1850.
9. Ghamkhari, M., Sadeghi-Mobarakeh, A., & Mohsenian-Rad, H. (2017). Strategic bidding for producers in nodal electricity markets: A convex relaxation approach. *IEEE Transactions on Power Systems*, *32*(3), 2324–2336.
10. Carrión, M., Arroyo, J. M., & Conejo, A. J. (2009). A bilevel stochastic programming approach for retailer futures market trading. *IEEE Transactions on Power Systems*, *24*(3), 1446–1456.
11. Carrion, M., Conejo, A. J., & Arroyo, J. M. (2007). Forward contracting and selling price determination for a retailer. *IEEE Transactions on Power Systems*, *22*(4), 2105–2114.

12. Nguyen, D. T., Nguyen, H. T., & Le, L. B. (2016). Dynamic pricing design for demand response integration in power distribution networks. *IEEE Transactions on Power Systems*, *31*(5), 3457–3472.
13. Li, R., Wang, Z., Chenghong, G., Li, F., & Hao, W. (2016). A novel time-of-use tariff design based on gaussian mixture model. *Applied Energy*, *162*, 1530–1536.
14. Yang, J., Zhao, J., Wen, F., & Dong, Z. (2019). A model of customizing electricity retail prices based on load profile clustering analysis. *IEEE Transactions on Smart Grid*, *10*(3), 3374–3386.
15. Yang, J., Zhao, J., Wen, F., & Dong, Z. Y. (2018). A framework of customizing electricity retail prices. *IEEE Transactions on Power Systems*, *33*(3), 2415–2428.
16. Yang, P., Tang, G., & Nehorai, A. (2012). A game-theoretic approach for optimal time-of-use electricity pricing. *IEEE Transactions on Power Systems*, *28*(2), 884–892.
17. Chapman, A. C., Verbič, G., & Hill, D. J. (2016). Algorithmic and strategic aspects to integrating demand-side aggregation and energy management methods. *IEEE Transactions on Smart Grid*, *7*(6), 2748–2760.
18. Samadi, P., Mohsenian-Rad, H., Schober, R., & Wong, V. W. S. (2012). Advanced demand side management for the future smart grid using mechanism design. *IEEE Transactions on Smart Grid*, *3*(3), 1170–1180.
19. Varian, H. R. (2010). *Intermediate microeconomics: A modern approach* (8th ed.). W.W. Norton Co.
20. Saez-Gallego, J., Morales, J. M., Zugno, M., & Madsen, H. (2016). A data-driven bidding model for a cluster of price-responsive consumers of electricity. *IEEE Transactions on Power Systems*, *31*(6), 5001–5011.
21. Ratliff, L. J., Dong, R., Ohlsson, H., & Sastry, S. S. (2014). Incentive design and utility learning via energy disaggregation. *IFAC Proceedings Volumes*, *47*(3), 3158 – 3163. 19th IFAC World Congress.
22. Chiu, T.-C., Shih, Y.-Y., Pang, A.-C., & Pai, C.-W. (2016). Optimized day-ahead pricing with renewable energy demand-side management for smart grids. *IEEE Internet of Things Journal*, *4*(2), 374–383.
23. García-Bertrand, R. (2013). Sale prices setting tool for retailers. *IEEE Transactions on Smart Grid*, *4*(4), 2028–2035.
24. Wang, Y., Chen, Q., Kang, C., Zhang, M., Wang, K., & Zhao, Y. (2015). Load profiling and its application to demand response: A review. *Tsinghua Science and Technology*, *20*, 117–129, 04.
25. Imamoto, A., & Tang, B. (2008). A recursive descent algorithm for finding the optimal minimax piecewise linear approximation of convex functions. In *Advances in Electrical and Electronics Engineering-IAENG Special Edition of the World Congress on Engineering and Computer Science 2008* (pp. 287–293). IEEE.
26. Price elasticity of demand. Technical report, Australian Energy Regulator, 2005.

Chapter 8
Socio-demographic Information Identification

Abstract This chapter investigates how such characteristics can be inferred from fine-grained smart meter data. A deep convolutional neural network (CNN) first automatically extracts features from massive load profiles. A support vector machine (SVM) then identifies the characteristics of the consumers. Comprehensive comparisons with state-of-the-art and advanced machine learning techniques are conducted. Case studies on an Irish dataset demonstrate the effectiveness of the proposed deep CNN-based method, which achieves higher accuracy in identifying the socio-demographic information about the consumers.

8.1 Introduction

A better understanding of the socio-demographic characteristics of their customers can help retailers provide more personalized services and make more reliable decisions on the targeting of demand response and energy efficiency programs [1, 2]. Leveraging smart meter data to obtain socio-demographic information can, therefore significantly enhance the competitiveness of retailers. In addition, some business models such as energy consulting can also benefit from the identification of socio-demographic characteristics. Take energy consulting as an example, the choice of consumers and the effectiveness of consulting can be largely improved with the socio-demographic information of consumers.

The socio-economic status of individual consumers influences their consumption behavior. Conversely, this socio-economic status can probably be inferred from their consumption behavior. Studies on the relationship between socio-demographic information and electricity consumption data can be divided into two types: estimating the load profile according to the socio-demographic information and identifying the socio-demographic information of consumers from smart meter data. Several authors have worked on inferring load profiles from socio-economic data. McLoughlin et al. [3] analyze the correlation between the electricity consumption of a dwelling and the socio-economic variables of its occupants to estimate load profiles. In [4], these authors then apply self-organizing maps (SOMs) to obtain a set of profile classes and use multi-nominal logistic regression to link the profile classes to household characteristics. Kavousian et al. [5] investigate how climate, building characteristics,

Y. Wang et al., *Smart Meter Data Analytics*,
https://doi.org/10.1007/978-981-15-2624-4_8

appliance stock, and occupants' behavior influence electricity consumption using the factor analysis regression method. Jin et al. [6] link unusual consumption patterns with consumers' socio-demographic characteristics and generate descriptive and predictive models to identify subgroups of consumers. Tong et al. [7] define an energy behavior correlation rate and an indicator dominance index to form a mapping relationship between different energy behavior groups of Irish people and their energy behavior indicators using wavelet analysis and X-means clustering. Vercamer et al. [8] address the issue of assigning new customers, for whom no advanced metering infrastructure (AMI) readings are available, to one of these load profiles based on spectral clustering, random forests, and stochastic boosting-based classification.

Other authors have worked on the identification of socio-demographic information from load profiles. Beckel et al. [9] propose a household characteristic estimation system called CLASS, where features selection and classification are conducted, and the accuracies of the majority of the household characteristic estimations are greater than 70%. In [10] these authors extend the classification work to regression and provide additional details on the consumption figures, ratios, temporal and statistical properties based on feature extraction. Hopf et al. [11] describe an extended system based on the CLASS tool [9], where a total of 88 features are designed, and a combined feature selection method is proposed for classification. Viegas et al. [12] use transparent fuzzy models to estimate the characteristics of consumers and extract knowledge from the fuzzy model rules. Zhong et al. [13] combine discrete Fourier transform (DFT) and a classification and regression tree (CART) to systematically divide the consumers into different groups. Wang et al. [14] apply non-negative sparse coding to extract partial usage patterns from load profiles and use SVM to identify the types of consumers.

As this review of the literature shows, the existing methods for identifying socio-demographic information about the consumers include three main stages: feature extraction to form a feature set, feature selection, and classification or regression. The majority of the works on feature extraction are implemented manually, such as the calculation of consumption, ratios, statistics, and temporal characteristics from load profiles. These manually extracted features may not effectively model the high variability and nonlinearity of individual load profiles. This chapter proposes an automatic feature extraction method based on deep learning techniques to learn features from a different dataset in a flexible manner.

Deep learning is an emerging technique that has advanced considerably since efficient optimization methods were proposed to train deep neural networks [15, 16]. Different types of deep neural networks have been proposed, including auto-encoder, convolutional neural networks (CNNs), recurrent neural networks (RNNs), restricted Boltzmann machine (RBM), and deep belief network (DBN) [17]. Networks that can effectively handle time series, such as deep RNN [18] and RBM [19], have been proposed for load forecasting. Auto-encoders have been applied to extract features from load profiles [20]. CNNs are an effective approach for generating useful and discriminative features from raw data and have broad applications in image recognition, speech recognition, and natural language processing [21]. In this chapter, a deep CNN is proposed to extract the highly nonlinear relationships between electricity consumption at different hours and on different days and the socio-demographic status of the consumer. To further improve the identification performance, a support vector machine (SVM) is used to replace the softmax classifier to identify the socio-demographic information of consumers based on the automatically extracted features.

8.2 Problem Definition

In this chapter, socio-demographic information is obtained from surveys of consumers and include sex, age, employment, social class, and residence. This survey consists mainly of multiple-choice questions that are easy for consumers to answer and can conveniently be encoded using a series of discrete numbers.

Formally, let $i \in I$ and $j \in J$ be the indices of consumers and labels, respectively; let the categorical variable $y_{i,j}$ denote the jth characteristic of the ith consumer; and let $\mathbf{c}_i$ denote the smart meter data of the ith consumer over a certain time period. In general, a feature extraction function G_j is used before classification to transform the original electricity consumption data into a form better suited for classification:

$$\mathbf{s}_{i,j} = G_j(\mathbf{c}_i, \mathbf{w}_{1,j}). \tag{8.1}$$

For the jth label, the classification model $F_j(\mathbf{s}_{i,j}, \mathbf{w}_{2,j})$ needs to be trained, where F_j denotes the mapping relationship from smart meter data to the jth label and $\mathbf{w}_{2,j}$ denotes the trained optimal parameters for classification. Thus, for given $\{\mathbf{s}_{i,j}, y_{i,j}\}$, the jth label of the ith consumer can be estimated:

$$\hat{y}_{i,j} = F_j(\mathbf{s}_{i,j}, \mathbf{w}_{2,j}). \tag{8.2}$$

A smooth function categorical cross entropy is used as the loss function to guide the training of the functions $G_j(\mathbf{c}_i, \mathbf{w}_{1,j})$ and $F_j(\mathbf{s}_{i,j}, \mathbf{w}_{2,j})$ when the total number of training samples is K_j:

$$L_j(\mathbf{w}_{1,j}, \mathbf{w}_{2,j}) = \frac{1}{K_j}\sum_{i=1}^{K_j}[y_{i,j} log \hat{y}_{i,j} - (1 - y_{i,j}) log(1 - \hat{y}_{i,j})]. \tag{8.3}$$

Since feature extraction and classification models are established for each label, the subscript j will be omitted for simplicity.

For the socio-demographic information identification problem, three issues should be addressed:

1. The determination of the feature extraction model G_j to obtain the input data $\mathbf{s}_{i,j}$.
2. The determination of the classification model F_j to produce the estimated label $\hat{y}_{i,j}$.
3. The determination of the training method to obtain $\mathbf{w}_{1,j}$ and $\mathbf{w}_{2,j}$ that achieves the optimal classification performance.

8.3 Method

This section first introduces the rationale for applying a CNN for feature selection and extraction rather than other machine learning techniques such as the least absolute shrinkage and selection operator (LASSO), principal component analysis (PCA), sparse coding. Then, it describes how the CNN architecture is constructed for feature extraction and classification. Finally, it proposes techniques to reduce overfitting and train optimal parameters.

8.3.1 Why Use a CNN?

8.3.1.1 Time Shift Invariance

Figure 8.1 shows the daily load profiles of a consumer over a week. Although some trends can be observed, such as higher consumption in the morning and at night and nearly zero consumption at midnight, considerable uncertainty exists regarding when and how much electricity is used. Time shifting is one of the main characteristics of residential load profiles. The peaks highlighted in the three red circles in Fig. 8.1 show that this consumer uses comparable amounts of electricity during adjacent time periods on different days. The load profiles are highly similar but slightly shifted. In a CNN, the filter weights are uniform for different regions. Thus, the features calculated in the convolutional layer are invariant to small shifts, which means that relatively stable features can be obtained from varying load profiles.

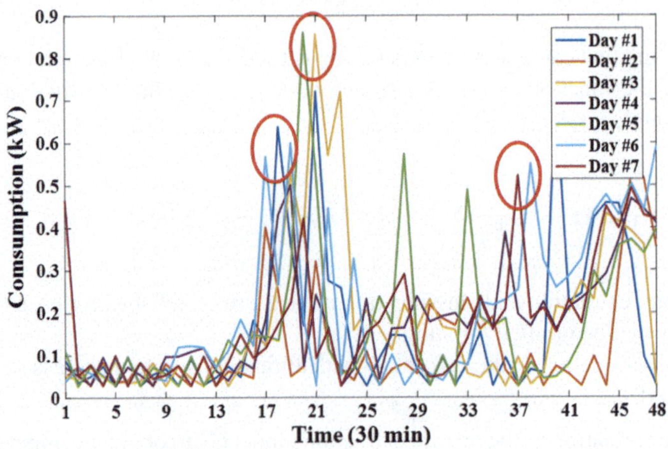

Fig. 8.1 Smart meter data of one consumer over one week

8.3.1.2 Nonlinear Relationship

Unlike the load profile for an entire power system, which is considerably more regular and has a relatively clear relationship with time and weather conditions, the residential load profiles are affected not only by the weather conditions and the type of day, but also by the socio-demographic status of the consumer, the house size, and other factors. The correlations between electricity consumption and these factors are highly nonlinear. Neural networks are able to model these highly nonlinear correlations, particularly networks with multiple layers. A deep CNN can rely on multiple convolutional and fully connected layers to learn the highly nonlinear relationships between load profiles and the socio-demographic information.

8.3.1.3 Data Visualization

The filters learned by the convolutional layers in a deep CNN can be visualized according to the learned weights. This visualization can show how the original profiles are transformed into other forms at different layers. Furthermore, the load profiles that produce the largest activations of neurons can be extracted.

8.3.2 Proposed Network Structure

Figure 8.2 shows the proposed deep CNN architecture. It consists of eight layers, three of which are convolutional layers, three layers are pooling layers, one is a fully connected layer, and the last one is an SVM layer.

Two factors are considered in determining the CNN network structure. The first factor is the characteristics of consumer electricity consumption behaviors. Since the load profiles are so variable, two CNN layers are applied to capture the hidden patterns to identify the socio-demographic information. Moreover, since the dimensions of the input data size are 7×24, which is quite small compared with what is used in image recognition problems, only one pooling layer is used. The second factor is the number of training samples. Since the number of samples is limited, to reduce

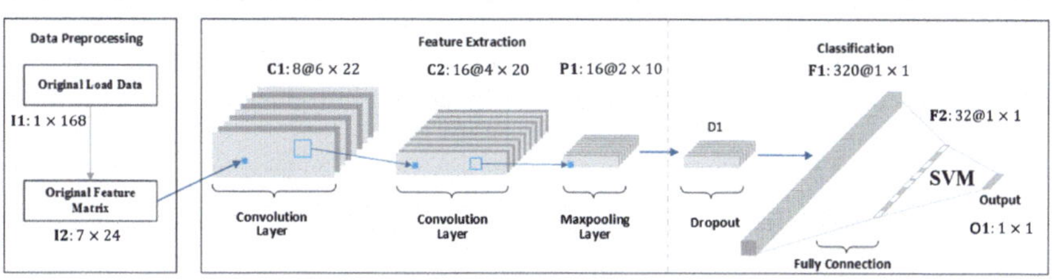

Fig. 8.2 Proposed deep CNN architecture

Table 8.1 Hyperparameters and parameters of the proposed deep CNN

Layer	Layer type	Hyperparameters	Number of parameters
C1	Convolution	Input size: 7 × 24 × 1	56
		Kernel size: 2 × 3	
		Kernel number: 8	
C2	Convolution	Input size: 6 × 22 × 8	160
		Kernel size: 3 × 3	
		Kernel number: 16	
P1	Max-Pooling	Input size: 4 × 20 × 16	None
Dr1	Dropout	None	None
F1	Flatten	None	None
D1	Dense	Input size: 320	10560
		Neuron number: 32	
D2	Softmax	Input size: 32	32
		Neuron number: 1	

the risk of overfitting, the network structure cannot be too complex. Thus, to reduce the parameters, the architecture consists of two convolutional layers, followed by a max-pooling layer and a dropout layer. Finally, the fully connected layer performs the final classification based on the flattened inputs from the previous layers.

The hyperparameters of the proposed deep CNN include the number of kernels, the kernel size of the CNN layers, the pool size of the max-pooling layer, the ratio of dropout, and the number of outputs of the last dense layer. These hyperparameters are obtained by grid search and cross-validation. Table 8.1 summarizes the hyperparameters and the number of parameters of the proposed deep CNN. A total of 10808 parameters must be trained.

8.3.3 *Description of the Layers*

Section 8.3.2 provides the overall structure of the proposed network. This subsection introduces how each layer works. Generally, for the lth layer with input $\mathbf{x}_l$, the learnt weight and bias are $\mathbf{W}_l$ and $\mathbf{b}_l$, respectively. g_l is the transformation function of the layer. The learnt features can be expressed as $g_l(\mathbf{W}_l, \mathbf{b}_l, \mathbf{x}_l)$. In the following, the exact expressions of g_l for different types of layers are introduced.

8.3.3.1 Activation

For each layer, the information is transferred along each neuron which is formulated as activation function. The activation of a neuron is a function from the input of the neuron to its output. Various activation functions have been designed, such as $g_{tanh}(x) = tanh(x)$ and $g_{sig}(x) = (1 + e^{-x})^{-1}$. Activation functions with saturating nonlinearities can significantly slow training with gradient descent or even block weight convergence, which is called vanishing gradient [22]. A non-saturating activation function named rectified linear unit (ReLU) is used in the proposed deep CNN [23], which has been proven to be several times faster than $tanh$ in deep CNN in [24].

$$g_{ReLU}(\mathbf{x}_l) = max(0, \mathbf{x}_l). \tag{8.4}$$

Note that a sigmoid activation function is used in the last layer for the classification tasks.

8.3.3.2 Convolutional Layers

Convolutional layers are the main layers for feature extraction in a deep CNN. Each convolutional layer has a certain number of feature filters. The number of filters in the lth layer is F_l. The f_lth feature filter has its own learnable parameters $\mathbf{W}_{l,f_l}$. Thus, the convolution results obtained by the f_lth filter can be expressed as follows:

$$g_{con}(\mathbf{x}_{l,f_l}) = \sum_{f_l=1}^{F_l} \mathbf{x}_{l,f_l} * \mathbf{W}_{l,f_l} + \mathbf{b}_{l,f_l}. \tag{8.5}$$

where $*$ is the convolution operation. Note that both $\mathbf{x}_{l,f_l}$ and $\mathbf{b}_{l,f_l}$ are matrices with the same size of filter $\mathbf{W}_{l,f_l}$.

8.3.3.3 Dense Layer

A dense layer is also called a fully connected layer. All the input features $\mathbf{x}_l$ are transmitted to the next layer by the weight $\mathbf{W}_l$:

$$g_{den}(\mathbf{x}_l) = \mathbf{W}_l \cdot \mathbf{x}_l + \mathbf{b}_l. \tag{8.6}$$

8.3.3.4 Pooling

A pooling stage is used to downsample and retain discriminant information. Pooling is conducted by transforming small windows into single values by averaging or maxing. The shift-invariance is thus further promoted because the features learned within the small window are similar even with small shifts in electricity consumption. Average-pooling and max-pooling return the average and max values of the activations in the small window since very small activations in the window may submerge the larger ones. Max-pooling retains the particular features, and experience indicates that max-pooling has better performance than average-pooling [25]. Thus, max-pooling is used in the pooling layers:

$$g_{mp}(\mathbf{x}_l) = max_{a \in A}\mathbf{x}_{l,a}. \tag{8.7}$$

8.3.3.5 Dropout Layer

The dropout layer randomly selects a fraction of inputs and sets them to 0. The random selection is assumed to have a Bernoulli distribution with a probability p:

$$\mathbf{r}_l \sim Bernoulli(p). \tag{8.8}$$

where r_l is a matrix of the same size as the input $\mathbf{x}_l$ and its elements are either 0 or 1 following a Bernoulli distribution. The dropout layer can be expressed as follows:

$$g_{do}(\mathbf{x}_l) = \mathbf{r}_l * \mathbf{x}_l. \tag{8.9}$$

8.3.3.6 Classification

Traditionally, softmax is used for classification in the last layer. Softmax is also a fully connected layer:

$$g_{sm}(\mathbf{x}_l) = \mathbf{W}_l \cdot \mathbf{x}_l + \mathbf{b}_l. \tag{8.10}$$

Thus, the probability of the mth class can be calculated using (8.11), and the predicted class is the class corresponding to the maximum probability.

$$P(y = m \mid \mathbf{x}) = \frac{e^{x_m}}{\sum_{m=1}^{M} e^{x_m}}. \tag{8.11}$$

Rather than applying softmax for classification, the proposed method uses an SVM to predict the class based on the learned features:

$$\hat{y} = g_{svm}(\mathbf{x}_l) = sgn(\mathbf{W}_{svm} \cdot \mathbf{x}_l + \mathbf{b}_l). \tag{8.12}$$

where $sgn(\cdot)$ is the sign function, which maps negative values to -1 and positive values to 1. The parameter $\mathbf{W}_l$ in the SVM layer is formulated as an optimization problem:

$$min \left\{ \lambda \parallel \mathbf{W}_{svm} \parallel + \left[\frac{1}{K_i} \sum_{i=1}^{K_i} max(0, 1 - y_i(\mathbf{W}_{svm} \cdot \mathbf{x}_{l,i} + \mathbf{b}_l)) \right] \right\}. \tag{8.13}$$

where $\|\cdot\|$ denotes the 2-norm; λ denotes the trade-off between increasing the margin size and confidence that it lies on the correct side of the margin [26].

8.3.3.7 Loss Function

The objective is to minimize the classification error, which is evaluated by cross-entropy as shown in (8.3).

8.3.4 Reducing Overfitting

Although a deep network with a large number of parameters is very powerful for feature extraction and classification, it can easily become over-fitted. In this case, the number of parameters to be trained is 10808. Changes are made in the inputs, in the model, and in the training method to reduce overfitting of the deep CNN.

8.3.4.1 Data Augmentation

Increasing the sample size is an effective way to reduce overfitting. Various data augmentation techniques, including noise injection, horizontal reflection, and random sampling, have been applied in CNN-based image classification to enlarge the size of the input. For the socio-demographic information identification problem, we use one-week smart meter data to refer to each socio-demographic information of the consumer. Even though the electricity consumption behavior of individual consumers can be affected by their socio-demographical status, weather condition, and even their mood, the previous study shows that each weekly load profile can more or less reveal the socio-demographic information of consumers [10]. Thus, data augmentation consists simply of using smart meter data of other weeks as training data. If the dataset contains Q weeks of smart meter data, then the training dataset can be enlarged Q times.

8.3.4.2 Dropout

Establishing a model with good generalization is important for the proposed deep CNN. Dropping units randomly from the neural network during training can prevent units from co-adapting too much [27] and make a neuron not rely on the presence of other specific neurons. Dropout is quite similar to the ensemble method by varying the hyperparameters to obtain a less correlated model at each epoch.

8.3.4.3 Weight Decay

Applying an appropriate training method is also useful for reducing overfitting. The weight decay term in (8.14) is essentially a regularizer that adds a penalty for weight update at each iteration. Regularization in stochastic gradient descent (SGD) reduces the risk of overfitting.

8.3.5 Training Method

The deep CNN model is trained using stochastic gradient descent with given batch size B, learning rate r, weight decay d, and momentum m. Iterations are implemented as follows [28]:

$$\mathbf{v}_{t+1} = m \cdot \mathbf{v}_t - d \cdot r \cdot \mathbf{W}_t - r \cdot \frac{\partial L}{\partial \mathbf{W}}|_{\mathbf{W}_t,\mathbf{B}_t}. \tag{8.14}$$

$$\mathbf{W}_{t+1} = \mathbf{W}_t + \mathbf{v}_{t+1}. \tag{8.15}$$

where $\mathbf{v}_t$ denotes the changes in the weights at the tth iteration, $\mathbf{W}_t$ denotes the learnt weights at the tth iteration, $m \cdot \mathbf{v}_t$ smooths the direction of gradient descent and accelerates the training process, $d \cdot r \cdot \mathbf{W}_t$ reduces the risk of overfitting, and $r \cdot \frac{\partial L}{\partial \mathbf{W}}|_{\mathbf{W}_t,\mathbf{B}_t}$ denotes the average value of the partial derivative of the loss function with respect to the weight of the tth batch data $\mathbf{B}_t$.

The weights in each layer are initialized by random sampling from a normal distribution with a mean of zero and a standard deviation of 0.01. Biases of all the neurons are initialized at a value of 1 to accelerate the early stage of learning because the inputs of ReLUs are positive in this case.

8.4 Performance Evaluation and Comparisons

This section discusses several evaluation criteria used to quantify the performance of the proposed method. Other methods proposed in the literature are also tested for comparison.

8.4.1 Performance Evaluation

For a classification problem with M classes, an $M \times M$ confusion matrix C can be statistically obtained, where $C_{m,n}$ denotes the number of samples of class m classified into class n. If $m = n$, then $C_{m,n}$ denotes the number of samples that are correctly classified, and vice versa. Thus, the **Accuracy** can be calculated as follows:

Table 8.2 Confusion matrix of a binary classification

		True	
		True positive	True negative
Predicted	Predicted positive	TP	FP
	Predicted negative	FN	TN

$$\textbf{Accuracy} = \frac{\sum_{m=1}^{M} C_{m,m}}{\sum_{m=1}^{M}\sum_{n=1}^{M} C_{m,n}}. \tag{8.16}$$

In particular, for a binary classification problem, we can obtain a confusion matrix, as shown in Table 8.2 according to the predicted and true label of the test samples. TP, FN, FP, and TN represent the number of samples that are correctly predicted as positive, incorrectly predicted as negative, incorrectly predicted as positive, and correctly predicted as negative. Based on these four indices, the $\mathbf{F}_1$ score (also called balanced F score) can be defined as follows to evaluate the performance on the imbalanced label dataset:

$$\mathbf{F}_1 = 2\frac{Pr \times Re}{Pr + Re}. \tag{8.17}$$

where Pr and Re denote precision and recall, respectively, and are calculated as follows:

$$\begin{aligned} Pr &= TP/(TP + FN) \\ Re &= TP/(TP + FP). \end{aligned} \tag{8.18}$$

8.4.2 Competing Methods

The seven methods that are compared with the method proposed in this chapter are briefly introduced in the following paragraphs.

8.4.2.1 Biased Guess (BG)

Since we have prior knowledge of the proportions of different classes in the training dataset, we can identify the socio-demographic information of consumers as the class with the largest proportions. The accuracy of this BG strategy is larger than that of a random guess and can be expressed as follows [10]:

$$\textbf{Accuracy}_{BG} = \sum_{m=1}^{M}\left(\frac{I_m}{I}\right)^2. \tag{8.19}$$

where I_m and I denote the number of samples of class m and the total number of samples. The accuracy of BG is used as a naive benchmark for other methods of identifying socio-demographic information.

8.4.2.2 Manual Feature Selection (MF)

Beckel et al. [10] proposed a consumer characteristic identification system where the majority of the features are extracted manually. The accuracies reported in [10] are compared in the case studies.

8.4.2.3 SVM

SVM is applied directly to predict the socio-demographic information based on the smart meter data of one week without any feature extraction or selection strategy.

8.4.2.4 L_1-Based Feature Selection+SVM (LS)

The linear model with an L_1 regularizer penalty can obtain sparse solutions, i.e., part of the coefficients corresponding to electricity consumptions at different time periods are set to zero. A linear SVM combined with an L_1 regularizer is first used for feature selection and to retain non-zero coefficients.

8.4.2.5 PCA+SVM (PS)

Principal component analysis (PCA) is a frequently used method for dimensionality reduction [29]. PCA is first applied to the original smart meter data to orthogonalize features, where each feature is a linear combination of the original data. The features are sorted in descending order according to variance. The first K transformed features are then used as input to the SVM for the identification task. The accuracies are different with a different number of features (or value of K). The highest accuracy is regarded as the accuracy of the PS method.

8.4.2.6 Sparse Coding+SVM (SS)

Sparse coding is a compressive sensing technique to map the original data into a higher-dimensional space, which is quite different from PCA. The basic idea of sparse coding is to generate redundant vectors such that the original data can be represented in terms of a linear combination of a limited number of vectors [14]. The coefficients learned by sparse coding are then fed into the SVM for socio-demographic information identification.

8.4.2.7 CNN+Softmax (CS)

Softmax in (8.11) rather than SVM is used in the last layer of the proposed deep CNN and is also compared with the proposed method.

8.5 Case Study

In this section, the case studies are implemented using Python 2.7.13 on a standard PC with an Intel® Core™ i7-4770MQ CPU running at 2.40 GHz and with 8.0 GB of RAM. The deep CNN architecture is constructed based on Tensorflow [30], and the interface between CNN and SVM is programmed using scikit-learn [31] and Keras [32].

8.5.1 Data Description

The dataset used in this section was provided by the Commission for Energy Regulation (CER), which is the regulator for the electricity and natural gas sectors in Ireland [33]. This dataset contains the smart meter data of 4232 residential consumers over 536 days at an interval of 30 min. Among the 536 days of smart meter data, the first 75 weeks (525 days) data were chosen to train, validate, and test the proposed deep CNN. More specifically, the consumers are first listed in increasing order according to the ID of the consumers. Then, the smart meter data of the first 80% consumers are used to train and validate the CNN model; the smart meter data of the rest 20% consumer are used to test the model. If there are null values or continuous zero values, the data for these weeks are removed. A total of 300,138 weeks of smart meter data are used. The training data is thus approximately 28 times the number of parameters to be estimated, which reduces the risk of overfitting.

The Irish dataset also contains two survey datasets (pre-trial and post-trial surveys) [33] which contain socio-demographic information about the consumers and are used as labels in the supervised learning task. For a fair comparison with the existing method, we identify the ten survey questions (socio-demographic information) in this section that are also investigated in the existing literature. Table 8.3 lists the socio-demographic information to be identified. To help readers easily find the corresponding survey questions, the question numbers in the survey are also provided in the second column. These questions cover information of the occupants of the house, the house itself, and the domestic appliances.

8.5.2 Basic Results

Figure 8.3 shows the accuracies and $\mathbf{F}_1$ scores of different socio-demographic information. Among these ten questions, the accuracies of #2 (chief income earner has

Table 8.3 Socio-demographic Information to be Identified

No.	Question No.	Socio-demographic information question	Answers	Number
1	300	Age of chief income earner	Young(<35)	436
			Medium(35 ~ 65)	2819
			Old(>65)	953
			Yes	1285
2	310	Chief income earner has retired or not	No	2947
			A or B	642
3	401	Social class of chief income earner	C1 or C2	1840
			D or E	1593
4	410	Have children or not	Yes	1229
			No	3003
5	450	House type	Detached or bungalow	2189
			Semi-detached or terraced	1964
6	453	Age of the house	Old(>30)	2151
			New(<30)	2077
7	460	Number of bedrooms	Very low(<3)	404
			Low(=3)	1884
			High(=4)	1470
			Very High(>4)	474
8	4704	Cooking facility type	Electrical	1272
			Not electrical	2960
9	4905	Energy-efficient light bulb proportion	Up to half	2041
			Three quarters or more	2191
10	6103	Floor area	Small(<100)	232
			Medium(>100 and <200)	1198
			Big(>200)	351

retired or not), #4 (have children or not), and #8 (cooking facility type) are higher than 75%; the accuracies of #7 (number of bedrooms) and #9 (energy-efficient light bulb proportion) are lower than 60%; and the accuracies of the remaining questions are between 60 and 75%. Note that the number of the two answers of #9 (up to half/three quarters or more) are 2041 and 2149, respectively. Although the accuracy of #9 is lower, its $\mathbf{F}_1$ score is not the lowest. Clearly, whether having children or not has a great influence on the daily life of consumers and significantly affects the load profiles. The type of cooking appliance and light bulbs directly determine electricity consumption. Thus, it is rather easy to identify these two factors from smart meter data. Compared with other information, the number of bedrooms has a weak relationship with the behavior of electricity consumption. The average accuracy and $\mathbf{F}_1$ score of these questions are 67.3% and 0.622, respectively. Three ways, data aug-

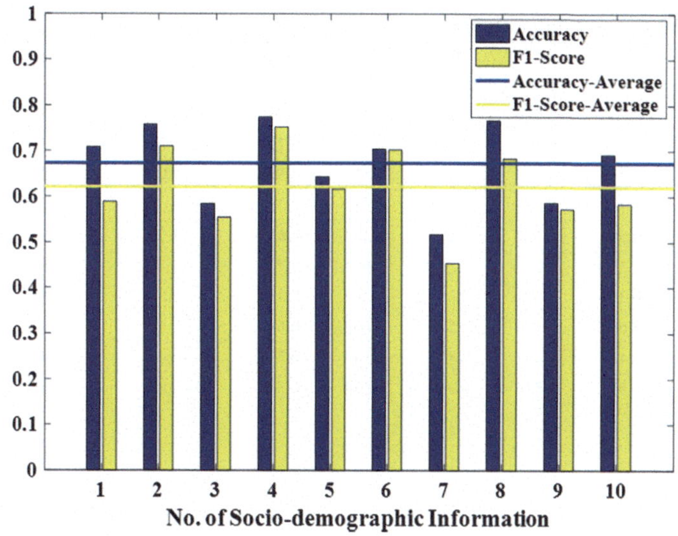

Fig. 8.3 Performance of the proposed method

Table 8.4 Accuracies of different methods

	BG	MF	SVM	LS	PS	SS	CS	Proposed	Improvement 1 (%)	Improvement 2 (%)	Improvement 3 (%)
1	0.511	0.59	0.648	0.664	0.667	0.666	0.688	**0.708**	3.15	**2.95**	29.47
2	0.577	0.73	0.697	0.702	0.696	0.693	0.748	**0.758**	2.47	**1.37**	24.39
3	0.382	0.53	0.506	0.498	0.513	0.5	0.55	**0.584**	3.77	**6.09**	37.66
4	0.588	0.73	0.709	0.715	0.73	0.714	0.748	**0.774**	2.47	**3.45**	24.39
5	0.501	0.59	0.572	0.567	0.567	0.531	0.621	**0.643**	5.25	**3.53**	16.65
6	0.5	0.64	0.564	0.577	0.582	0.566	0.665	**0.703**	3.91	**5.69**	22.77
7	0.416	0.39	0.472	0.476	0.49	0.467	0.501	**0.517**	2.24	**3.17**	13.77
8	0.579	0.71	0.687	0.675	0.694	0.698	0.739	**0.766**	4.08	**3.70**	22.60
9	0.501	0.55	0.511	0.492	0.528	0.525	0.565	**0.586**	2.73	**3.70**	7.13
10	0.508	0.5	0.643	0.649	0.644	0.639	0.658	**0.691**	1.39	**4.96**	24.41

mentation, dropout, and weight decay, are used to reduce the overfitting risk. We also conduct numerical experiments without dropout layer and weight decay. The average accuracies of the methods without dropout layer and weight decay are 61.2% and 64.3%, respectively; the average $\mathbf{F}_1$ scores are 0.597 and 0.608, respectively. It can be seen that the network with dropout layer and weight decay has better performance which verifies the effectiveness of the two ways to reduce overfitting.

8.5.3 Comparative Analysis

Tables 8.4 and 8.5 compare the accuracies and $\mathbf{F}_1$ scores of the proposed and competing methods. The column Improvement 1 shows the relative improvements of CS

Table 8.5 $\mathbf{F}_1$ scores of different methods

	SVM	LS	PS	SS	CS	Proposed	Improvement 1 (%)	Improvement 2 (%)
1	0.562	0.563	0.539	0.533	0.571	**0.589**	1.42	**3.15**
2	0.652	0.659	0.602	0.569	0.687	**0.71**	4.25	**3.35**
3	0.474	0.458	0.47	0.451	0.512	**0.554**	8.02	**8.20**
4	0.709	0.711	0.687	0.615	0.737	**0.752**	3.66	**2.04**
5	0.446	0.563	0.562	0.451	0.584	**0.616**	3.73	**5.48**
6	0.488	0.576	0.52	0.519	0.661	**0.702**	14.76	**6.20**
7	0.418	0.389	0.42	0.361	0.432	**0.454**	2.86	**5.09**
8	0.584	0.605	0.574	0.574	0.652	**0.683**	7.77	**4.75**
9	0.446	0.454	0.491	0.409	0.547	**0.572**	11.41	**4.57**
10	0.539	0.538	0.516	0.499	0.552	**0.583**	2.41	**5.62**

(traditional CNN) compared with the best performer among the other six competing methods. The column Improvement 2 shows the relative improvements of the proposed CNN+SVM method compared with the CS method. Improvement 3 shows the relative improvements of the average performance of MF, SVM, LS, PS, SS, CS, and the proposed method compared with BG method. The accuracies of the BG and MF methods are provided in [10], the $\mathbf{F}_1$ scores of which are not provided. Note that the accuracies of the MF method are obtained by running the classification over the entire 75 weeks and majority voting.

It is clear that all classification models outperform the BG method. If the electricity consumption of residents has no relationship with one specific socio-demographic information, the feature extraction and classification process may not be able to improve the identification accuracy compared with BG method. In other words, the improvements of feature extraction and classification methods over BG method (Improvement 3 in Table 8.4) can more or less indicate how much the socio-demographic information can affect the electricity consumption behaviors of the consumers. The improvement of #9 is the lowest, which means that the energy-efficient light bulb proportion has very little influence on electricity consumption behavior.

The performance of Lasso-based SVM (LS) has better but very comparable performance with SVM in terms of average accuracy and $\mathbf{F}_1$ score. This result means that the Lasso-based feature extraction method has a very small effect on the performance of the SVM classifier. However, after PCA or sparse coding-based feature extraction, the classifiers have better performance than the Lasso-based method or simple SVM. CNN-based deep learning network (CS) has distinct advantages over SVM, LS, PS, and SS, which means that the proposed method can extract highly nonlinear relationships hidden in these massive load profiles. By replacing softmax with SVM, the performance can be further improved in terms of both accuracy and $\mathbf{F}_1$ score.

8.6 Conclusions

This chapter proposes a CNN-based deep learning method for identifying consumer socio-demographic information. CNN can take into consideration the correlations between different hours of the day and different days. The proposed method automatically extracts the hidden usage patterns from massive and varying smart meter data to improve the accuracy of socio-demographic information identification. Case studies on an Irish dataset show the superiority of CNN over other feature extractions methods.

References

1. Keerthisinghe, C., Verbič, G., & Chapman, A. C. (2016). A fast technique for smart home management: ADP with temporal difference learning. *IEEE Transactions on Smart Grid*, *9*(4), 3291–3303.
2. Sun, Siyang, Yang, Qiang, & Yan, Wenjun. (2017). Optimal temporal-spatial pev charging scheduling in active power distribution networks. *Protection and Control of Modern Power Systems*, *2*(1), 34.
3. McLoughlin, Fintan, Duffy, Aidan, & Conlon, Michael. (2012). Characterising domestic electricity consumption patterns by dwelling and occupant socio-economic variables: An irish case study. *Energy and Buildings*, *48*, 240–248.
4. McLoughlin, Fintan, Duffy, Aidan, & Conlon, Michael. (2015). A clustering approach to domestic electricity load profile characterisation using smart metering data. *Applied Energy*, *141*, 190–199.
5. Kavousian, Amir, Rajagopal, Ram, & Fischer, Martin. (2013). Determinants of residential electricity consumption: Using smart meter data to examine the effect of climate, building characteristics, appliance stock, and occupants' behavior. *Energy*, *55*, 184–194.
6. Jin, Nanlin, Flach, Peter, Wilcox, Tom, Sellman, Royston, Thumim, Joshua, & Knobbe, Arno. (2014). Subgroup discovery in smart electricity meter data. *IEEE Transactions on Industrial Informatics*, *10*(2), 1327–1336.
7. Tong, Xing, Li, Ran, Li, Furong, & Kang, Chongqing. (2016). Cross-domain feature selection and coding for household energy behavior. *Energy*, *107*, 9–16.
8. Vercamer, Dauwe, Steurtewagen, Bram, Van den Poel, Dirk, & Vermeulen, Frank. (2015). Predicting consumer load profiles using commercial and open data. *IEEE Transactions on Power Systems*, *31*(5), 3693–3701.
9. Beckel, C., Sadamori, L., & Santini, S. (2013). Automatic socio-economic classification of households using electricity consumption data. In *Proceedings of the Fourth International Conference on Future Energy Systems* (pp. 75–86). ACM.
10. Beckel, Christian, Sadamori, Leyna, Staake, Thorsten, & Santini, Silvia. (2014). Revealing household characteristics from smart meter data. *Energy*, *78*, 397–410.
11. Hopf, Konstantin, Sodenkamp, Mariya, Kozlovkiy, Ilya, & Staake, Thorsten. (2016). Feature extraction and filtering for household classification based on smart electricity meter data. *Computer Science-Research and Development*, *31*(3), 141–148.
12. Viegas, J. L., Vieira, S. M., & Sousa, J. M. C. (2016). Mining consumer characteristics from smart metering data through fuzzy modelling. In *International Conference on Information Processing and Management of Uncertainty in Knowledge-Based Systems* (pp. 562–573) Springer.
13. Zhong, Shiyin, & Tam, Kwa-Sur. (2015). Hierarchical classification of load profiles based on their characteristic attributes in frequency domain. *IEEE Transactions on Power Systems*, *30*(5), 2434–2441.

14. Wang, Yi, Chen, Qixin, Kang, Chongqing, Xia, Qing, & Luo, Min. (2016). Sparse and redundant representation-based smart meter data compression and pattern extraction. *IEEE Transactions on Power Systems*, *32*(3), 2142–2151.
15. LeCun, Yann, Bengio, Yoshua, & Hinton, Geoffrey. (2015). Deep learning. *Nature*, *521*(7553), 436–444.
16. Hinton, G. E., & Salakhutdinov, R. R. (2006). Reducing the dimensionality of data with neural networks. *Science*, *313*(5786), 504–507.
17. Schmidhuber, Jürgen. (2015). Deep learning in neural networks: An overview. *Neural Networks*, *61*, 85–117.
18. Shi, H., Minghao, X., & Li, R. (2017). Deep learning for household load forecasting—a novel pooling deep rnn. *IEEE Transactions on Smart Grid*, *9*(5), 5271–5280.
19. Mocanu, E., Nguyen, P. H., Gibescu, M., & Kling, W. L. (2016). Deep learning for estimating building energy consumption. *Sustainable Energy, Grids and Networks*, *6*, 91–99.
20. Varga, E. D., Beretka, S. F., Noce, C., & Sapienza, G. (2015). Robust real-time load profile encoding and classification framework for efficient power systems operation. *IEEE Transactions on Power Systems*, *30*(4), 1897–1904.
21. Sharif Razavian, A., Azizpour, H., Sullivan, J., & Carlsson, S. (2014). Cnn features off-the-shelf: an astounding baseline for recognition. In *Proceedings of the IEEE Conference on Computer Vision and Pattern Recognition Workshops* (pp. 806–813).
22. He, K., Zhang, X., Ren, S., & Sun, J. (2015). Delving deep into rectifiers: Surpassing human-level performance on imagenet classification. In *Proceedings of the IEEE International Conference on Computer Vision* (pp. 1026–1034).
23. Nair, V., & Hinton, G. E. (2010). Rectified linear units improve restricted boltzmann machines. In *Proceedings of the 27th International Conference on Machine Learning (ICML-10)* (pp. 807–814).
24. Krizhevsky, A., Sutskever, I., & Hinton, G. E. (2012). Imagenet classification with deep convolutional neural networks. In *Advances in Neural Information Processing Systems* (pp. 1097–1105).
25. Boureau, Y. L., Ponce, J., & LeCun, Y. (2010). A theoretical analysis of feature pooling in visual recognition. In *Proceedings of the 27th International Conference on Machine Learning (ICML-10)* (pp. 111–118).
26. Furey, T. S., Cristianini, N., Duffy, N., Bednarski, D. W., Schummer, M., & Haussler, D. (2000). Support vector machine classification and validation of cancer tissue samples using microarray expression data. *Bioinformatics*, *16*(10), 906–914.
27. Srivastava, Nitish, Hinton, Geoffrey, Krizhevsky, Alex, Sutskever, Ilya, & Salakhutdinov, Ruslan. (2014). Dropout: A simple way to prevent neural networks from overfitting. *The Journal of Machine Learning Research*, *15*(1), 1929–1958.
28. Bengio, Y. (2012). Practical recommendations for gradient-based training of deep architectures. In *Neural Networks: Tricks of the Trade* (pp. 437–478) Springer.
29. Wold, Svante, Esbensen, Kim, & Geladi, Paul. (1987). Principal component analysis. *Chemometrics and Intelligent Laboratory Systems*, *2*(1–3), 37–52.
30. Abadi, M., Agarwal, A., Barham, P., Brevdo, E., Chen, Z., Citro, C., Corrado, G. S., Davis, A., Dean, J., Devin, M. et al. (2016). Tensorflow: Large-scale machine learning on heterogeneous distributed systems. arXiv:1603.04467.
31. Pedregosa, F., Varoquaux, G., Gramfort, A., Michel, V., Thirion, B., Grisel, O., Blondel, M., Prettenhofer, P., Weiss, R., Dubourg, V. et al. (2011). Scikit-learn: Machine learning in python. *Journal of Machine Learning Research*, *12*(Oct), 2825–2830.
32. Chollet, F. et al. (2015). Keras.
33. Irish Social Science Data Archive. (2012). Commission for Energy Regulation (CER) Smart Metering Project. http://www.ucd.ie/issda/data/commissionforenergyregulationcer/.

Chapter 9
Coding for Household Energy Behavior

Abstract Household energy behavior is a key factor that dictates energy consumption, efficiency, and conservation. In the past, household energy behavior was typically unknown because conventional meters only recorded the total amount of energy consumed by a household over a significant period of time. The rollout of smart meters enables real-time household energy consumption data to be recorded and analyzed. This chapter uses smart meter readings from more than 5000 Irish households to identify energy behavior indicators through a cross-domain feature selection and coding approach. The idea is to extract and connect customers' features from the energy domain and demography domain, i.e., smart meter data and household information. Smart meter data are characterized by typical energy spectral patterns, whereas household information is encoded as the energy behavior indicator. The results show that employment status and internet usage are highly correlated with household energy behavior in Ireland because employment status and internet usage have an important effect on lifestyle, including when to work, play, and rest, and hence yield a difference in electricity use style. The proposed approach offers a simple, transparent and effective alternative to a challenging cross-domain matching problem with massive smart meter data and energy behavior indicators.

9.1 Introduction

With the development of household-level low carbon technologies such as PVs (photovoltaics), EVs (electric vehicles) and HPs (heat pumps), households have taken a more active role in the energy system. Understanding their energy behaviors would be beneficial to decreasing energy loss, improving system efficiency, and enhancing sustainable energy integration. In the past, little information on household energy has been collected because conventional meters only record the total amount of energy consumed for a household over a significant period of time. Smart meters record the consumption of electric energy in intervals of an hour or less and communicate that information back to the utility for monitoring and billing.

The recent rollout of smart meters has brought opportunities to provide insight into household energy behavior because energy behavior dictates the shape and magnitude

Y. Wang et al., *Smart Meter Data Analytics*,
https://doi.org/10.1007/978-981-15-2624-4_9

of the electrical load, which can be captured by smart meter data. With increasingly extensive and massive smart meter data, the energy industry has embraced big data [1], in which a series of data-mining-related methods in different aspects, such as classification [2–4], regression [5–7], and clustering [8–10], are applied to mine and connect data in the energy domain. Based on these methods, several smart meter mining applications have emerged, including load profiling [11–13], customer segmentation [14–16], load forecasting [17], and NILM (non-intrusive load monitoring) [18].

Reference [19] proposes a load prediction method for industrial and commercial consumers based on socioeconomic factors. The socioeconomic factors considered include the population, crime rate, building size, available area, turnover rate, number of employees, and etc. The basic idea is to obtain consumers' typical load profiles using spectral clustering method, and then to establish the adaptive stochastic boosting and random forests classification model between socioeconomic factors and consumers' typical load profiles. The method has been applied to more than 6000 Belgian industrial and commercial consumers. The results show high prediction accuracy. Except for this work, few studies have linked data in the demography domain with smart meter data. Smart meter data represent human energy behavior, which is affected by data in the demography domain. The idea is to extract and connect human features from the energy domain to the demography domain, i.e., smart meter data and household information. Smart meter data are characterized by typical energy spectral patterns representing energy behavior, whereas household information is encoded as an energy behavior indicator.

Motivated by this idea, this chapter proposes a cross-domain feature selection and coding method for household energy behavior. Human-feature-related data in the demography domain is typically recorded in a questionnaire composed of a series of specially designed questions, such as age, occupancy, employment status, income, and energy usage habits. Each question can have a range of answers which can be designated as A, B, C, D, etc. In this method, the questionnaire answers constitute a label sequence, and the subset of labels that has the most significant effect on energy behavior represented by the typical energy spectral pattern is identified as energy behavior indicators, making it possible to construct a connection between data in the energy domain and the demography domain. Through this approach, energy patterns and energy behavior indicators are connected, providing a deep understanding of why people behave differently, illustrating the underlying factors that are responsible for differing human energy behavior, and showing how energy behavior might change in the future if people's status changes.

9.2 Basic Idea and Framework

This chapter presents a socioeconomic factors-based consumer load profile prediction method to link the energy domain and the demography domain. The basic idea is shown in Fig. 9.1: since the shape of consumers' load profiles vary greatly, it is necessary to cluster consumers into different classes according to the shape of

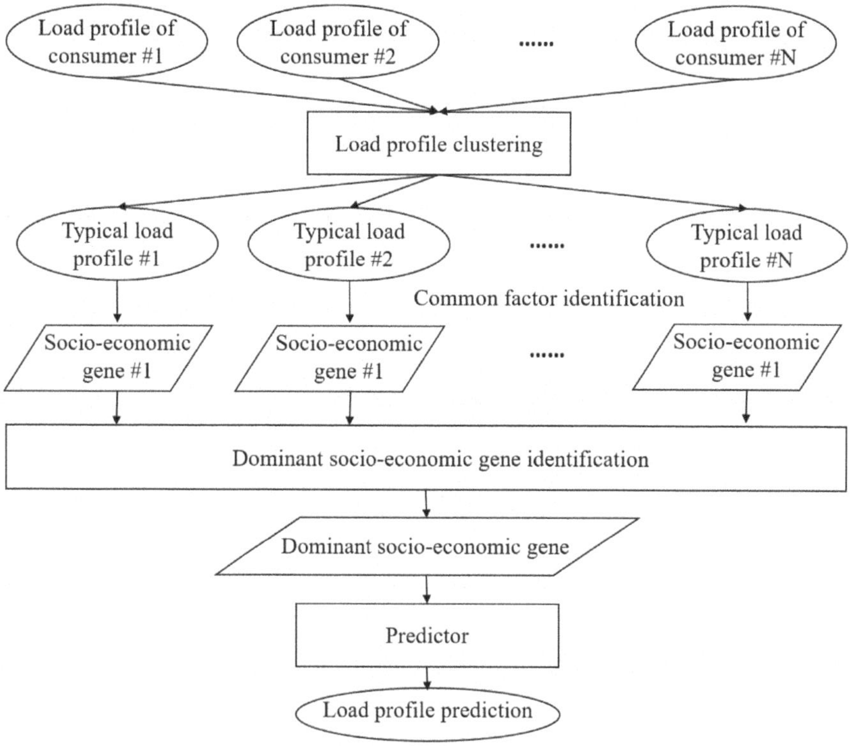

Fig. 9.1 Framework of the proposed method

load profiles first. On this basis, for each class of consumers with similar load profile shapes, their common socioeconomic factors can be identified. These identified factors are viewed as influential socioeconomic factors. They are defined as "socioeconomic genes". Then, for different types of consumers, the differences of their "socioeconomic genes" will be analyzed, and the "socioeconomic genes" with significant differences among different classes will be identified as the socioeconomic related factors that affect consumers' electricity consumption behavior. It is defined as "dominant socioeconomic gene". Finally, the dominant socioeconomic gene is used as the input of the consumer load profile predictor to realize the prediction process.

9.3 Load Profile Clustering

For the consumer typical load profile extraction, a Gaussian Mixture Model (GMM)- [20] and X-mean- [21] based two-stage clustering approach is proposed. The load profiles of different consumers are very different. Even the load profiles on different

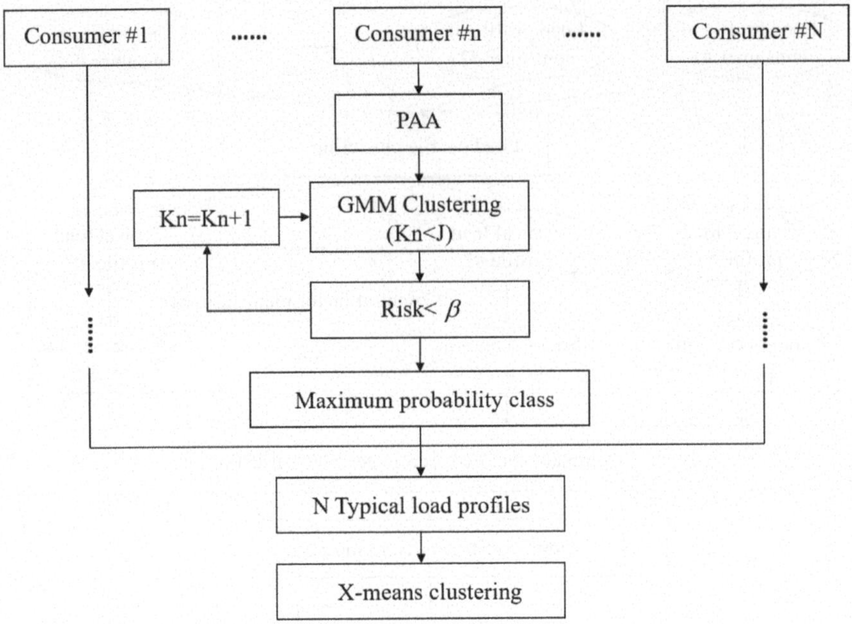

Fig. 9.2 GMM and X-means-based two-stage load profile clustering

days of one consumer are also very different. It is necessary to extract a typical load profile for each consumer first to characterize the electricity consumption behavior of this consumer. Then clustering can be conducted on different consumers based on their typical load profiles.

In the process of typical load profile extraction for each consumer, the GMM clustering method is directly applied to extract several types of load profiles. Then, the typical load curves of multiple consumers are clustered by X-means clustering method. The overall steps are described in Fig. 9.2. Details are provided in the following subsections.

9.3.1 GMM-Based Typical Load Profile Extraction

A simple and efficient method for consumer typical load profile extraction is proposed here, which can be divided into two steps: piecewise average approximation (PAA) and GMM clustering.

(1) Piecewise Average Approximation

A fixed window width w is used to divide the original load profile into several segments, and each segment is approximated by the average load within the window. In this way, the fluctuation of the original load profile can be reduced. We denote the

N daily load profiles as $\mathbf{P}_n = [P_{n,1}\ P_{n,2}\ \cdots\ P_{n,T}]$, where $n = 1, 2, \ldots, N$; T is the number of total time periods per day; w is an integer and divisible by T. Thus, the approximated load profile is $\mathbf{P}'_n = [P'_{n,1}\ P'_{n,2}\ \cdots\ P'_{n,T/w}]$, where

$$\begin{aligned} P'_{n,1} &= \frac{1}{w}\sum_{i=1}^{w} P_{n,i}, \\ P'_{n,2} &= \frac{1}{w}\sum_{i=1}^{w} P_{n,w+i}, \\ &\ldots, \\ P'_{n,T/w} &= \frac{1}{w}\sum_{i=1}^{w} P_{n,T-w+i}. \end{aligned} \tag{9.1}$$

(2) GMM Clustering

GMM model is a mixed distribution model, which can be used to model the data consist of different distributions. The probability density function (PDF) of the mixed distribution can be expressed as the weighted sum of a series of known finite PDFs (usually normal distribution or other distributions). For a GMM model with finite distributions, the probability density of load profiles appearing in the model is as follows.

$$f(\mathbf{P}'_n; \Psi) = \sum_{k=1}^{K} \lambda_k f_k(\mathbf{P}'_n; \theta_k) \tag{9.2}$$

where λ_k denotes the weight of thc finite distribution f_k; Ψ denotes all the parameters of the GMM model; θ_k denotes the parameters of the kth PDF; $\lambda_k f_k(\mathbf{P}'_n; \theta_k)$ denotes the weighted PDF. Since the integral of the mixed probability density distributions is 1, the sum of weights should be equal to 1:

$$\sum_{k=1}^{K} \lambda_k = 1 \tag{9.3}$$

The GMM model can be solved by Expectation Maximum (EM) algorithm with historical load profiles $\mathbf{P}'_n$, finite distribution f_k, and the number of distributions K. Then, according to the weighted probability density that $\mathbf{P}'_n$ belongs to each finite distribution, the posterior probability of each finite distribution f_k that $\mathbf{P}'_n$ belongs to can be obtained by Bayesian theorem.

Then we choose the class k with the greatest posterior probability as the classification of $\mathbf{P}'_n$. The term (1−posterior probability that $\mathbf{P}'_n$ belongs to class k) is defined as the risk. The optimal number of distributions K can be searched by gradually increasing the value of K from $K = 1$, and the minimum value of K where the total risk of all historical days is less than the threshold β is selected as the optimal value.

For all classes of historical load profiles, the load profile with the maximum posterior probability is selected as the typical load profile of this class. On this basis, and the typical load profile covering the most historical days is taken as the typical load profile of the consumer.

9.3.2 X-Means-Based Load Profile Clustering

X-means clustering method is an extension of k-means clustering method, which provides an effective approach to estimate the number of clusters. This method overcomes the main shortcomings of k-means clustering method. It searches for a series of different values of k and chooses the optimal k value according to Bayesian information criterion (BIC), In this way, it is unnecessary to determine the number of clusters before clustering. Its search strategy is as follows:

(1) For the given k clusters, the cluster center and the corresponding BIC are determined.
(2) Some existing classes are split into two clusters by a k-means clustering method ($k = 2$), where whether the cluster should be split depends on whether it can help increase the value of BIC.
(3) Repeat the above steps until the number of clusters k reaches the pre-set maximum value, and select k that corresponds to the optimum value of BIC as the final number of clusters.

9.4 Socioeconomic Genes Identification Method

9.4.1 Socioeconomic Information Classification

In the questionnaire recording Irish household information, each question's answer list is represented by the labels A, B, C, D, etc. These different labels record different household information. The following is a sample question in the Irish smart meter trials questionnaire.

Question 3: What is the employment status of the chief income earner in your household, is he/she

Option A *An employee*
Option B *Self-employed (with employees)*
Option C *Self-employed (with no employees)*
Option D *Unemployed (actively seeking work)*
Option E *Unemployed (not actively seeking work)*
Option F *Retired*
Option G *Career: Looking after relative or family*

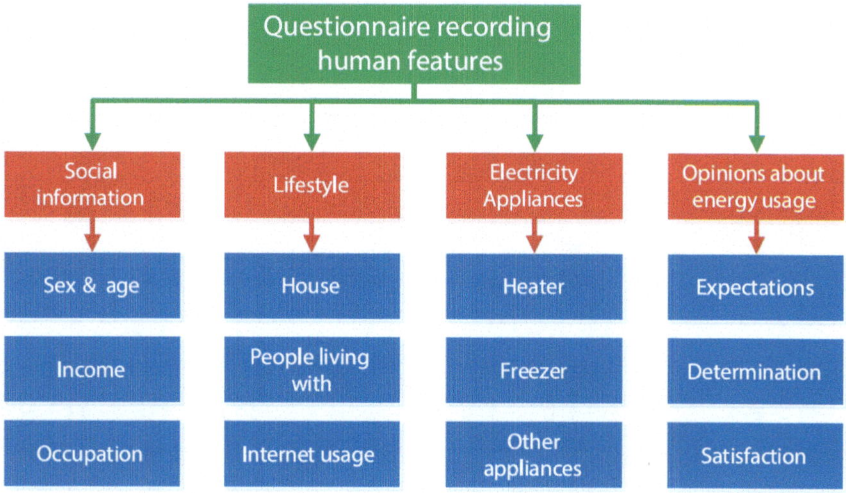

Fig. 9.3 The classification of the Irish people's household information released by SEAI

In the Irish smart meter trial data released by Electric Ireland and SEAI (Sustainable Energy Authority of Ireland) [22], the questionnaire recording Irish people's household information comprises 144 questions, and these questions can be divided into four major categories and 12 minor categories, as shown in Fig. 9.3. The four major categories of questions address social information, lifestyle, electrical appliances and opinions about energy usage. These four major categories of questions can then be classified into 16 minor categories, which are sex & age, income, occupation, house, people living with, internet usage, heater, freezer, other appliances, expectations, determination, and satisfaction.

To illustrate the questionnaire information clearly, the most representative question in each category is listed in Table 9.1. For example, in the sex & age category, the representative question is "please record sex from voice", with the answer A corresponding to Male, and B corresponding to Female. In the occupation category, according to the NRS (National Readership Survey) social grades system [23], the customers are classified into six grades, which are AB, C1, C2, DE, F, and refused. These grades are represented by answer A to F, respectively. Note that there are 12 categories of questions, and for each category, one representative question and its corresponding answers are shown in this table. Some answer options, such as options C, D, E, and F in the sex & age category, are blank, indicating that there is no corresponding answer to the options.

Table 9.1 Representative questions in the behavior questionnaire released by SEAI

Category	Question	A	B	C	D	E	F
Sex & age	Please record sex from voice	Male	Female				
Income	Can you state which of the following broad categories best represents the yearly household income BEFORE TAX?	Less than 15,000 Euros	15,000–30,000 Euros	30,000–50,000 Euros	50,000–75,000 Euros	75,000 or more Euros	Refused
Occupation	SOCIAL CLASS Interviewer, Respondent said that occupation of chief income earner was <CLASS>Please code	AB	C1	C2	DE	F	Refused
House	Do you own or rent your home?	Rent (from a private landlord)	Rent (from a local authority)	Own outright (not mortgaged)	Own with mortgage etc.	Other	
People living with	What best describes the people you live with?	I live alone	All people in my home are over 15 years of age	Both adults and children under 15 years of age live in my home			
Internet usage	Do you use the internet regularly yourself?	Yes	No				
Heater	Do you have a timer to control when your heating comes on and goes off?	Yes	No				
Freezer	Have any of standalone freezers ever applied to you?	Yes	No				
Non-heating appliances	Number of washing machine	None	1	2	More than 2		
Expectations	I would now like to ask you about your expectations about: Learn how to reduce my energy usage	Yes	No				
Determination	My household may decide to make major changes to the way we use electricity	Strongly agree	Agree	Neutral	Disagree	Strongly disagree	
Satisfaction	The overall cost of electricity	Very Satisfied	Satisfied	Neutral	Disagree	Very Dissatisfied	

9.4.2 The Concept of Socioeconomic Genes

For each consumer, the answer labels of all the questions in the questionnaire can be arranged in the order of questions to form a label sequence, which represents the socioeconomic characteristics of residents, only a part of which has an impact on residents' electricity consumption behavior. Coincidentally, hereditary molecular DNA is composed of four bases, A, T, C, and G, arranged in an orderly manner. Most of the base sequences of DNA are not expressed, but only those that can express and affect biological traits are called genes. Inspired by the similarity between DNA carrying biological traits and socioeconomic information labeling sequences affecting consumers' electricity consumption behavior, the following concepts are vividly defined:

Socioeconomic DNA: An orderly sequence of labels representing users' socioeconomic information. The labels of different locations represent different social and economic characteristics of different aspects, and the labels of the same location represent different social and economic characteristics of the same aspect.

Socioeconomic genes: labels that have a dominant impact on the consumer's electricity consumption profiles.

Socioeconomic gene loci: The sequence of socioeconomic genes in socioeconomic DNA.

Socioeconomic gene profiles: Maps of socioeconomic genes and their loci are represented on consumers' socioeconomic DNA.

9.4.3 Socioeconomic Genes Evaluation Indicators

(1) Genes Entropy

It is impossible to determine which segments of socioeconomic DNA belong to socioeconomic genes based on single consumer's socioeconomic DNA and their electricity profiles. However, if we analyze a class of consumers with similar load profiles, we can determine the socioeconomic genes and their loci by calculating the gene entropy of such consumers.

Gene Entropy: A measure of gene purity. 0 represents complete homogeneity, and closer to 1 means more confusion.

For a group of consumers, if they contain c different labels at a certain locus, and the proportion of the jth label is p_j, then the gene entropy S_g of the locus can be calculated as follows:

$$S_g = \sum_{j=1}^{c} -p_j \log_c(p_j) \tag{9.4}$$

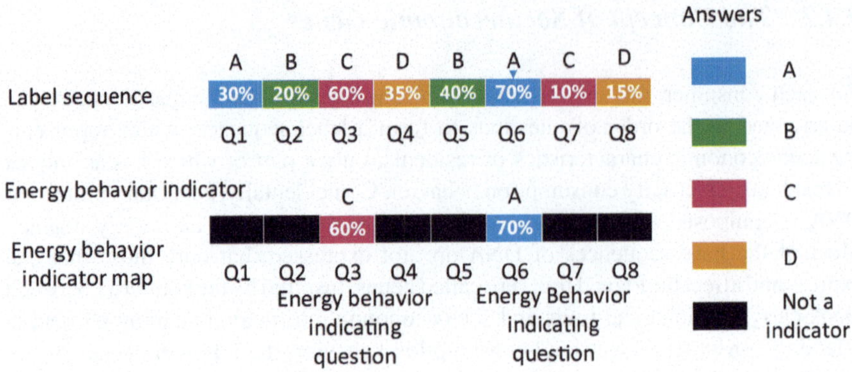

Fig. 9.4 Label sequence, energy behavior indicating question, energy behavior indicator and energy behavior indicator map for a person

By setting a certain threshold value ϵ_1, we can judge whether the entropy value S_g is less than ϵ_1 or not. If $S_g < \epsilon_1$, the locus belongs to the gene locus, and the label with the highest proportion of the locus is the socioeconomic gene.

As shown in Fig. 9.4, The four different answers A, B, C, and D are marked with different colors. The sample household label sequence is a sequence of answers for eight questions in the questionnaire. The sample questionnaire is composed of eight different questions marked with Q1 Q8, and the sample household answers are A, B, C, D, B, A, C, and D, respectively. Hence, the sample household label sequence is composed of eight sequential answers, which are A, B, C, D, B, A, C, and D. The number contained in each small rectangle shows the rate of the corresponding answer in the group of the sample household. For example, the answer rates for Q3 and Q6 are 60% and 70%, respectively; for the remainder, the answer rates are all below 50%. For this sample household, only the answer to Q3 and Q6 are the most frequent in the group, and the corresponding EBCRs are greater than the preset threshold. Hence Q3 and Q6 are energy behavior indicating questions, and the corresponding energy behavior indicators are C (Q3) and A (Q6). From the energy behavior indicator map of the household, it is obvious that only question 3 and 6 in the energy behavior sequence are identified as energy behavior indicating questions, which means their answers C (Q3) and A (Q6) are energy behavior indicators and are correlated with the household energy behavior.

(2) Classification Entropy

In biology, genes can be divided into dominant genes and recessive genes. Compared with recessive genes, dominant genes play a significant role in biological traits. When dominant genes and recessive genes coexist, recessive genes will be "shielded" by dominant genes, and thus have no effect on biological traits. In socioeconomic genes, high gene entropy characterizes that a class of consumers contain almost the same gene at the corresponding locus. However, there may be many types of consumers with the same gene at the same time. It is impossible to judge which kind of consumers belong to according to the gene. Thus, we define the dominant socioeconomic genes which have strong classification ability for consumers, while the recessive socioeconomic genes which have weak classification ability for consumers. The so-called consumer classification ability of socioeconomic gene is the ability to distinguish which kind of consumer belongs to according to the socioeconomic gene. Generally speaking, the more uniformly the socioeconomic genes among the consumers are distributed, the worse their classification ability is., the worse their classification ability; the more concentrated they are among the minority consumers, the stronger their classification ability is. In order to distinguish dominant genes from recessive genes, it is necessary to reflect the classification ability of genes. The classification entropy index of genes defined here is as follows:

Classification Entropy: A measure of the ability of a gene for consumer classification. 0 means that the gene is simply distributed in one class; it is closer to 1 when more genes are distributed in many classes.

For a gene, if it distributes in m clusters and the proportion of the ith class is q_i, the classification entropy of the gene can be calculated as follows:

$$S_c = \sum_{i=1}^{m} -q_i \log_m(q_i) \tag{9.5}$$

By setting a certain threshold value ϵ_2, we can judge whether the entropy value S_c is less than ϵ_2 or not. If $S_c < \epsilon_2$, the gene is the dominant gene.

The IGD represents the uniqueness of the energy behavior indicator. Different groups of customers may have the same energy behavior indicator. The more groups the energy behavior indicator is shared by, the lower the IGD is. A unique energy behavior indicator means a higher dominance on energy behavior, and a common energy behavior indicator probably has no decisive effects on energy behavior. There is also a threshold for the IGD index to determine whether an energy behavior indicator is dominant or recessive. As shown in Fig. 9.5, there are three groups of customers that have different energy behaviors. The energy behavior indicators of group 1 include Q3 (C) and Q8 (D). The energy behavior indicators of group 2 include Q3 (C), Q6 (A), and Q8 (D). The energy behavior indicators of group 3 include Q3 (B) and Q8 (D). The energy behavior indicator Q8 (D) is shared by all groups; hence, it has little ability to differentiate energy behavior among different groups. The energy behavior indicator Q3 (B) only exists in group 3, and Q6 (A) only exists in group 2. Therefore, these indicators are more beneficial in identifying the energy behavior for groups 2 and 3 than any other energy behavior indicators. The energy behavior indicator Q3 (C) is shared by two groups - group 1 and group 2, so Q3 (C) is more unique than Q8 (D), which is shared by three groups. The IGD fully represents these energy behavior indicators' uniqueness and dominance on energy behavior. By calculating the IGD value and setting the threshold as 50%, the energy behavior indicators Q3 (B), Q3 (C), and Q6 (A) can be identified as dominant indicators, whereas Q8 (D) is recessive.

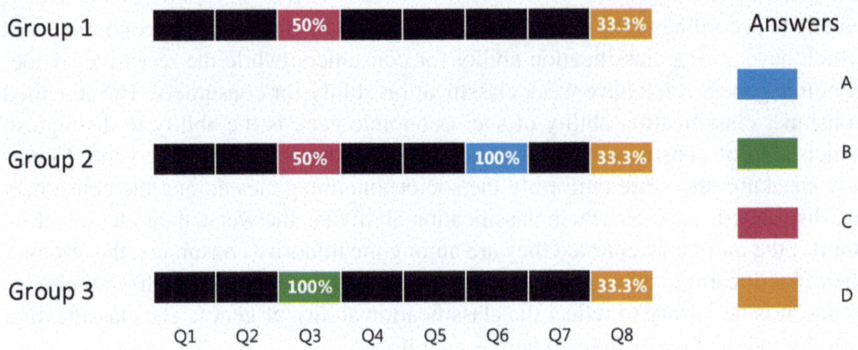

Fig. 9.5 The dominance value represents the uniqueness and dominance of the energy behavior indicator

9.4.4 *Socioeconomic Gene Search Method*

Assume that the number of consumers is M, the socioeconomic questionnaire contains Q questions (which is the same as the number of socioeconomic gene loci). As stated above, the socioeconomic gene search method is as follows:

(1) Consumer Load Profile Clustering

Classify the M consumers into k classes according to the clustering method described in Sect. 9.3.

(2) Socioeconomic Gene Identification

(2.1) Traverse the consumer classes $i = 1 \sim k$, the gene locus $j = 1 \sim Q$ and the jth locus label $x = 1 - q(j)$;

(2.2) For the ith class of consumers, calculate the gene entropy $S^c_{i,j,x}$ of the jth locus labeled x. If the gene entropy $S^c_{i,j,x}$ exceeds the threshold of the gene entropy ϵ_1, the label (i, j, x) will be labeled as a socioeconomic gene and go to Step 2.3, otherwise, return to Step 2.1.

(2.3) For socioeconomic gene (i, j, x), calculate the distribution proportion in the classes of consumers $i = 1 \sim k$ and calculate the classification entropy $S^g_{i,j,x}$. If $S^g_{i,j,x}$ exceeds the threshold of classification entropy ϵ_2, it will be labeled as dominant socioeconomic genes and return to Step 2.1.

9.5 Load Profile Prediction

Naive Bayesian classifier [24] is used in this chapter to predict the consumer load profile. The inputs of the classifier are the dominant socioeconomic genes. This method assumes that all dominant socioeconomic genes have the same importance and independence. The posterior probability of each consumer load profile is calculated based on the training data using Bayesian rule. According to the values of socioeconomic genes of test data, the consumer load profile can be finally predicted as the load profile with the maximum posterior probability. Assume that there are n dominant socioeconomic gene loci, the labels of which are $g_1 \sim g_n$. The value of each gene for the typical load profiles of M consumers is $C_1 \sim C_M$. The load profile prediction process can be divided into the following three steps:

(1) Prior probability calculation

According to the training data, calculate the label values of the n dominant socioeconomic genes $P(g_1) \sim P(g_n)$.

Calculate the occurrence probability of the ith ($i = 1 \sim M$) class of typical load profile $P(C = C_i)$.

Calculate the values of dominant socioeconomic gene loci $P(g_1|C = C_i)$, $P(g_2|C = C_i), \ldots, P(g_n|C = C_i)$ for all typical load profiles $C = C_i$.

(2) Posterior probability calculation

For test data, according to Bayesian theorem and dominant socioeconomic independent hypothesis, when the label value of n dominant socioeconomic gene locus is $g_1 \sim g_n$, the posterior probability of consumer load profiles occurrence is calculated as follows:

$$P(C = C_i|g_1 \bigcap g_2 \bigcap \cdots \bigcap g_n) = P(C{=}C_i)\frac{P(g_1|C{=}C_i)P(g_2|C{=}C_i)\cdots P(g_n|C{=}C_i)}{P(g_1)P(g_2)\cdots P(g_n)} \tag{9.6}$$

(3) Maximum posterior probability prediction

Compare the conditional probability of $C = C_1 \sim C_M$, and then select C_j with the maximum conditional probability $P(C = C_j|g_1 \bigcap g_2 \bigcap \cdots \bigcap g_n) = \max P(C = C_i|g_1 \bigcap g_2 \bigcap \cdots \bigcap g_n)$ as the predicted load profile.

9.6 Case Studies

The smart meter data recording energy behavior and the demographic data recording household information used in this chapter, are from Electric Ireland and SEAI. SEAI released online fully anonymized data sets from smart meter trials for electricity customers. The smart meter trials occurred during 2009 and 2010, with more than 5000 Irish households and businesses participating. The participating households were all carefully recruited to ensure that they were representative of the national population. 5375 residential participants were initially recruited with a return rate of 78.7%, which means 4232 participants returned the pre-trial questionnaires [24]. Of participants who return the questionnaires, only 3487 have a record of smart meter data. Hence, the final number of participants adopted in this study is 3487. These participants' data were mainly composed of two parts: (1) the smart meter data

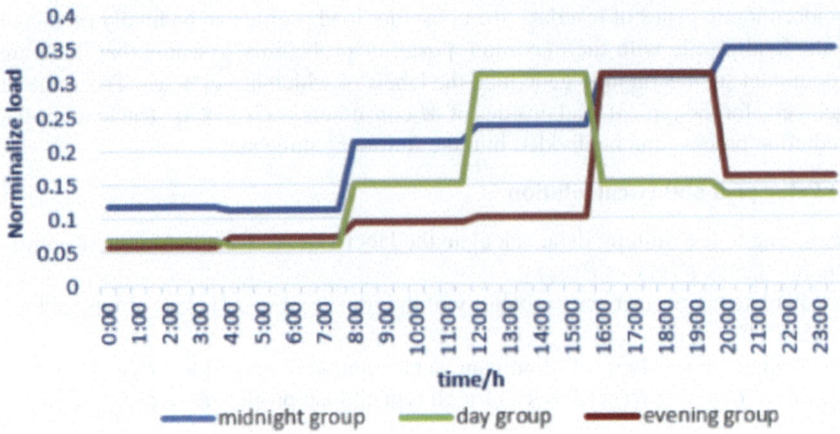

Fig. 9.6 The typical energy spectral patterns of three groups of customers

which recorded the daily customer energy consumptions at a 30 min interval; (2) the demographic data in the form of a questionnaire, comprising of 144 questions. These questions and their answers include household information, such as customers' age, employment status, social class, electrical appliances, and energy usage habits.

9.6.1 *Consumer Load Profile Classification*

In this step, according to smart meter data, Irish people are clustered into three groups through x-means clustering. X-means clustering is an extended k-means method with an efficient estimation of the number of clusters. This method overcomes the main shortcoming of k-means method that the number of clusters k has to be predetermined. Applying the technique to cluster the Irish smart meter data, three typical energy spectral patterns are identified and labeled as "day group", "evening group", and "midnight group". As shown in Fig. 9.6, for the day group, the major energy usage occurs in the afternoon from 12:00 to 16:00, which means the group uses electricity mainly during the day. For the evening group, the main consumption is from 16:00 to 20:00, which means these people use electricity mainly in the evening. For the midnight group, the load gradually increases throughout the day, with the lowest load in the morning and the highest load late in the midnight.

9.6.2 *Socioeconomic Gene Search Result*

Through the energy behavior indicator searching method, 91 energy behavior indicators are found, of which 74 (81%) are recessive indicators and 17 (19%) are dominant indicators, as shown in Fig. 9.7. Most of the behavior indicators are recessive, which

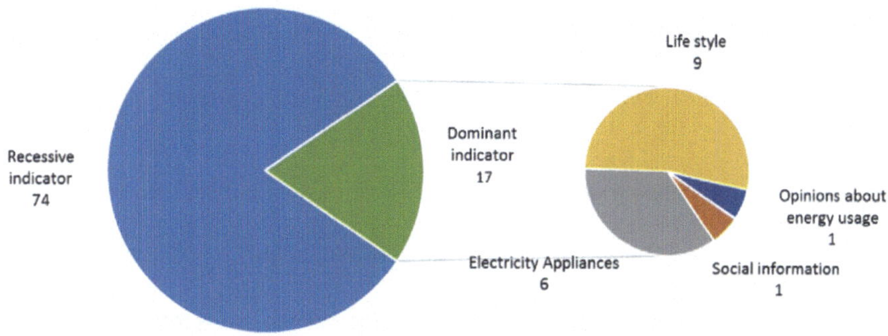

Fig. 9.7 Composition of the energy behavior indicators found in Irish people

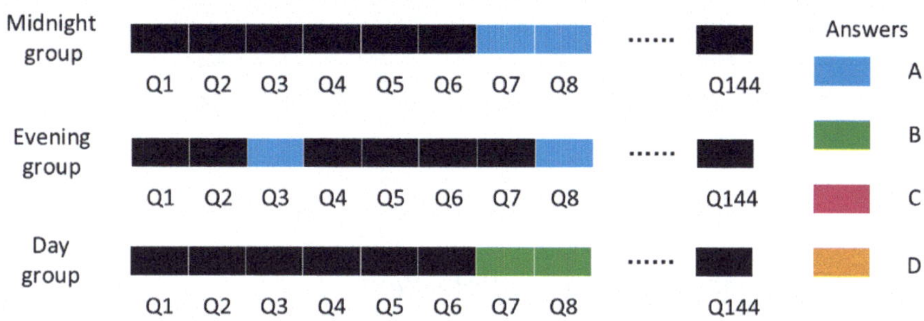

Fig. 9.8 The energy behavior indicator maps for dominant indicators of the Irish people

indicates that although abundant behavior indicators are found, only a small proportion of them are dominant, and contribute to the difference in energy behavior.

According to the classification results of the questions, the composition of dominant indicators can also be divided into four major categories. Of the 17 dominant indicators, nine behavior indicators belong to the lifestyle category, six are in the electricity appliances category, one belongs to the category of social information, and one is in the category of opinions about energy usage. This statistical result shows that human features in the lifestyle category have the greatest effect on Irish people's energy behavior.

All of the energy behavior indicators in the lifestyle category are related to internet usage habits. Superficially, internet usage is entirely unrelated to energy behavior; however, they are highly associated according to the energy behavior indicator results. To understand how the internet-usage-related behavior indicators affect energy behavior in-depth, the energy behavior indicator map of the three groups are shown in Fig. 9.8. The energy behavior indicator map is composed of 144 questions. In this energy behavior indicator map, four of the internet-usage-related indicators and another important employment-related indicator are plotted. The internet-usage-related indicators include Q7 (A), Q7 (B), Q8 (A) and Q8 (B), and the employment-related indicator is Q3 (A). From the figure, these energy behavior indicators' IGD values are greater than the IGD threshold; hence, they are all dominant.

Table 9.2 Consumer load profile prediction accuracy

	Day group (95)	Evening group (438)	Midnight group (164)
Prediction accuracy (%)	66.3	74.2	68.3

Table 9.3 The energy behavior indicating questions and corresponding energy behavior indicators in three groups

Id	Energy behavior indicating question	Day group	Evening group	Midnight group
3	What is the employment status of the chief income earner in your household, is he/she	Nag	An employee	Nag
7	Do you use the internet regularly yourself?	No	Nag	Yes
8	Are there other people in your household that use the internet regularly?	No	Yes	Yes

9.6.3 Consumer Load Profile Prediction

A naive Bayesian classifier was constructed based on 17 dominant socio-economic genes. In order to verify the prediction accuracy, 3487 users were divided into two groups: training group (2790 families, 80%) and testing group (697, 20%). The training group and the test group were divided into random sampling methods. The training group was used to train Bayesian classifiers, while the test group's results were used to characterize the accuracy of dominant socio-economic gene prediction. Table 9.2 shows the predicted results. It can be seen that the number of households using daytime electricity, evening electricity, and midnight electricity are 95, 438 and 164, respectively. In the evening, most users use electricity because the vast majority of the residents in society belong to "working people", followed by midnight users, and at least daytime users, because the elderly living alone accounts for a small proportion of society. In the evening, the prediction accuracy of the power consumption model is the highest, 74.2%, followed by the midnight power consumption model, 68.3%, and the lowest is the day power consumption model, 66.3%.

Table 9.3 shows the socioeconomic characteristics described by three of the 17 dominant genes where Nag means "not an energy behavior indicating question". Among the three dominant genes, two are related to the frequency of Internet use and one is employment status. According to the dominant genes shown in Table 9.3, the load curves of different types of residents are matched with their socioeconomic characteristics, as shown in Fig. 9.9.

The dominant indicators, and the energy behavior indicating questions the indicators are located in, are illustrated in Table 9.3. The dominant indicators are in three

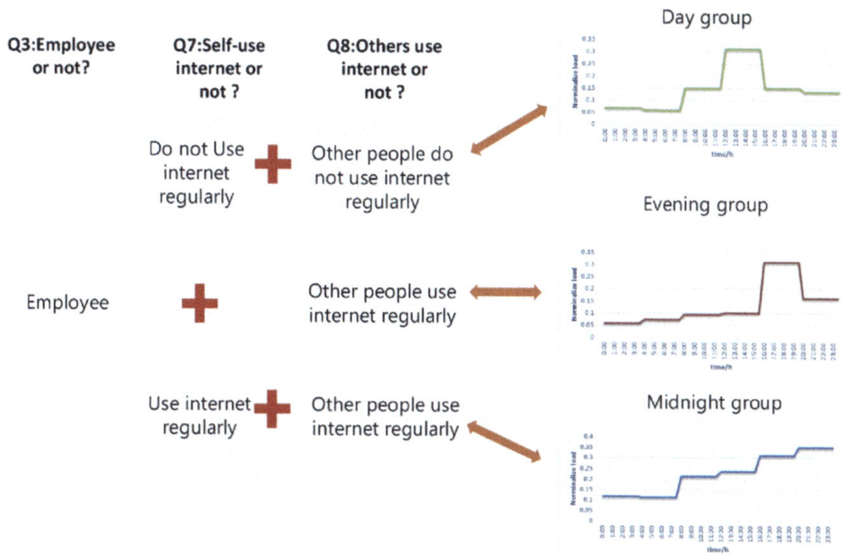

Fig. 9.9 The mapping relationship between different energy behavior groups of the Irish people and their energy behavior indicators

different behavior indicating questions. Of the three questions, two address the frequency of internet usage, and one considers employment status. Using the energy behavior indicators provided by Table 9.3, a mapping between Irish people's energy behavior indicator and energy behavior is performed in Fig. 9.9. From this figure, it can be seen that the Irish people in the day group do not use the internet. This is most likely to describe an elderly group that is retired and also tends not to use internet. The people in this group also tend to stay at home; hence, their daytime electricity usage is the highest. For the midnight group, questions 7 and 8 show that this group uses the internet regularly, suggesting this group is largely composed of young people in a shared home or students. Young people often have a habit of joining parties, going to pubs and clubs, or resting late. Therefore, their energy behavior peaks near midnight. In the evening group, people are typically employed, which can explain why people in this group use electricity mainly in the early evenings.

The Irish people's internet usage habits are related to energy behavior. This phenomenon is mainly because internet usage habits represent human's preference in lifestyle to some degree, including when to work, play, and rest. The lifestyle information plays an important role in people's energy consumption patterns; therefore, internet and electricity usage are highly correlated.

9.7 Conclusions

In this chapter, the relationship between customers' energy behaviors and their household information is first extracted and analyzed through the proposed cross-domain feature selection and coding method. This enables access to disaggregating smart meter data into a range of energy behaviors. Each energy behavior can then be uniquely traced by a set of energy behavior indicators, thus offering a simple, transparent and effective alternative to a challenging matching problem with massive smart meter data and a huge range of possible indicators.

Energy behavior indicators explain household energy behavior with a deeper view of the rationale behind differing energy behavior patterns, the underlying factors influencing human's energy consumption and are able to predict people's future energy behavior if their behavior status changes. In the energy industry, according to the Ireland case study, household energy behavior is highly correlated with the features of employment status and internet usage. Through this finding, household energy behavior can be forecasted based on several features, saving the investment of installing millions of smart meters in Ireland. If other household load profiles in Ireland are known, this information can also be used to infer household information, thus creating value-added services and products for energy utilities and energy service providers. The correlation between household energy behavior and human features of employment status and internet usage in Ireland may not be appropriate for other countries. Because the socioeconomic status (developed, developing or under-developed) and geographical location (European, Asian, African) of countries differ, different countries' people may have different energy behavior patterns and also different energy behavior indicators. Although different countries' results may differ, the method validated by the Ireland case can be applied universally.

References

1. Brown, B., Chui, M., & Manyika, J. (2011). Are you ready for the era of 'big data'. *McKinsey Quarterly*, *4*(1), 24–35.
2. Williams, C. K. I., & Barber, D. (1998). Bayesian classification with gaussian processes. *IEEE Transactions on Pattern Analysis and Machine Intelligence*, *20*(12), 1342–1351.
3. Suykens, J. A. K., & Vandewalle, J. (1999). Least squares support vector machine classifiers. *Neural Processing Letters*, *9*(3), 293–300.
4. Friedl, M. A., & Brodley, C. E. (1997). Decision tree classification of land cover from remotely sensed data. *Remote Sensing of Environment*, *61*(3), 399–409.
5. Kutner, M. H., Nachtsheim, C. J., Neter, J., Li, W., & et al. (2005). *Applied linear statistical models* (Vol. 5). Boston: McGraw-Hill Irwin.
6. Park, D. C., El-Sharkawi, M. A., Marks, R. J., Atlas, L. E., & Damborg, M. J. (1991). Electric load forecasting using an artificial neural network. *IEEE Transactions on Power Systems*, *6*(2), 442–449.
7. Smola, A. J., & Schölkopf, B. (2004). A tutorial on support vector regression. *Statistics and Computing*, *14*(3), 199–222.

8. Grygorash, O., Zhou, Y., & Jorgensen, Z. (2006). Minimum spanning tree based clustering algorithms. *2006 18th IEEE International Conference on Tools with Artificial Intelligence (ICTAI'06)* (pp. 73–81). IEEE.
9. Langfelder, P., Zhang, B., & Horvath, S. (2007). Defining clusters from a hierarchical cluster tree: The dynamic tree cut package for r. *Bioinformatics, 24*(5), 719–720.
10. Hartigan, J. A., & Wong, M. A. (1979). Algorithm as 136: A k-means clustering algorithm. *Journal of the Royal Statistical Society. Series C (Applied Statistics), 28*(1), 100–108.
11. Wang, Y., Chen, Q., Kang, C., Zhang, M., Wang, K., & Zhao, Y. (2015). Load profiling and its application to demand response: A review. *Tsinghua Science and Technology, 20*(2), 117–129.
12. Chuan, L., & Ukil, A. (2014). Modeling and validation of electrical load profiling in residential buildings in singapore. *IEEE Transactions on Power Systems, 30*(5), 2800–2809.
13. Stephen, B., Mutanen, A. J., Galloway, S., Burt, G., & Järventausta, P. (2013). Enhanced load profiling for residential network customers. *IEEE Transactions on Power Delivery, 29*(1), 88–96.
14. Chicco, G., & Akilimali, J. S. (2010). Renyi entropy-based classification of daily electrical load patterns. *IET Generation, Transmission & Distribution, 4*(6), 736–745.
15. Tsekouras, G. J., Hatziargyriou, N. D., & Dialynas, E. N. (2007). Two-stage pattern recognition of load curves for classification of electricity customers. *IEEE Transactions on Power Systems, 22*(3), 1120–1128.
16. Chicco, G., Ionel, O.-M., & Porumb, R. (2012). Electrical load pattern grouping based on centroid model with ant colony clustering. *IEEE Transactions on Power Systems, 28*(2), 1706–1715.
17. Espinoza, M., Joye, C., Belmans, R., & De Moor, B. (2005). Short-term load forecasting, profile identification, and customer segmentation: a methodology based on periodic time series. *IEEE Transactions on Power Systems, 20*(3), 1622–1630.
18. Zeifman, M., & Roth, K. (2011). Nonintrusive appliance load monitoring: Review and outlook. *IEEE Transactions on Consumer Electronics, 57*(1), 76–84.
19. Vercamer, D., Steurtewagen, B., Van den Poel, D., & Vermeulen, F. (2015). Predicting consumer load profiles using commercial and open data. *IEEE Transactions on Power Systems, 31*(5), 3693–3701.
20. Rasmussen, C. E. (2000). The infinite gaussian mixture model. *Advances in Neural Information Processing Systems* (pp. 554–560).
21. Pelleg, D., Moore, A. W., & et al. (2000). X-means: Extending k-means with efficient estimation of the number of clusters. *Icml* (Vol. 1, pp. 727–734).
22. Irish Social Science Data Archive. (2012). Commission for energy regulation (cer) smart metering project. http://www.ucd.ie/issda/data/commissionforenergyregulationcer/.
23. Meier, E., & Moy, C. (2004). Social grading and the census. *International Journal of Market Research, 46*(2), 141–170.
24. McCallum, A., Nigam, K., & et al. (1998). A comparison of event models for naive bayes text classification. *AAAI-98 Workshop on Learning for Text Categorization* (Vol. 752, pp. 41–48). Citeseer.

Chapter 10
Clustering of Consumption Behavior Dynamics

Abstract In a competitive retail market, large volumes of smart meter data provide opportunities for load-serving entities (LSEs) to enhance their knowledge of customers' electricity consumption behaviors via load profiling. Instead of focusing on the shape of the load curves, this chapter proposes a novel approach for the clustering of electricity consumption behavior dynamics, where "dynamics" refer to transitions and relations between consumption behaviors, or rather consumption levels, in adjacent periods. First, for each individual customer, symbolic aggregate approximation (SAX) is performed to reduce the scale of the data set, and a time-based Markov model is applied to model the dynamic of electricity consumption, transforming the large data set of load curves to several state transition matrices. Second, a clustering technique by Fast Search and Find of Density Peaks (CFSFDP) is primarily carried out to obtain the typical dynamics of consumer behavior, with the difference between any two consumption patterns measured by the Kullback–Liebler (K–L) distance, and to classify the customers into several clusters. To tackle the challenges of big data, the CFSFDP technique is integrated into a divide-and-conquer approach toward big data applications. A numerical case verifies the effectiveness of the proposed models and approaches.

10.1 Introduction

The existing studies on load profiling mainly focus on individual large industrial/commercial customer, medium or low voltage feeder, or a combination of small customers, load profiles of which shows much more regularity [1]. It should be noted that although these dynamic characteristics are always "deluged" in a combination of customers, they could be described by several typical load patterns. However, with regard to residential customers, at least two new challenges will be faced. One challenge is the great variety and variability of the load patterns. As indicated in Fig. 10.1, there are clear differences in the electricity consumption patterns of the two residents. Peak loads have different amplitudes and occur at different times of day, for example. Electricity consumption patterns also vary daily even for the same customer. In this case, several typical daily load patterns are not fine enough to reveal the actual consumption behaviors. The daily profile should be decomposed

Y. Wang et al., *Smart Meter Data Analytics*,
https://doi.org/10.1007/978-981-15-2624-4_10

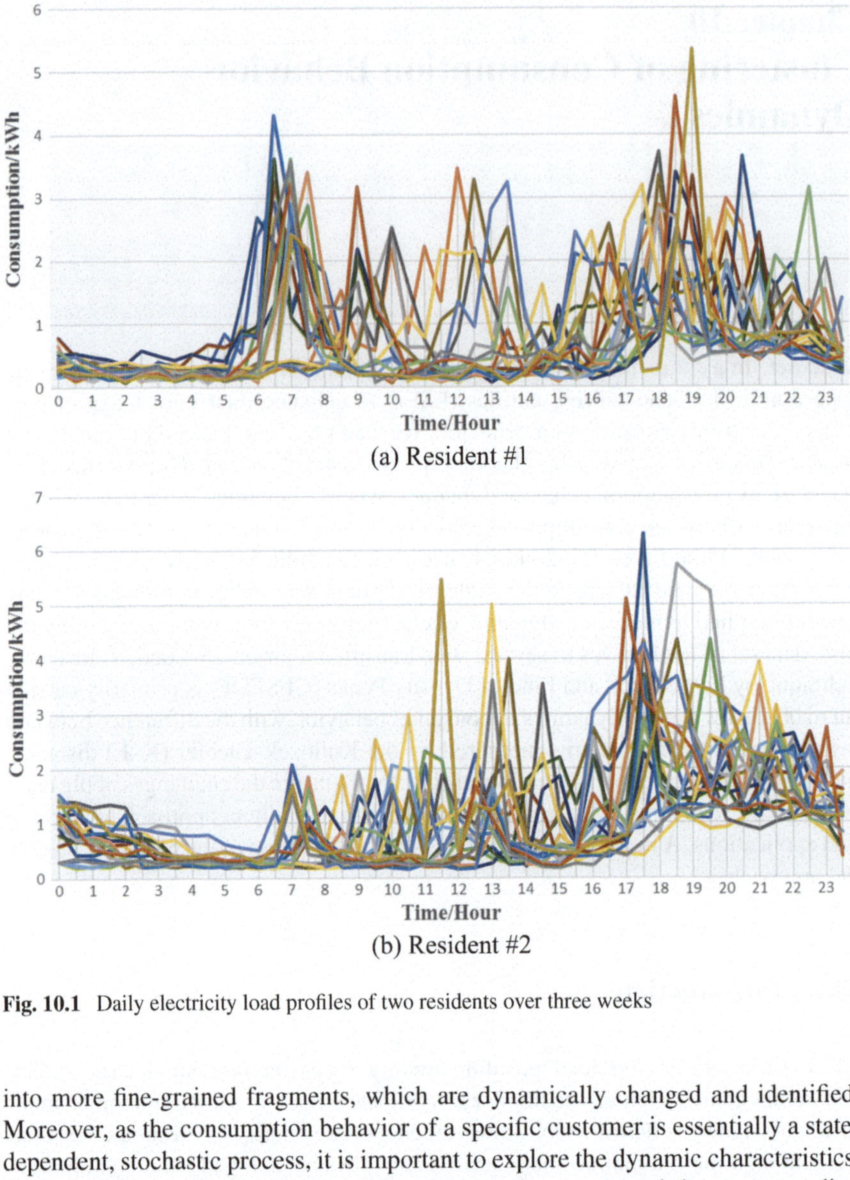

Fig. 10.1 Daily electricity load profiles of two residents over three weeks

into more fine-grained fragments, which are dynamically changed and identified. Moreover, as the consumption behavior of a specific customer is essentially a state-dependent, stochastic process, it is important to explore the dynamic characteristics, e.g., switching and maintaining, of the consumption states and the corresponding probabilities. The other challenge is that of "big data". Considering the high frequency and dimensionality of the data contained in the load curves, data sets in the multi-petabyte range will be analyzed [2]. Traditional clustering techniques are tricky to be executed in a "big data world".

To tackle these two challenges, this chapter implements a time-based Markov model to formulate the dynamics of customers' electricity consumption behaviors, considering the state-dependent characteristics, which indicates that future consumption behaviors would be related to the current states. This assumption is reasonable as various electricity consumption behaviors would last for different periods before being capable of changes, as could be abstracted from historical performances. The transitions and relations between consumption behaviors, or rather consumption levels, in adjacent periods are referred to as "dynamics" in this chapter. These dynamics have been modeled by the Markov model in several works [3]. However, few papers consider the dynamics as a factor for clustering. Profiling of the dynamics could provide useful information for understanding the consumption patterns of customers, forecasting the consumption trends in short periods, and identifying the potential demand response targets. Moreover, this approach formulates the large data set of load curves as several state transition matrices, greatly reducing the dimensionality and scale.

In addition to the Markov model, this chapter tries to address the "data deluge" issue in three other ways. First, applying SAX to transform the load curves into a symbolic string to reduce the storage space and ease the communication traffic between smart meters and data centers. Second, a recently reported effective clustering technique by Fast Search and Find of Density Peaks (CFSFDP) is first utilized to profile the electricity consumption behaviors, which has the advantages of low time complexity and robustness to noise points [4]. The dynamics of electricity consumption are described by the differences between every two consumption patterns, as measured by the Kullback–Liebler (K–L) distance [5]. Third, to tackle the challenges of big and dispersed data, the CFSFDP technique is integrated into a divide-and-conquer approach to further improve the efficiency of data processing, where adaptive k-means is applied to obtain the representative customers at the local sites and a modified CFSFDP method is performed at the global sites. The approach could be further applied to big data applications.

Finally, the potential applications of the proposed method to demand response targeting, abnormal consumption behavior detecting and load forecasting are analyzed and discussed. Especially, entropy analysis is conducted based on the clustering results to evaluate the variability of consumption behavior for each cluster, which can be used to quantify the potential of price-based and incentive-based demand response.

10.2 Basic Methodology

The proposed methodology for the dynamic discovery of electricity consumption can be divided into six stages, as shown in Fig. 10.2. The first stage conducts some load data preparations, including data cleaning and load curve normalization. The second stage reduces the dimensionality of the load profiles using SAX. The third stage formulates the electricity consumption dynamics of each individual customer utilizing a time-based Markov model. The K–L distance is applied to measure the difference between any two Markov models to obtain the distance matrix in the fourth stage. The fifth stage performs a modified CFSFDP clustering algorithm to discover

Fig. 10.2 Processes of clustering of electricity consumption behavior dynamics

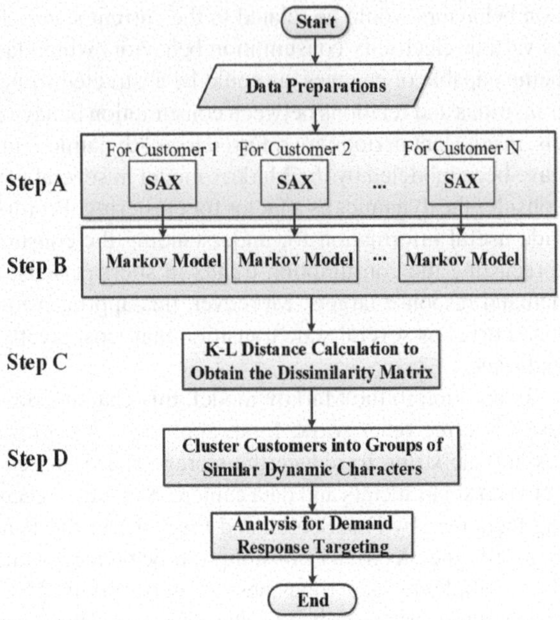

the typical dynamics of electricity consumption. Finally, the results analysis of the demand response targeting is conducted in the sixth stage. The details of the first five stages will be introduced in the following, and the demand response targeting analysis part will be further explained in the case studies.

10.2.1 Data Normalization

Data preparations including data cleaning is not the subject of this chapter and will not be discussed. To make the load profiles comparable, the normalization process transforms the consumption data of arbitrary value $x = \{x_1, x_2, \ldots x_H\}$ to the range of (0, 1), as shown in (10.1).

$$x_i' = \frac{x_i - x_{\min}}{x_{\max} - x_{\min}} \tag{10.1}$$

where, x_i and x'_i denote the actual and normalized electricity consumption at time i; $x_{\min}$ and $x_{\max}$ denote the minimum and maximum consumption over H periods respectively.

It should be noted that the normalization is performed daily instead of over entire periods. This strategy is chosen for at least three reasons. First, it can weaken the impact of anomalous days with critical peaks or bad data injections. Second, it can provide load shapes which maximum values are less affected by the daily or seasonal changes. Third, it can filter out the baseload, which has little effect on demand response and reserve, in favor of the fluctuant part, which shows greater potential in demand response.

10.2.2 SAX for Load Curves

SAX is a powerful technique for the dimensional reduction and representation of time series data with the lower bounding of the Euclidean distance [6]. SAX discretizes numeric time series into symbolic strings by two steps: transforming the load data into a piecewise aggregate approximation (PAA) representation and then symbolizing the PAA representation into a discrete string.

The basic idea of PAA is intuitive and simple, replacing the amplitude values falling in the same time interval with their mean values, as shown in (10.2).

$$\bar{x}_i' = \frac{1}{k_i - k_{i-1}} \sum_{j=k_{i-1}+1}^{k_i} x'_j \tag{10.2}$$

where j is the index of the normalized load data; i is the index of the transformed PAA load data; k_i is the ith time domain breakpoint; and $\bar{x}_i'$ is the average value of the ith segment [6].

The averaging of the PAA can smooth out large, short-duration "spikes" of load profiles. It has been proven that PAA has all the pruning power of the Haar-based DWT and can be defined for arbitrary length queries with lower computation cost [6].

The transformed PAA time series data are then referred by the SAX algorithm to obtain a discrete symbolic representation. The amplitude axis is partitioned into N intervals, and each univocal representation w_p corresponds to an amplitude range $[\beta_{p-1}, \beta_p]$. On this basis, the mapping from a PAA approximation $\bar{x}_i'$ to a word w_p is obtained as follows:

$$\alpha_i = w_p \quad \text{if } \beta_{p-1} < \bar{x}_i' < \beta_p \tag{10.3}$$

Hence, the load curves can be represented by a symbolic string α. For example, Fig. 10.3 shows the normalized electricity consumption data collected from customer #1512 over one week (168 hours) at a frequency of 30 min. The time axis is divided into four periods each day. These data can be represented as "**abcabbcaaacabaabbaccbbbcbabb**", with three symbols and a total of 28 periods. For traditional SAX, the time domain is divided into regular intervals, and inside each interval, the average of the amplitude values is calculated.

The main concern of SAX is the determination of the time domain breakpoint k_i and the amplitude breakpoint β_p. Generally, the time domain is partitioned uniformly, and the amplitude axis is partitioned based on the normal distribution hypothesis [7].

To make the breakpoints clear in physics meaning, this chapter adopts a non-regular interval on both time domain and amplitude. Specifically, the time domain

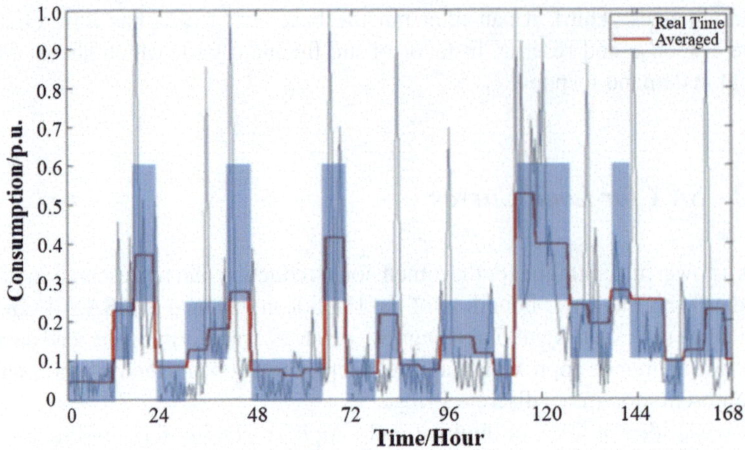

Fig. 10.3 Electricity consumption data of customer #1512 over one week and its SAX representation

breakpoint k_i is determined by comprehensively taking the implementation of the ToU tariff and the regular routine of customers into consideration. For example, the time domain can be divided into three intervals according to the definition of the peak, flat, valley periods, during which the electricity prices are the same respectively under ToU tariff. For another example, four-time periods named overnight period, breakfast period, daytime period, and evening period can be approximately chosen through some statistics [8].

Each word transformed from normalized load profiles by SAX corresponds to a discrete state of the Markov model in the next step. In general, the states are equally probable for optimal use of the Markov models [9]. Thus, the amplitude breakpoint β_p is determined by the quantiles of the statistical distribution of the amplitudes in the whole data set. For example, if we want to simplify the electricity consumption by three states, the consumption levels correspond to 33.33 and 66.67% of the cumulative distribution function (CDF). Another question is how many states are needed in the Markov model. The dissimilarity between transformed symbolic string by SAX and original load profiles is gradually reduced by increasing the number of states [8]. However, many more states may result in large size and sparsity of the transition probability matrix, which may result in the meaninglessness of transition probability matrix and "curse of dimensionality" in the step of clustering. Thus, the number of states is a tradeoff between the information loss by SAX and size if the transition probability matrix.

10.2.3 Time-Based Markov Model

If we want to predict the trend or level of electricity consumption for each customer, we may make full use of their past and present states. If the future consumption level or state depends only on the present state, it is called a Markov property and can be modeled by a Markov chain. Various Markov models have been applied to load forecasting [10].

For a symbolic string with N symbols, discrete Markov model with N corresponding states can be applied to model the dynamic characteristics of their consumption levels. However, customers have different dynamic characteristics at different periods for their regular routines every day. Therefore, time-based Markov model is applied to formulate the characteristics. For each adjacent period, a Markov chain can be modeled. Then, the one-step transition number matrix F^t at period t can be calculated. From F^t, the transition probability matrix P^t at period t can be further estimated according to (10.4).

$$\hat{p}_{ij}^t = \begin{cases} \dfrac{f_{ij}^t}{\sum\limits_{k=1}^{n} f_{kj}^t} & \text{if } \sum\limits_{k=1}^{n} f_{kj}^t \neq 0 \\ 0 & \text{otherwise} \end{cases} \quad (10.4)$$

where, $F^t = \begin{bmatrix} f_{11}^t & f_{12}^t & \cdots & f_{1n}^t \\ f_{21}^t & f_{22}^t & \cdots & f_{2n}^t \\ \cdots & \cdots & \cdots & \cdots \\ f_{n1}^t & f_{n2}^t & \cdots & f_{nn}^t \end{bmatrix}$; $\hat{P}^t = \begin{bmatrix} \hat{p}_{11}^t & \hat{p}_{12}^t & \cdots & \hat{p}_{1n}^t \\ \hat{p}_{21}^t & \hat{p}_{22}^t & \cdots & \hat{p}_{2n}^t \\ \cdots & \cdots & \cdots & \cdots \\ \hat{p}_{n1}^t & \hat{p}_{n2}^t & \cdots & \hat{p}_{nn}^t \end{bmatrix}$; $\hat{p}_{ij}^t$ denotes the estimated one-step transition probability from state j to state i at time t.

It has been proved that $\hat{P}^t$ is the unbiased estimation of transition probability matrix P^t [11].

In the following, a test of the Markov property of electricity consumption should be conducted to validate the Markov hypothesis. Following is the test theorem of the Markov property proposed in [12]

$$\chi^2 = 2\sum_{i=1}^{N}\sum_{j=1}^{N} f_{ij}\left|\log\frac{p_{ij}}{p_{\bullet j}}\right| \quad (10.5)$$

where $p_{\bullet j} = \sum\limits_{i=1}^{N} f_{ij} / \sum\limits_{i=1}^{N}\sum\limits_{k=1}^{N} f_{ik}$, N is the number of states.

Given a significance level α, if $\chi^2 \geq \chi_\alpha^2((N-1)^2)$ holds, we can be reasonably confident that the electricity consumption of customers has a Markov property.

10.2.4 Distance Calculation

The dissimilarity/distance measurement is a fundamental problem in clustering. There exist many ways to compute the distances between two matrices, such as 1-norm distance and 2-norm distance (Euclidean distance). However, different from general matrices, a $N \times N$ state transition probability matrix essentially consists of N probability distributions, where each row (e.g., the ith row) corresponds to a probabilistic distribution of the state of the next period at the current state (e.g., the ith state). K–L distance is an effective way to quantify the dissimilarity between two probabilistic distributions [5]. Thus, for discrimination between two Markov model with the state transition matrices P_i^t and P_j^t, the K–L distance is defined as [13]

$$KLD(P_i^t, P_j^t) = \frac{1}{N}\sum_{m=1}^{N}\sum_{n=1}^{N} p_{imn}^t \log \frac{p_{imn}^t}{p_{jmn}^t} \tag{10.6}$$

Note that $KLD(P_i^t, P_j^t) = KLD(P_j^t, P_i^t)$ is not guaranteed to hold; that is to say, the K–L distance is unsymmetrical. For the convenience of clustering, we define the symmetric K–L distance D_{ij}^t of two Markov model at period t as [14]

$$D_{ij}^t = \frac{KLD(P_i^t, P_j^t) + KLD(P_j^t, P_i^t)}{2} \tag{10.7}$$

Each customer is modeled by T Markov model for T periods of the day. We further extend the K–L distance to T periods as follows:

$$D_{ij} = \sum_{t=1}^{T} D_{ij}^t \tag{10.8}$$

The dissimilarity matrix is derived by calculating the K–L distance among all customers according to (10.8).

10.2.5 CFSFDP Algorithm

CFSFDP is a recently reported clustering algorithm that can effectively recognize clusters regardless of their shape with a reasonable assumption that the cluster centers must have a higher local density and relatively larger distance to the points of higher density [4].

For a data set, the neighbors can be recognized by a soft threshold like the Gaussian kernel function or a hard threshold as defined in (10.9). To reduce the computation complexity for big data sets, we employ the hard threshold to calculate the local density

$$\rho_i = \sum_{j=1}^{N} \chi(D_{ij} - d_c) \tag{10.9}$$

where $\chi(x) = \begin{cases} 1 & x < 0 \\ 0 & otherwise \end{cases}$, D_{ij} is the dissimilarity/distance between objects i and j; and d_c is the cutoff distance chosen by the principle proposed in [4] or experience.

The minimum distance δ_i between object i and any other object j of higher density is calculated as follows:

$$\delta_i = \min_{j:\rho_j > \rho_i} (D_{ij}) \tag{10.10}$$

For the point with the highest local density, the minimum distance $\delta_i = \max_j(D_{ij})$. Thus, the object with much larger δ_i has the maximum density in the local or global area.

Hence, each object or point has two important quantities: local density ρ_i and distance δ_i. We can plot all the points $A_i(\rho_i, \delta_i)$ on a two-dimensional plane, which is called the decision graph. The points of higher local density and a larger distance than the thresholds (ρ_0, δ_0) can be identified as density peaks or cluster centers. After these density peaks are found, other remaining points are assigned to the same cluster as its nearest neighbor of higher density.

As stated above, the proposed clustering method has the following advantages so that we adopt in our study.

First, CFSFDP is so elegant and simple that fewer parameters are needed with low time complexity, and it has shown high performance in classifying several data sets. After finding the density peaks, the assignment of each object can be performed in a single step without iteration, in contrast with many other clustering methods like k-means.

Second, CFSFDP as density-based clustering technique can effectively detect non-spherically distributed data and be robust to noise points, which is verified in our case studies.

Third, the distribution of objects on the decision graph reveals much information. For example, it is easy to detect the outliers or bad data injections with a small ρ_i and large δ_i, and find the objects around the edge of the cluster with both small ρ_i and δ_i. The number of clusters can be adjusted elastically according to the distribution of objects by setting different thresholds for ρ_i and δ_i.

10.3 Distributed Algorithm for Large Data Sets

The electricity consumption data skyrocketing for population-level customers is challenging the storage, communication, and analysis of the data. Although SAX and time-based Markov model have largely reduced the dimensionality of the load profiles, the centralized clustering technique is not effective in dealing with big data challenges. On the one hand, the electricity consumption data are collected and distributed on different sites. The electricity consumption data of customers are collected and stored on different substations they belong to. It is costly and time-consuming to transmit whole data from each distributed site to a central site. On the other hand, the analysis and clustering of large data sets gathered from each distributed site need a very long time and memory overhead. When applying the CFSFDP, the dissimilarity matrix of all the customers should first be obtained, which accounts for most of the computation time. Both the time and space complexity of the CFSFDP are $O(N^2)$.

In fact, there exist many works on parallel clustering for big data applications [15, 16]. For these algorithms, the whole data set should reside on the same data center and then be distributed to different clients like map-and-reduce in Hadoop. It is not satisfied with the practical situation of electricity consumption data collecting and storing. Besides, some fully distributed clustering algorithms are also proposed [17] by aggregating the information of local data and then sending it to a central site for central analysis. However, these algorithms do not have the advantages as CFSFDP. Thus, this section is proposed to design a fully distributed instead of parallel clustering algorithm to ease the communication and computation burden as well as retain the advantages of the CFSFDP by a divide-and-conquer framework.

10.3.1 Framework

Figure 10.4 gives a divide-and-conquer framework for distributed clustering, where L_i denotes the original data on the ith distributed local site; M_i denotes the representative objects selected from the ith distributed local site; and R denotes the global clustering results. Each object corresponds to a customer described by transition probability matrices. The proposed algorithm consists of three steps:

Step 1: The SAX and time-based Markov model for individual customers are handled separately. Divide the big data set into k parts, each marked as L_i. Note that the data on one distributed site can be further partitioned to make the size of the data sets on each site more even.

Step 2: An adaptive k-means method is performed for each individual part to obtain a certain number of cluster centers. Each cluster center can represent all the objects belonging to this cluster with a small error. All these cluster centers of L_i are selected as the representative objects M_i, which are defined as a local model.

Step 3: A modified CFSFDP method is applied to all the representative objects (local models) that are centralized and gathered to classify them into several groups R, which are defined as a global model. Then, according to the final clustering result, the cluster label of each local site would be updated.

It is worth noting that the adaptive k-means and modified CFSFDP are not interchangeable in Step 2 and Step 3. k-means as a partitioning based clustering algorithm tries to minimize the within-class distance of all the clusters, which is consistent with the object of Step 2, i.e., selecting the objects that can represent the remaining objects around them for each individual part. While modified CFSFDP applied in Step 3 can inherit its advantages by global clustering. The adaptive k-means method and the modified CFSFDP method will be described in detail in the next two parts.

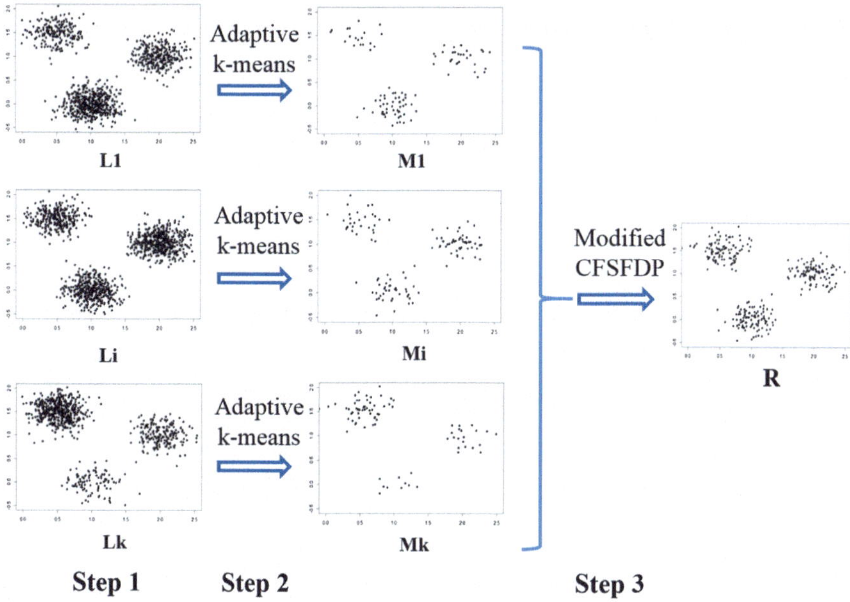

Fig. 10.4 Divide-and-conquer framework for distributed clustering

10.3.2 Local Modeling-Adaptive k-Means

A set of clustering centers will be obtained by k-means, where the sum of the squared distances between each object is minimized. These centroids can be used as a "code book": each object can be represented by the corresponding centroid with the least error. This is called vector quantization (VQ). We try to establish a local model by finding the "code book" that guarantees that the distortion of each object by VQ satisfies the threshold condition according to (10.11)

$$E_k = \sum_{t=1}^{T}\sum_{j=1}^{N}\sum_{i=1}^{N}(p_{ij}^t - C_{kij}^t)^2 \leq \theta \sum_{t=1}^{T}\sum_{j=1}^{N}\sum_{i=1}^{N} {C_{kij}^t}^2 \tag{10.11}$$

where C_{kij}^t denotes the kth centroid; and θ denotes the distortion threshold.

Traditional k-means needs a given number of centers, which makes it difficult to guarantee that (10.11) holds. In this chapter, an adaptive k-means is adopted to dynamically adjust the number of centers following a simple rule: if an object violates the threshold condition, 2-means (i.e. k-means for $k = 2$) will be applied to partition this cluster further and add a new center to the "codebook" [18].

The Fig. 10.5 shows the detailed procedures of the adaptive k-means method. The distortion threshold θ varies depending on different needs. The smaller threshold corresponds to higher clustering accuracy and a larger number of local represen-

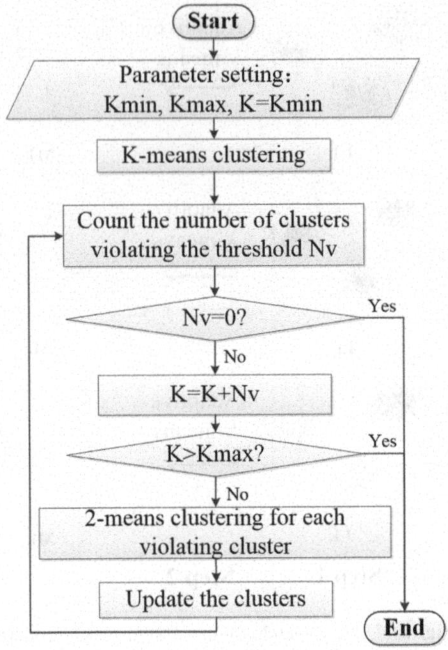

Fig. 10.5 Adaptive *k*-means for local modeling based on threshold

tative objects, and vice versa. As a supplement of distortion threshold and another terminating condition of the iteration, the parameters, K_{min} and K_{max}, are given to limit the size of the "codebook". The value of K_{min} and K_{max} can be determined according to the data transmission limits. Especially, if ensuring certain precision is the priority, the adaptive k-means can start from $K_{min} = 2$ until the (10.11) holds by setting the value of K_{max} to positive infinity. The proposed adaptive k-means distinguishes from traditional k-means in at least four aspects:

First, for adaptive k-means, the number of clusters adjusts dynamically depending on whether the distortion threshold condition is satisfied, in contrast to traditional k-means, where it should be pre-determined.

Second, the convergence condition of adaptive k-means is given by (10.11) and K_{max}. While traditional k-means converges until the sum of the squared distances between each object no longer decreases.

Third, the proposed algorithm is capable of retaining the information of outliers on each site because these outliers will become separate clusters.

Fourth, this algorithm applies 2-means to the violating cluster separately. Thus it has a small computational burden, and parallel computation potential makes it applicable to large data sets. While traditional k-means is conducted on the whole data sets.

10.3.3 Global Modeling-Modified CFSFDP

The original CFSFDP algorithm considers the clustered objects equally. However, in a two-level clustering framework, the selected representative models from different local sites might represent "samples" of different populations. It would be reasonable to consider the representativeness of the local models in the centralized clustering. Thus, a modified CFSFDP method is proposed, which introduces a weight factor to differentiate the representativeness of the local models. Without loss of generality, the weight factor, C_j, is added to the local density calculation

$$\rho_i = \sum_{j=1}^{N} C_j \chi \left(D_{ij} - d_c \right) \tag{10.12}$$

where C_j refers to the weight of the representative points of each cluster in Mi, which is equal to the number of objects belonging to the cluster. The calculation of δ_i is the same as (10). Similarly, based on the calculated ρ_i and δ_i, a decision graph can be drawn to find the density peaks that have a higher local density ρ_i and larger distance δ_i as cluster centers. After the determination of the cluster centers, each of the other objects is assigned to the same cluster as its nearest neighbor of higher density [4].

Now that each representative object from the distributed site has its own cluster label, the objects on the distributed sites will be relabeled according to the cluster label of the representative object. If two centroids ended up in the same cluster, then all their objects will belong to the same cluster.

10.4 Case Studies

10.4.1 Description of the Data Set

The data set used in this chapter was provided by Research Perspective, Ltd. and contains the electricity consumption of 6,445 customers (4,511 residents, 391 industries, and 1533 unknown) over one and a half years (537 days) at a granularity of 30 min [19]. The whole data set consists of a total of 3.46 million (6445 $\times$ 537) daily load profiles. The bad load profiles are roughly identified by detecting the load profiles with missing values or all zeroes. Among these massive load data, we eliminate 6187 bad load profiles, which is a very small sample (approximately 0.18%) of the whole data set.

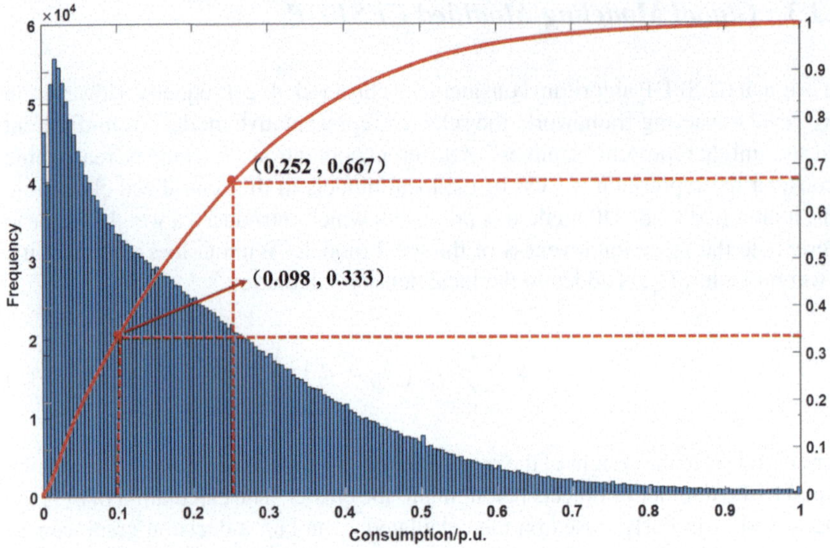

Fig. 10.6 Histogram and CDF of PAA representations of the whole data sets

10.4.2 Modeling Consumption Dynamics for Each Customer

According to the regular routine of electrical customers, we reasonably divide a day into four periods: Period 1 (00:00–06:30, 22:00–24:00, overnight period), Period 2 (06:30–11:30, morning period), Period 3 (11:30–17:00, daytime period), and Period 4 (17:00–22:00, night period). On this basis, the load data are transformed into PAA representations which also vary from 0 to 1. Figure 10.6 shows the histogram and CDF of PAA representations of the whole data sets. It can be seen that the higher the consumption, the lower the density.

For the number of Markov states, we change it from 1 to 6, and then calculate the average recovery error for each case, as shown in Fig. 10.7. The average error drops rapidly as the number of states increases. However, it changes little when the number of states is greater than 3. The breakpoints are approximately valued as 0.1 (one-tenth of the maximum consumption) and 0.25 (one-fourth of the maximum consumption) corresponding to 0.333 and 0.667 in CDF respectively. Thus, we divide the amplitude into three parts: symbol a for 0 0.1; symbol b for 0.1 0.25; and symbol c for 0.25 1.0. These three states can be defined as absence, passive occupancy, and active occupancy [3]. Then, the electricity consumption data of each individual customer can be represented as a symbolic string, like the case in Fig. 10.3. Then, four Markov models of each customer are modeled for the four periods of the day.

We calculated the χ^2 test statistic according to (10.5) for 6445 customers over four periods. Given the significance level $\alpha = 0.05$, $\chi^2_{\alpha}((N-1)^2) = \chi^2_{0.05}(4) = 9.488$. The results show that the electricity consumption of over 99% (6387) of the customers have much larger χ^2 test statistic and show a significant Markov property.

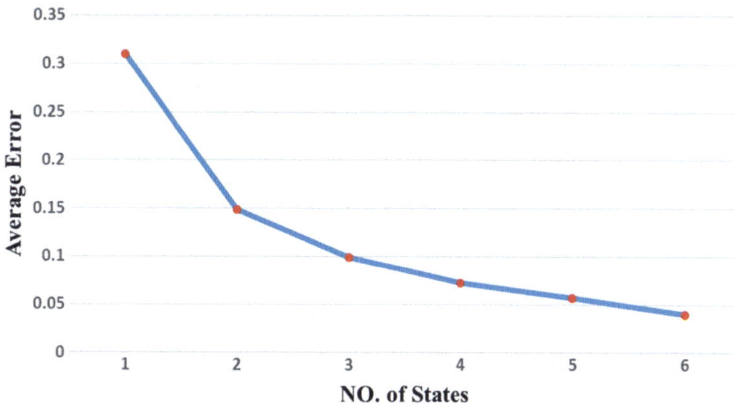

Fig. 10.7 The average error with different number of Markov states

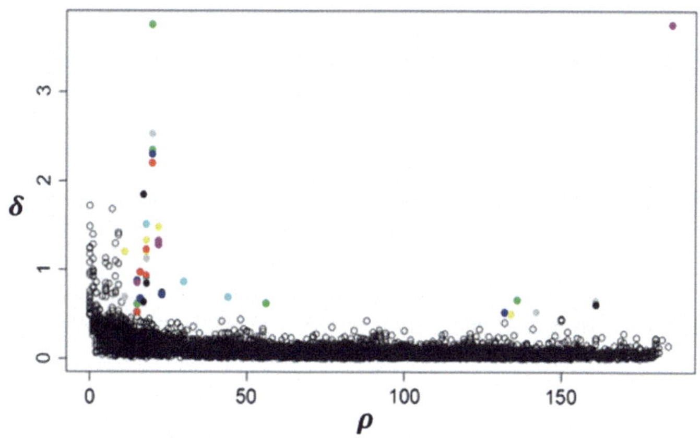

Fig. 10.8 Decision graph to find density peaks for full periods

10.4.3 Clustering for Full Periods

To obtain the typical dynamic characteristics of electricity consumption and to segment customers into several groups, CFSFDP is first applied to the full periods. After calculating the dissimilarity matrix following (10.8), we plot the local density ρ and distance δ of each customer, calculated according to (10.9) and (10.10), respectively, in the decision graph, as shown in Fig. 10.8. We choose the density peak with $\rho > 10$ and $\delta > 0.5$, where a total of 40 clusters can be obtained, which have been marked with different colors in Fig. 10.8.

To show the distribution of the 6445 customers, we mapped the customers into a 2-D plane according to their dissimilarity matrix by multidimensional scaling (MDS) [20], as shown in Fig. 10.9. MDS is a very effective dimensional reduction way for visualizing the level of similarity among different objects of a data set. It tries to

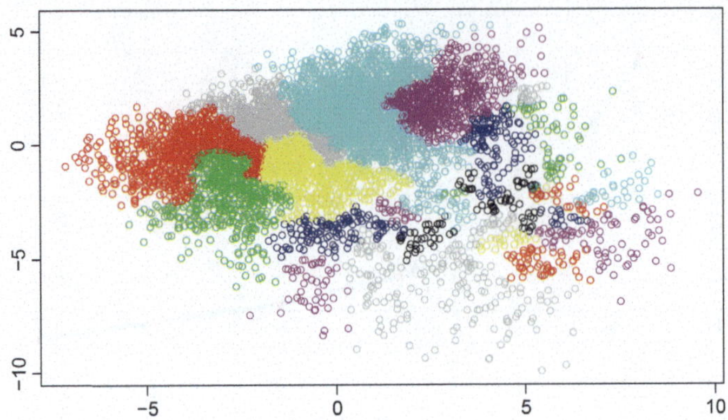

Fig. 10.9 2-D plane mapping for full periods of 6445 customers by MDS according to their K–L distance

place each object in N-dimensional space such that the between-object distances are preserved as closely as possible. Each point in the plane stands for a customer. Points in the same cluster are marked with the same color. It can be seen that the customers of different clusters are unevenly distributed. Approximately 90% of the customers belong to the 10 larger clusters, whereas the other 10% are distributed in the other 30 clusters. In this way, these 6445 customers are segmented into different groups according to their electricity consumption dynamic characteristics for full periods. Note that the customers in the same cluster have similar electricity consumption behavior dynamics over a certain period instead of similar shape in load profiles.

10.4.4 Clustering for Each Adjacent Periods

Sometimes, we may not be concerned with the dynamic characteristics of full periods and instead concentrate on a certain period. For example, to evaluate the demand response potential in noon peak shaving of each customer, the dynamics from Period 1 to Period 2 are much more important; to measure the potential to follow the change of wind power at midnight, the dynamics from Period 4 to Period 1 should be emphasized. Thus, it is necessary to conduct customer segmentation for different adjacent periods. Figure 10.10 illustrates the decision graph and 2-D plane mapping of customers for the four adjacent periods.

It can be seen that the distributions of the customers of the four adjacent periods are shaped like bells, and the proposed clustering technique can effectively address the non-spherically distributed data. Unsurprisingly, the dynamics from Period 2 to Period 3 and from Period 3 to Period 4 show more diversity because people become more active during the day, whereas the dynamics from Period 1 to Period 2 and from Period 4 to Period 1 show less diversity because most people are off duty and go to sleep with less electricity consumption. Taking the dynamics from Period

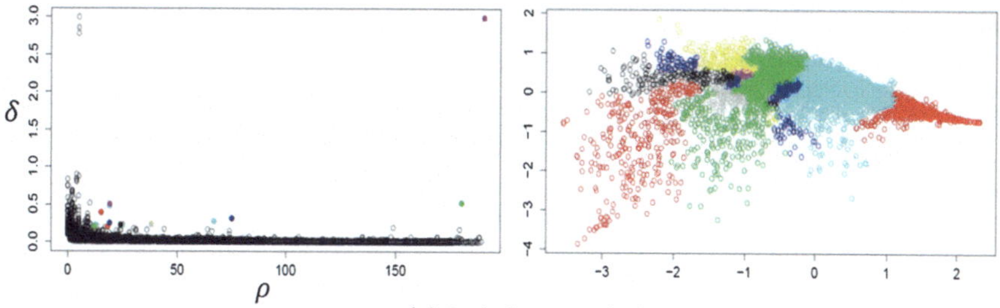

(a) Period 1 to Period 2

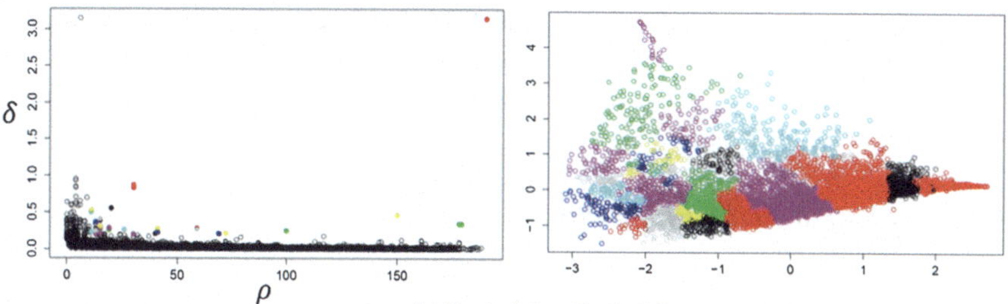

(b) Period 2 to Period 3

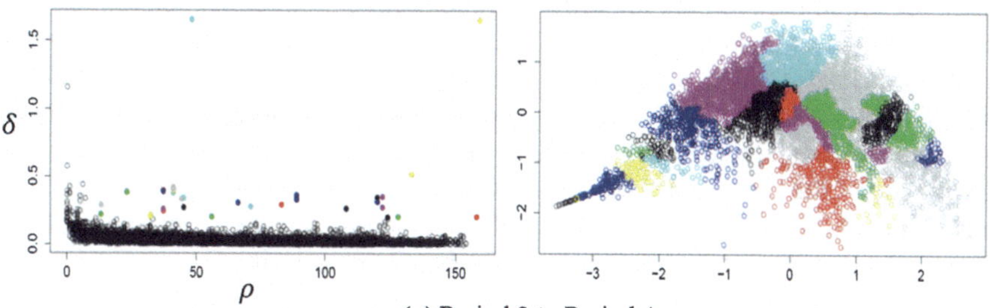

(c) Period 3 to Period 4

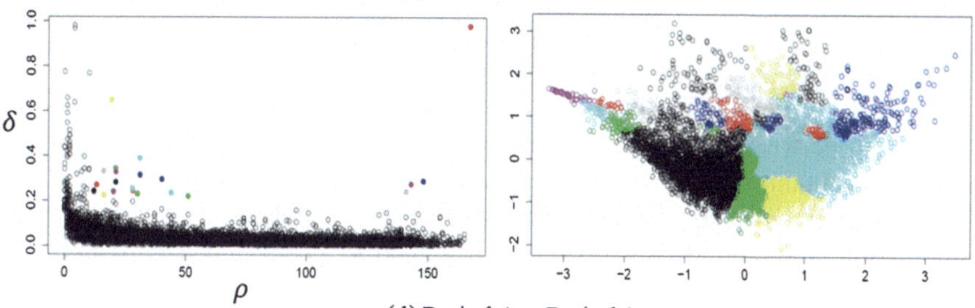

(d) Period 4 to Period 1

Fig. 10.10 Decision graph and 2-D plane mapping of customers for different adjacent periods

2 to Period 3 as an example, the six most typical dynamic patterns are shown in Fig. 10.11. The percent in each matrix stands for the percentage of customers who belong to the cluster. For example, approximately 37% of the customers have very similar electricity consumption dynamics to that of Type_1.

Fig. 10.11 The six most typical dynamic patterns from Period 2 to Period 3

Type_1 **37%**

$$\begin{bmatrix} 0.283 & 0.434 & 0.283 \\ 0.120 & 0.526 & 0.354 \\ 0.200 & 0.400 & 0.400 \end{bmatrix}$$

Type_2 **30%**

$$\begin{bmatrix} 0.194 & 0.424 & 0.382 \\ 0.114 & 0.438 & 0.448 \\ 0.100 & 0.450 & 0.450 \end{bmatrix}$$

Type_3 **10%**

$$\begin{bmatrix} 0.334 & 0.333 & 0.333 \\ 0.111 & 0.333 & 0.556 \\ 0.002 & 0.004 & 0.994 \end{bmatrix}$$

Type_4 **5.3%**

$$\begin{bmatrix} 0.735 & 0.183 & 0.082 \\ 0.531 & 0.318 & 0.151 \\ 0.345 & 0.379 & 0.276 \end{bmatrix}$$

Type_5 **2.4%**

$$\begin{bmatrix} 0.375 & 0.375 & 0.250 \\ 0.333 & 0.333 & 0.334 \\ 0.037 & 0.037 & 0.926 \end{bmatrix}$$

Type_6 **1.1%**

$$\begin{bmatrix} 0.716 & 0.199 & 0.085 \\ 0.709 & 0.213 & 0.078 \\ 0.585 & 0.293 & 0.123 \end{bmatrix}$$

10.4.5 Distributed Clustering

To verify the proposed distributed clustering algorithm, we divide the 6445 customers into three equal parts. Then, the distortion threshold θ is carefully selected for the adaptive k-means method, as a larger threshold leads to poor accuracy, whereas a smaller one leads to little compression. We run 100 cases by varying θ from 0.0025 to 0.25 with steps of 0.0025 and calculate the average compression ratio (CR) of the three distributed sites for each case. The CR is defined as the ratio between the volume of the compressed data and the volume of the original data. Especially, the compressed data refers to local models obtained by adaptive k-means, and the original data refers to the whole objects distributed on each site:

$$\text{CR} = \frac{\text{No. of local models}}{\text{No. of the whole objects}} \tag{10.13}$$

The lower the CR, the better the compression effect. Figure 10.12 shows the relationship between the average compression ratio and the threshold of different periods. To obtain a lower compression ratio and guarantee clustering quality, we choose "knee point" **A** as a balance, where θ is approximately 0.025 and the average compression ratio is approximately 0.065. K_{min} and K_{max} are valued as 10 and 1000 respectively.

To evaluate the performance of the proposed algorithm, we run both the centralized and distributed clustering processes. The high consistency indicates the good performance of the distributed algorithm. As shown in Table 10.1, the matching rate of the algorithm with centralized algorithm can be as high as 96.47%. This indicates that the proposed algorithm has a higher clustering quality with a lower CR. In addition, the time and space complexity of the modified CFSFDP in global modeling is $O((CR \cdot N)^2)$. This means that the efficiency of the global clustering has increased by $(1/CR)^2$ times, where CR < 1 holds. In this case, the efficiency has been boosted to approximately $(1/0.065)^2 \approx 235$ times.

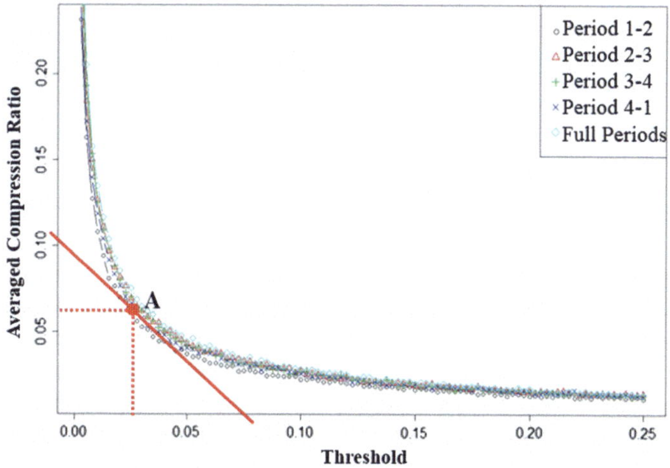

Fig. 10.12 The relationship between averaged compression ratio and threshold for Markov model of different periods

Table 10.1 Matching matrix of centralized clustering with three clusters for gull periods

		Centralized clustering		
		Cluster 1	Cluster 2	Cluster 3
Distributed clustering	Cluster 1	**2417**	15	143
	Cluster 2	46	**991**	0
	Cluster 3	22	3	**2808**

We implement the proposed distributed clustering algorithm by Matlab R2015a on a standard PC, with an Intel ®Core^{TM} i7-4770MQ CPU @ 2.40 GHz, and 8.0 GB RAM. The centralized clustering takes 60.058 s for 6445 customers. For the distributed clustering algorithm, the times needed for adaptive k-means on distributed sites range from 0.415 to 0.542 s, with an average of 0.472 s; the times needed for global modeling is only 0.226 s. Distance calculation consumes most of the time at the global modeling stage. The overall computation time reduced greatly. Note that the time consumed by adaptive k-means is greater than that of CFSFDP because many iterations are needed to satisfy the threshold condition proposed by (10.11) in contrast to CFSFDP.

10.5 Potential Applications

Different from the traditional load profiling methods which mainly focus on the shape of load profiles, this chapter tries to perform clustering on the load consumption change extents and possibilities in adjacent periods, indicating dynamic features of customer consumption behaviors. The proposed modeling method has many poten-

Table 10.2 Entropies of different types of Markov model in Fig. 10.11

Type	1	2	3	4	5	6
Entropy	3.092	2.967	2.076	2.818	2.496	2.473

tial applications. For example, on the decision graph obtained by CFSFDP such as Figs. 10.8 and 10.10, we can easily find the objects with small ρ_i and large δ_i, which can be considered as outlier. That is to say, this customer shows the great difference in electricity consumption behavior dynamics. However, customers of similar social eco-backgrounds are more likely to have similar electricity consumption behavior dynamics. Thus, we can detect abnormal or suspicious electricity consumption behavior quickly through the decision graph. For another example, future consumption can be simulated through Monte-Carlo from the angle of statistics and probability if the state transition probability matrix is known. Based on the simulated electricity consumption, optimal ToU tariff can be designed. Moreover, entropy-based demand response targeting will be further analyzed in this section as an illustration of the applications.

It is believed that customers of less variability and heavier consumption are suitable for incentive-based demand response programs like direct load control (DLC) for their predictability for control, whereas customers of greater variability and heavier consumption are suitable for price-based demand response programs, like ToU pricing, for their flexibility to modify their consumption. Note that a $N \times N$ state transition probability matrix is essentially a combination of N probability distributions as mentioned before. Obviously, though the dynamic characteristics have been abstracted into 3×3 matrices as shown in Fig. 10.11, we can make intuitive evaluations on the customers toward demand response targeting by introducing the approach of entropy evaluation to further extract information from the matrices. The variability could be quantified by the Shannon entropy [21] of the state transition matrix:

$$\text{Entropy} = -\sum_{t=1}^{T}\sum_{i=1}^{N}\sum_{j=1}^{N} p_{ij}^{t} \log p_{ij}^{t} \tag{10.14}$$

Table 10.2 shows the entropies of the Markov model in Fig. 10.11. It can be seen that Type_3 shows the minimum entropy. The 0.994 in the Type_3 matrix means that the Type_3 customers have a greater opportunity to remain unchanged in state c, i.e., the higher consumption level, and are easier to predict. Thus, customers of Type_3 may have a greater potential for an incentive-based demand response during Period 3. However, Type_1 and Type_2 show much higher entropies and have a relatively higher consumption level than Type_3, which makes them much more suitable for price-based demand response. For example, the Type_1 and Type_2 customers have almost the same probability of switching from state c to state b and state c, which is hard to predict, and have more flexibility to adjust their consumption behaviors.

10.6 Conclusions

In this chapter, a novel approach for the clustering of electricity consumption behavior dynamics toward large data sets has been proposed. Different from traditional load profiling from a static perspective, SAX and the time-based Markov model are utilized to model the electricity consumption dynamic characteristics of each customer. A density-based clustering technique, CFSFDP, is performed to discover the typical dynamics of electricity consumption and segment customers into different groups. Finally, a time-domain analysis and entropy evaluation are conducted on the result of the dynamic clustering to identify the demand response potential of each group's customers. The challenges of massive high-dimensional electricity consumption data are addressed in three ways. First, SAX can reduce and discretize the numerical consumption data to ease the cost of data communication and storage. Second, the Markov model is utilized to transform long-term data to several transition matrices. Third, a distributed clustering algorithm is proposed for distributed big data sets.

References

1. Notaristefano, A., Chicco, G., & Piglione, F. (2013). Data size reduction with symbolic aggregate approximation for electrical load pattern grouping. *IET Generation, Transmission & Distribution*, *7*(2), 108–117.
2. Rodriguez, M., González, I., & Zalama, E. (2014). Identification of electrical devices applying big data and machine learning techniques to power consumption data. In *International Technology Robotics Applications*, pp. 37–46. Springer.
3. Torriti, J. (2014). A review of time use models of residential electricity demand. *Renewable and Sustainable Energy Reviews*, *37*, 265–272.
4. Rodriguez, A., & Laio, A. (2014). Clustering by fast search and find of density peaks. *Science*, *344*(6191), 1492–1496.
5. Kullback, S., & Leibler, R. A. (1951). On information and sufficiency. *The Annals of Mathematical Statistics*, *22*(1), 79–86.
6. Lin, J., Keogh, E., Lonardi, S., & Chiu, B. (2003). A symbolic representation of time series, with implications for streaming algorithms. In *Proceedings of the 8th ACM SIGMOD workshop on Research issues in data mining and knowledge discovery*, pp. 2–11. ACM.
7. Lin, J., Keogh, E., Wei, L., & Lonardi, S. (2007). Experiencing SAX: A novel symbolic representation of time series. *Data Mining and Knowledge Discovery*, *15*(2), 107–144.
8. Haben, S., Singleton, C., & Grindrod, P. (2015). Analysis and clustering of residential customers energy behavioral demand using smart meter data. *IEEE Transactions on Smart Grid*, *7*(1), 136–144.
9. Labeeuw, W., & Deconinck, G. (2013). Residential electrical load model based on mixture model clustering and Markov models. *IEEE Transactions on Industrial Informatics*, *9*(3), 1561–1569.
10. Niu, D., Shi, H., Li, J., & Xu, C. (2010). Research on power load forecasting based on combined model of Markov and BP neural networks. In *2010 8th World Congress on Intelligent Control and Automation*, pp. 4372–4375. IEEE.
11. Yang, Y., Wang, Z., Zhang, Q., & Yang, Y. (2010). A time based Markov model for automatic position-dependent services in smart home. In *2010 Chinese Control and Decision Conference*, pp. 2771–2776. IEEE.

12. Zhang, Y., Zhang, Q., & Yu, R. (2010). Markov property of Markov chains and its test. In *2010 International Conference on Machine Learning and Cybernetics*, vol. 4, pp. 1864–1867.
13. Liao, T. W. (2005). Clustering of time series data—a survey. *Pattern Recognition*, *38*(11), 1857–1874.
14. Tabibian, S., Akbari, A., & Nasersharif, B. (2015). Speech enhancement using a wavelet thresholding method based on symmetric Kullback-Leibler divergence. *Signal Processing*, *106*, 184–197.
15. Zhao, W., Ma, H., & He, Q. (2009). Parallel k-means clustering based on mapreduce. In *IEEE International Conference on Cloud Computing*, pp. 674–679. Springer.
16. Sun, Z., Fox, G., Weidong, G., & Li, Z. (2014). A parallel clustering method combined information bottleneck theory and centroid-based clustering. *The Journal of Supercomputing*, *69*(1), 452–467.
17. Januzaj, E., Kriegel, H.-P., & Pfeifle, M. (2004). Dbdc: Density based distributed clustering. In *International Conference on Extending Database Technology*, pp. 88–105. Springer.
18. Kwac, J., Flora, J., & Rajagopal, R. (2014). Household energy consumption segmentation using hourly data. *IEEE Transactions on Smart Grid*, *5*(1), 420–430.
19. Commission for Energy Regulation (CER). (2012). CER Smart Metering Project - electricity customer behaviour trial, 2009-2010. Irish Social Science Data Archive. SN: 0012-00.
20. de Leeuw, J., & Heiser, W. (1982). 13 theory of multidimensional scaling. *Handbook of Statistics*, *2*, 285–316.
21. Lin, J. (1991). Divergence measures based on the shannon entropy. *IEEE Transactions on Information Theory*, *37*(1), 145–151.

Chapter 11
Probabilistic Residential Load Forecasting

Abstract The installation of smart meters enables the collection of massive fine-grained electricity consumption data and makes individual consumer level load forecasting possible. Compared to aggregated loads, load forecasting for individual consumers is prone to non-stationary and stochastic features. In this chapter, a probabilistic load forecasting method for individual consumers is proposed to handle the variability and uncertainty of future load profiles. Specifically, a deep neural network, long short-term memory (LSTM), is used to model both the long-term and short-term dependencies within the load profiles. Pinball loss, instead of the mean square error (MSE), is used to guide the training of the parameters. In this way, traditional LSTM-based point forecasting is extended to probabilistic forecasting in the form of quantiles. Numerical experiments are conducted on an open dataset from Ireland. Forecasting for both residential and commercial consumers is tested. Results show that the proposed method has superior performance over traditional methods.

11.1 Introduction

Electrical load forecasting is the basis of power system planning and operation. It is of great significance to provide a load forecast that strikes a balance between supply and demand, thus allowing for more efficient planning and dispatch of the energy and minimization of energy waste [1]. Traditional load forecasting mainly focuses on system-level or bus-level loads. However, the wide installation of smart meters enables the collection of massive amounts of fine-grained electricity consumption data, making it possible to implement load forecasting for individual consumers. Individual consumer load forecasting acts as the operation data source for demand response implementation [2], energy home management [3], transactive energy [4], etc.

In recent years, an increasing amount of research has been carried out on individual consumer load forecasting. Loads of individual households show greater volatility compared with aggregated load [2]. Different machine learning techniques, such as linear regression, feed-forward NNs, SVR, and least squares support vector machine (LS-SVM), were applied to four house loads. The results show that these methods perform poorly on individual loads, with LS-SVM giving the best performance. A least absolute shrinkage and selection operator (Lasso) panelized linear regression

Y. Wang et al., *Smart Meter Data Analytics*,
https://doi.org/10.1007/978-981-15-2624-4_11

model was proposed in [5] to capture the sparsity of individual household load profiles. This linear Lasso regression has low computational burden, and the results are more interpretable compared with nonlinear machine learning models. A clustering-based approach considered in [6] took into account the daily load profile as a segment and directly forecasted the whole segment instead of the load at each time point separately. A clustering algorithm was also used in [7] for household load prediction where the transition of load shape between two days was characterized by a Markov model. Shape-based clustering can also consider the time drift of the load profiles. Recurrent neural networks (RNNs), the most widely used deep-learning architecture for time series modeling, were applied for household load forecasting in [8]. Case studies on 920 customers from Ireland show that an RNN-based method outperforms traditional forecasting models, including an autoregressive integrated moving average model (ARIMA) and SVR. A long short-term memory (LSTM) RNN was also used in [9, 10]. The difference between the work done in [10] and that in [9] is that more detailed information about appliance-level consumption was known. Sparse coding was applied to model the household load profiles in [11], and different forecasting methods including ARIMA, Holt-Winters, and ridge regression were then performed on the extracted features. Compressive sensing techniques were also applied in [12] to explore the spatiotemporal sparsity within the residential load profiles. Reference [13] investigated how calendar variables, forecasting granularity, and the length of the training set influenced the forecasting performance. Comprehensive case studies were carried out with various regression models. The performance was evaluated using root mean square error (RMSE) and normalized RMSE.

Since the individual load is highly volatile and may be very close to zero for some time periods, traditional forecasting error metrics, such as the mean absolute percentage error (MAPE), are not suitable for quantifying the performance of different methods. Thus, in addition to the establishment of new forecasting models, the metrics of forecasting error have also been studied. A novel adjusted p-norm error was proposed in [14] to reduce the "double penalty" effect caused by the time drift of residential load profiles. This metric is quite similar to dynamic time warping (DTW). To tackle the challenges of near-zero values and outliers for MAPE, another metric, the mean arctangent absolute percentage error (MAAPE), was proposed in [15]. MAAPE, a variation of MAPE, is defined as a slope and an angle, the tangent of which equals the ratio between the absolute error and the real value, i.e., the absolute percentage error (APE).

Traditional point load forecasting can only provide the expected values of future loads. One of the recent advances in load forecasting has been probabilistic load forecasting, which is presented in the form of density, quantiles, or intervals. Density load forecasts were obtained by Gaussian process quantile regression in [16]. The proposed Gaussian process quantile regression belongs to a nonparametric method. Different quantile regression methods have also been applied to net load forecasting [17] and modeling the effect of temperature on load [18]. Quantile regression averaging was applied to multiple point sister forecasts in [19]. The quantile regression averaging bridges point forecasts and probabilistic forecasts. A comprehensive review of probabilistic load forecasting can be found in [20].

Most probabilistic load forecasting that exists in the literature is conducted on an aggregation level or on system-level loads. There are very few works on probabilistic load forecasting for individual load profiles, which have large uncertainties and are less predictable. Conditional kernel density (CKD) estimation methods were used to forecast the uncertainty of smart meter data with different lead times (from 30 min to one week) in [21]. A boosting additive quantile regression method was proposed in [22] for probabilistic household load forecasting, where the base-learners of the additive model include linear and P-spline models. The proposed quantile regression model outperforms three other benchmarks in terms of continuous ranked probability score (CRPS). Gaussian and log-normal processes were used in [23] for residential load forecasting. Both point and probabilistic forecasting evaluation metrics were used, including MAE, RMSE, the prediction interval normalized average width (PINAW) and the prediction interval coverage probability (PICP). Results showed that the log-normal process has better performance than traditional Gaussian processes for residential loads. An auto-regressive integrated moving average (ARIMA) model-based probabilistic load forecasting was proposed in [24] which takes full consideration of the probabilistic hierarchical EVs' parking lot demand modeling.

Aggregated load forecasts can be used as the inputs of power system operation optimization model. While individual load forecasting for residents and SME has at least two potential applications in future smart grid. The first one is for home energy management (HEM). For the house with distributed energy storage, the HEM needs to control the charge/discharge of storage according to the forecasted loads to minimize the total cost. Probabilistic forecasts can help HEM make stochastic optimal decisions. The second one is for demand response, especially for incentive-based demand response. If the aggregator or retailer wants to control the loads of each house directly, the forecast can help them select suitable customers for demand response.

In this chapter, we aim to produce quantile probabilistic forecasts for individual loads, given that the deep neural network LSTM is an effective network to model both the long-term and short-term dependencies in the time series. This model has been proven to have good performance in [8, 9]. We would like to answer a very intuitive question: *Is it possible to exploit the power of LSTM to enhance the probabilistic forecasts for individual load profiles?* To fulfill this task, we use pinball loss instead of MSE to guide the training of LSTM networks. The proposed method combines LSTM and pinball loss to formulate a novel quantile probabilistic forecasting model.

11.2 Pinball Loss Guided LSTM

This section introduces the pinball loss guided LSTM regression for probabilistic residential load forecasting. The main idea is to combine the strength of LSTM with quantile regression: the former is able to capture the long- and short-term dependencies within the load data, and the latter is able to provide the future uncertainty information using predefined quantiles.

11.2.1 LSTM

LSTM is an efficient RNN architecture for time series modeling and forecasting. Traditional neural networks try to learn the correspondence between inputs and outputs from a static perspective. However, when the input data are a time series, the information will be lost if these data are independently trained as inputs and outputs of the neural network. Compared with traditional neural networks, RNNs make a link between each two "input-output" pair. Figure 11.1 shows the basic topology of a simple RNN, where X and Y denote the input and output data; h denotes the hidden state; W_{hx}, W_{yh}, and W_{hh} denote the weight matrices describing the relationship between X and h, h and Y, and h and h. The output $\mathbf{y}_t$ is not only determined by the input $\mathbf{X}_t$ but also by the last hidden state $\mathbf{h}_{t-1}$. The hidden state $\mathbf{h}_t$ is the key component to keep the temporal dependences within the time series.

However, the simple RNN has only a single hidden state $\mathbf{h}$, which is sensitive to short-term input. To capture long-term dependencies within the time series, an LSTM unit contains two hidden states $\mathbf{h}_t$ and $\mathbf{c}_t$, which are designed for keeping short-term information and long-term information, respectively. The inner structure of an LSTM unit is presented in Fig. 11.2.

The hidden state $\mathbf{c}$ contains an extra mechanism for strategically forgetting unrelated information corresponding to the current time. To retain the long-term information, three control gates are introduced in the LSTM unit, as shown in Fig. 11.2. These are the forget gate, the input gate, and the output gate. The control gates essentially fully connect the layers (denoted as σ).

The first gate in the LSTM unit is the forget gate $\mathbf{f}_t$, which determines how much information is kept from the last state $\mathbf{c}_{t-1}$. The forget state at time t is formulated as:

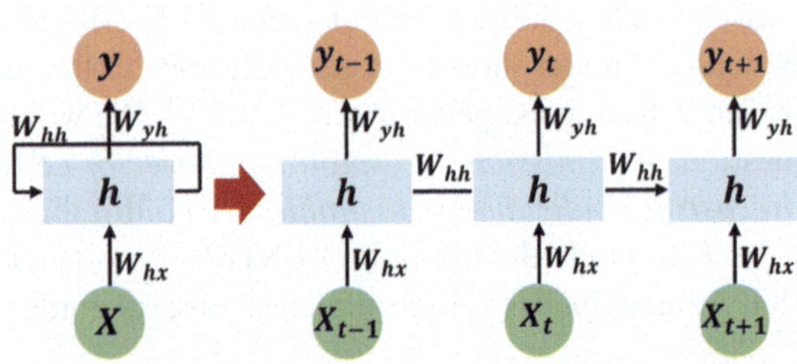

Fig. 11.1 The structure of LSTM

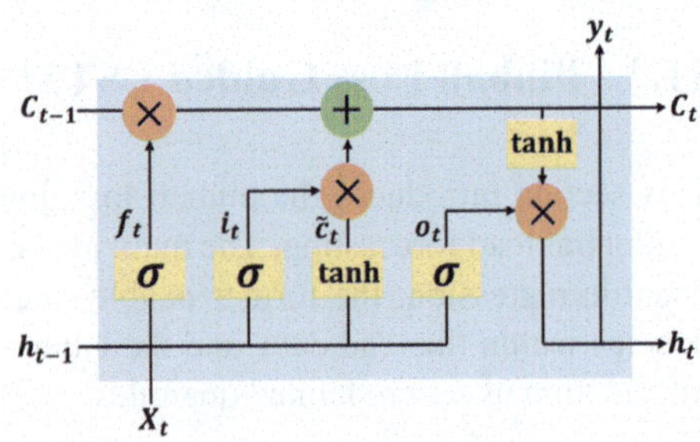

Fig. 11.2 Inner structure of an LSTM unit

$$\mathbf{f}_t = \sigma\left(\mathbf{W}_f \cdot [\mathbf{h}_{t-1}, \mathbf{X}_t] + \mathbf{b}_f\right), \tag{11.1}$$

where $\sigma(\cdot)$ denotes the sigmoid activation function; $\mathbf{X}_t$ is the input vector for the regression model, which mainly include historical load data, calendar data, and external factors; $\mathbf{f}_t$, $\mathbf{h}_{t-1}$, and $\mathbf{b}_f$ stand for the forget gate vector at time t, the output vector (also the state-h vector) at time $t-1$, and the bias of the forget gate at time t, respectively; $\mathbf{W}_f$ is the weight matrix of the forget gate; and $[\cdot]$ is the concatenating operator for vectors.

The second gate is the input gate $\mathbf{i}_t$, which determines how much current information should be treated as input to generate the current state $\mathbf{c}_t$. $\mathbf{i}_t$ is calculated by:

$$\mathbf{i}_t = \sigma\left(\mathbf{W}_i \cdot [\mathbf{h}_{t-1}, \mathbf{X}_t] + \mathbf{b}_i\right), \tag{11.2}$$

where $\mathbf{W}_i$ and $\mathbf{b}_i$ denote the weight matrix and bias of the input gate, respectively. It can be seen that $\mathbf{i}_t$ has a similar formulation to $\mathbf{f}_t$. Both gates are determined by $\mathbf{h}_{t-1}$ and $\mathbf{X}_t$.

The current hidden state $\mathbf{c}_t$ is determined by adding the parts of information they control. The long-term information is controlled by $\mathbf{f}_t$, and the short-term information is controlled by $\mathbf{i}_t$:

$$\tilde{\mathbf{c}}_t = tanh\left(\mathbf{W}_c \cdot [\mathbf{h}_{t-1}, \mathbf{x}_t] + \mathbf{b}_c\right), \tag{11.3}$$

$$\mathbf{c}_t = \mathbf{f}_t * \mathbf{c}_{t-1} + \mathbf{i}_t * \tilde{\mathbf{c}}_t, \tag{11.4}$$

where $tanh(\cdot)$ denotes the tanh activation function; $\mathbf{W}_c$ and $\mathbf{b}_c$ denote the weight matrix and the bias of the current gate, respectively; and the operator $*$ stands for the element-wise product.

The last phase of the LSTM unit is to calculate how much information can eventually be treated as the output. Another control gate is chosen as the output gate $\mathbf{o}_t$:

$$\mathbf{o}_t = \sigma\left(\mathbf{W}_o \cdot [\mathbf{h}_{t-1}, \mathbf{x}_t] + \mathbf{b}_o\right). \tag{11.5}$$

Since gates control the information flow by performing an element-wise product, the final output of LSTM $\mathbf{h}_t$ is defined by:

$$\mathbf{h}_t = \mathbf{o}_t * tanh(\mathbf{c}_t). \tag{11.6}$$

11.2.2 Pinball Loss

For traditional LSTM, the loss function is the MSE:

$$L_{MSE} = \frac{1}{T}\sum_{t=1}^{T}(y_t - \hat{y}_t^E)^2, \tag{11.7}$$

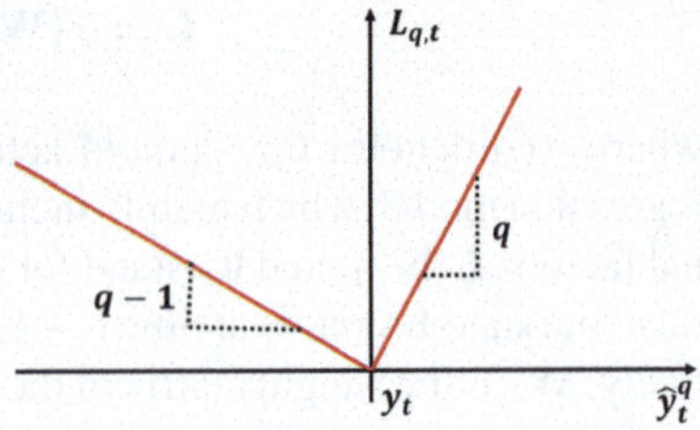

Fig. 11.3 Illustration of pinball loss

where y_t and $\hat{y}_t^E$ denote the measured and predicted load at time t, respectively, and T denotes the total prediction time period.

Traditional LSTM can only provide the expected value of the future load. To provide more information about future uncertainties, we replace the loss function MSE by the pinball loss, also called the quantile loss, to guide the training of the LSTM network. The pinball loss is calculated as follows:

$$L_{q,t}(y_t, \hat{y}_t^q) = \begin{cases} (1-q)(\hat{y}_t^q - y_t) & \hat{y}_t^q \geq y_t \\ q(y_t - \hat{y}_t^q) & \hat{y}_t^q < y_t. \end{cases} \tag{11.8}$$

where q denotes the targeted quantile, $\hat{y}_t^q$ denotes the estimated qth quantile at time t, and $L_{q,t}$ denotes the pinball loss for the qth quantile at time t. Figure 11.3 gives an illustration of pinball loss, which is asymmetric. When the forecasted quantile is higher than the real value, the penalty will be multiplied by $(1-q)$, and when the forecasted quantile is lower than the real value, the penalty will be multiplied by q.

There are at least two advantages to choosing pinball loss as the loss function:

1. Under the guidance of pinball loss, the trained LSTM network provides the targeted quantile value instead of the expected value. By varying the value of the quantile, we can obtain a series of quantiles to represent the uncertainties. The whole training process is non-parametric and requires no presumption about the distributions.
2. The probabilistic forecasts are usually evaluated using three aspects: reliability, sharpness, and calibration. Pinball loss is a comprehensive index for these three criteria, which means that the pinball loss can guarantee the performance of the final probabilistic forecasts.

11.2.3 Overall Networks

As introduced above, the pinball loss guided LSTM is a combination of LSTM and pinball loss. The overall pinball loss guided LSTM network is shown in Fig. 11.4.

Concretely, the proposed pinball loss guided LSTM in this chapter consists of three phases.

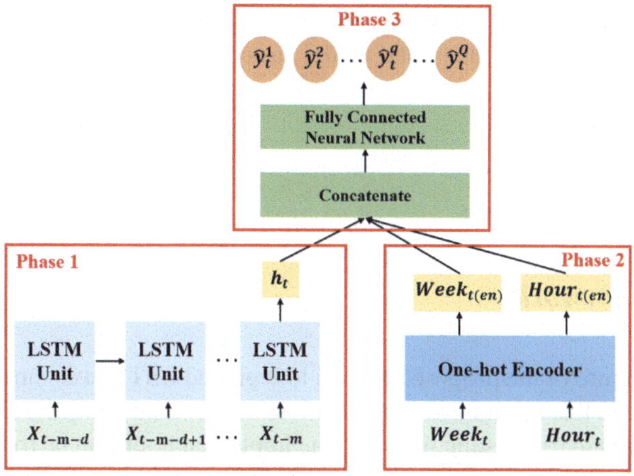

Fig. 11.4 Overall structure of pinball loss guided LSTM

The first phase is stacked by LSTM units where the inputs are the sequential loads at different time stamps, and the output is the hidden state h_t at the last timestamp, corresponding to the encoded features learned from the historical load. In this figure, m denotes the number of time periods ahead, and d denotes the number of time periods that are considered as the inputs in the forecasting model.

The second phase is a one-hot encoder, converting numerical time variables W_t and H_t into encoded vectors, where $Week_t$ and $Hour_t$ denote the day of the week and the hour of the day of the forecasted load y_t, respectively. $Week_{t(en)}$ and $Hour_{t(en)}$ denote the encoded vectors corresponding to the week and hour variables, respectively.

The third phase is a fully-connected (FC) network, where the inputs are the concatenated feature vectors generated from the two phases mentioned above, and the outputs are the forecasted quantiles.

In traditional quantile regression, the model would be trained for each quantile individually. The training objective is to minimize the average loss function L_q for the qth quantile, which is described as:

$$\min \quad L_q = \frac{1}{T}\sum_{t=1}^{T} L_{q,t}(y_t, \hat{y}_t^q). \tag{11.9}$$

There is a high computational burden when many quantiles need to be trained. To ease this computational burden, we design multiple outputs in the third phase, and the loss function is taken as the average pinball loss L for all quantiles:

$$\min \quad L = \frac{1}{Q \times T}\sum_{q=1}^{Q}\sum_{t=1}^{T} L_{q,t}(y_t, \hat{y}_t^q). \tag{11.10}$$

In this way, the LSTM network needs to be trained only once. Our numerical experiments show that the integrated model has comparable performance with multiple individual models.

11.3 Implementations

11.3.1 Framework

The basic structure of the proposed pinball loss guided LSTM was introduced in the above section. In this section, we provide more details on the implementation of the whole probabilistic forecasting process. The implementation can be roughly divided into three stages: data preparation, model training, and probabilistic forecasting. These are shown in Fig. 11.5.

11.3.2 Data Preparation

In the data preparation stage, we only use the historical load data since the weather data are not available in our dataset. We first clean the load dataset as follows. Any not-a-number (NAN) data are simply replaced by the average of the load data at the same time period from one day ahead and one data later. The input data of the regression model include the historical load data and the calendar variables, such as the current hour of the day and day of a week. After formulating the input and output dataset, we split the dataset into three parts for model training (S1), validation (S2), and testing (S3).

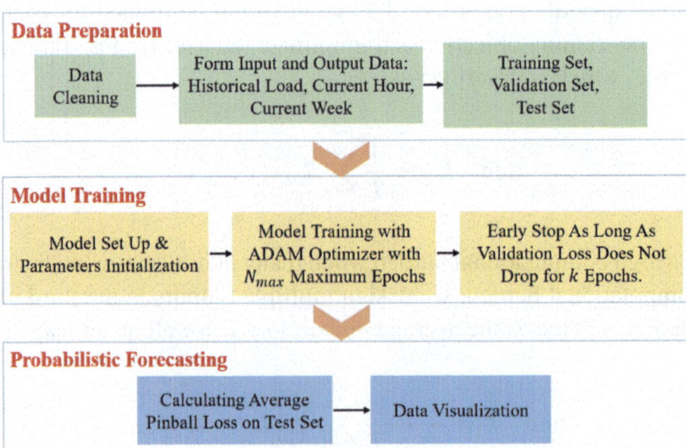

Fig. 11.5 Implementation flowchart for probabilistic individual load forecasting

11.3.3 Model Training

The first step is the setup of the neural network. A static computing graph is generated with TensorFlow [25] according to Fig. 11.4. Then, the parameters are initialized. The parameters can be divided into weights and biases. All weights in the three phases are initialized with values sampled from a truncated normal distribution with a mean of 0 and a standard deviation of 0.01. All biases are initialized to 0. Such initialization can, to some extent, prevent the neural network from becoming stuck in a local minimum. After that, the loss function of the neural network is optimized using a gradient-descent-based method with an adequate learning rate—Adam [26]. We define the maximum training epoch as N_{max}, and an early stopping mechanism is utilized to prevent the model from overfitting. Concretely, if the monitored validation loss does not drop for k epochs, the training process is terminated.

One requirement of the application of Adam is that the loss function should be differentiable so that the neural network can be trained using gradient descent. However, the pinball loss is not differentiable everywhere. In this chapter, we introduce the Huber norm [27] to the loss function, with very little approximation, in order to make the loss function differentiable everywhere. The Huber norm can be viewed as a combination of the L1- and L2-norms:

$$H(y_t, \hat{y}_t^q) = \begin{cases} \dfrac{(\hat{y}_t^q - y_t)^2}{2\varepsilon} & 0 \leq |\hat{y}_t^q - y_t| \leq \varepsilon \\ |\hat{y}_t^q - y_t| - \dfrac{\varepsilon}{2} & |\hat{y}_t^q - y_t| > \varepsilon, \end{cases} \tag{11.11}$$

where ε denotes the threshold magnitude for the L1- and L2-norms. When the forecast error $|\hat{y}_t^q - y_t|$ is below the threshold, the Huber norm is the L2-norm; when the forecast error is larger than the threshold, the Huber norm is the L1-norm.

We then substitute $(\hat{y}_t^q - y_t)$ into Eq. (11.8) with the Huber norm, and the approximated pinball loss can be calculated as:

$$L_{q,t}(y_t, \hat{y}_t^q) = \begin{cases} (1-q)h(y_t, \hat{y}_t^q) & \hat{y}_t^q \geq y_t \\ qh(y_t, \hat{y}_t^q) & \hat{y}_t^q < y_t. \end{cases} \tag{11.12}$$

Compared with standard pinball loss, the approximated pinball loss is differentiable when the forecast error is zero, i.e., $\hat{y}_t^q = y_t$. The gradient of the approximated pinball loss is equal to that of the standard pinball loss when the forecast error is larger than the threshold, and there is very little difference when the forecast error is below the threshold.

11.3.4 Probabilistic Forecasting

The performance of the probabilistic forecasts is evaluated by the average of the total pinball loss:

$$L = \frac{1}{Q \times |S3|} \sum_{q=1}^{Q} \sum_{t \in S3} L_{q,t}(y_t, \hat{y}_t^q), \tag{11.13}$$

where $|S3|$ denotes the length of the test dataset.

In addition to calculating the average pinball loss, the forecasted quantiles can also be plotted. In this way, we can visualize how these quantiles cover the real values at different time periods.

11.4 Benchmarks

In this section, we introduce three probabilistic load forecasting methods as the benchmarks in our case study.

11.4.1 QRNN

The quantile regression neural network (QRNN) is a nonlinear quantile regression model for probabilistic load forecasting. To capture the effect of calendar variables, the week and hour variables of the time period are also coded by the one-hot encoder. The overall structure of the QRNN is shown in Fig. 11.6. Compared with the proposed pinball loss guided LSTM, there are no LSTM units for the input variables.

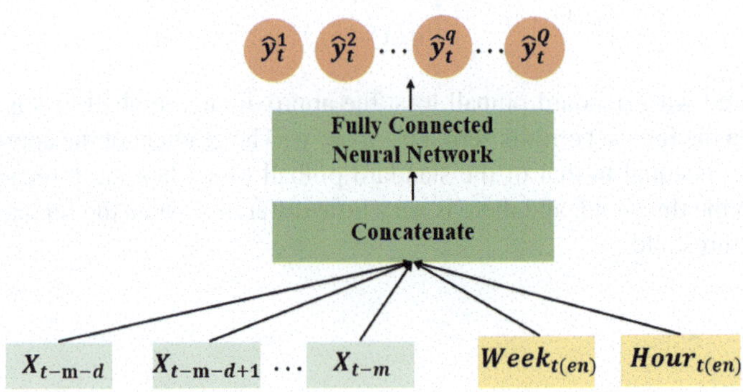

Fig. 11.6 Structure of QRNN

11.4.2 QGBRT

Gradient boosting regression tree (GBRT) is a powerful point forecasting method that has been applied in various competitions and achieves high rankings, including in load forecasting. Quantile gradient boosting regression tree (QGBRT) is an improvement to GBRT where the loss function is the pinball loss, producing quantile forecasts instead of expected values.

11.4.3 LSTM+E

In addition to two frequently used nonlinear quantile regression models, the probabilistic forecasts can also be obtained by the statistics of the point forecast errors. To make a fair comparison, the point forecasts are produced based on a traditional LSTM with the same structure as the proposed pinball loss guided LSTM. We simply assume that the errors follow Gaussian distributions. Since the distribution of the errors varies for different time periods, the variances for different time periods are calculated individually. Then, the quantiles can be calculated based on the corresponding variances.

11.5 Case Studies

The proposed pinball loss guided LSTM, QRNN, and traditional LSTM are implemented using TensorFlow [25]. QGBRT is implemented using the GBRT package in Scikit-Learn [28]. The model training is supported by CUDA8.0 and an Nvidia GPU, TITAN X (Pascal). The GPU is also applied for parallel computation to accelerate the model training process. For the implementation of QGBRT, a total of Q parallel processes are opened for the individual training of Q quantile regression models.

The hyperparameters of the proposed pinball loss guided LSTM (denoted as QLSTM in the following), and the competing methods (QRNN, QGBRT, and LSTM+E) are illustrated in Table 11.1. The structures of QLSTM and LSTM+E

Table 11.1 Hyperparameter settings for different models

Models	Parameters
QLSTM/LSTM+E	LSTM-unit:16
	FC-unit:16
	FC-layer:3
QRNN	FC-unit:16
	FC-layer:3
QGBRT	N_estimators:500
	min_samples_split = 2
	max_depth = 3
	samples_leaf = 1

are the same except for the loss function, allowing us to make a fair comparison. The full connection layer of QRNN is the same as that of Phase 3 in QLSTM. The number of estimators of QGBRT is set to 500, which makes the GBRT an adequately strong learning model.

11.5.1 Data Description

The dataset used in the case studies was collected from the Smart Metering Electricity Customer Behavior Trials (CBTs) proposed by the Commission for Energy Regulation (CER) in Ireland. It contains over 6000 residential load profiles and small and medium enterprises (SME) load profiles for approximately one and a half years (from the 1st of July 2009 to the 31st of December 2010). These load profiles were collected at 30-min intervals. There are a total of 26,000 data points for each individual consumer. We use the first 22,000 load data points for model training and validation (S1 and S2) and apply the following 2000 points for model testing (S3). We implement the case studies on the load profiles of 100 randomly selected residential and SME consumers. We provide comprehensive model testing by using several forecast lead times, 30 min, one hour, two hours, and four hours.

11.5.2 Residential Load Forecasting Results

Table 11.2 presents the performance of the proposed QLSTM and three other competing methods measured by the average pinball loss for all 100 residential consumers. The proposed QLSTM has the lowest pinball loss for the four different lead times. In Table 11.2, I_QRNN, I_QGBRT, and I_LSTM+E denote the relative improvements of the proposed QLTSM model compared with QRNN, QGBRT, and LSTM+E. Except for QLSTM, QRNN has better performance than QGBRT and LSTM+E, while LSTM+E has the worst performance. A possible reason for this is the unreasonable assumption that forecasting errors follow a Gaussian distribution. Compared with QRNN, the relative improvements of QLSTM are 3.46, 2.76, 2.18, and 2.19%.

Table 11.2 Overall performance of different methods for residential consumers

	Pinball loss (kW)				Relative improvement (%)		
	QLSTM	QRNN	QGBRT	LSTM+E	I_QRNN	I_QGBRT	I_LSTM+E
30 min	**0.0837**	0.0867	0.0886	0.0905	3.46	5.50	7.52
1 h	**0.0963**	0.0990	0.1030	0.1020	2.76	6.48	5.62
2 h	**0.1018**	0.1040	0.1077	0.1061	2.18	5.50	4.13
3 h	**0.1031**	0.1054	0.1090	0.1077	2.19	5.40	4.27

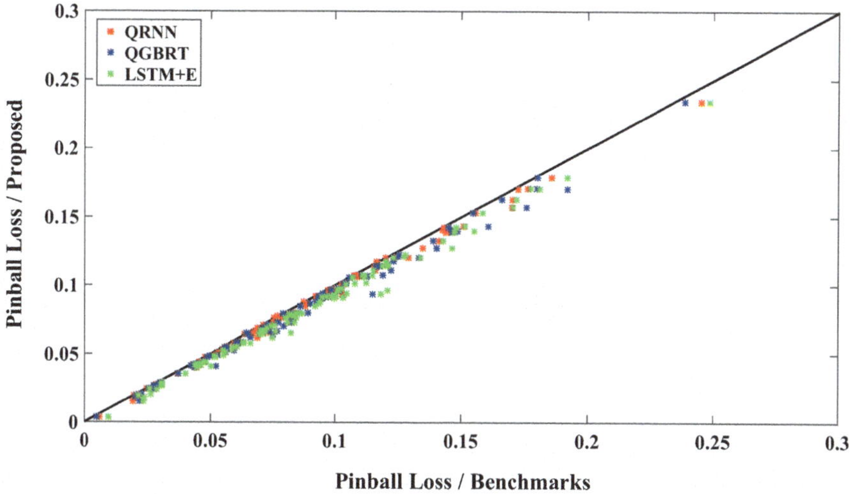

(a) 30 min

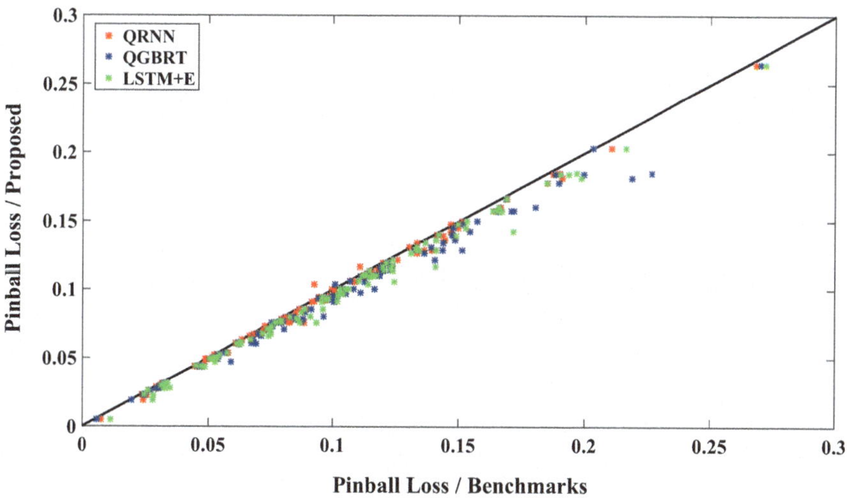

(b) One Hour

Fig. 11.7 Performance comparison between QLSTM and three benchmarks for all residential consumers

In addition, the averaged pinball loss gets larger with longer lead time, especially for the lead time from 30 min to one hour. However, the pinball loss stays relatively stable when the lead time is longer than one hour. The reason is that the residential load profiles are so stochastic that only 30 min or one hour ahead time data can effectively help to capture future trends.

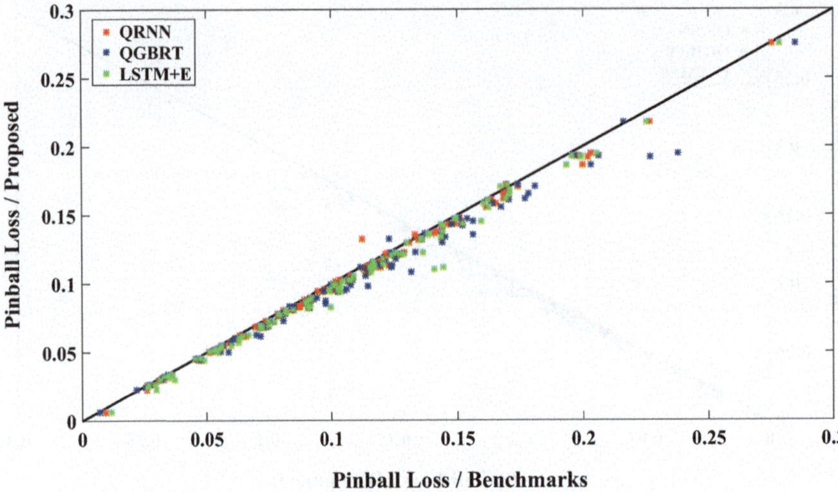

(c) Two Hours

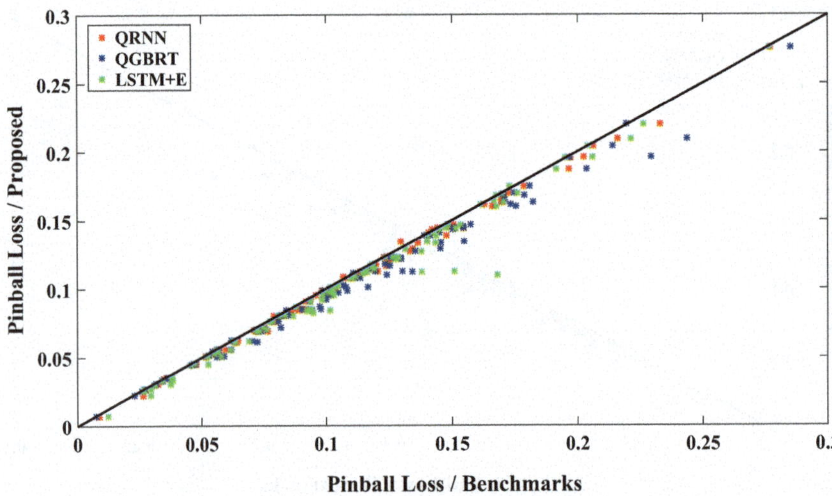

(d) Four Hours

Fig. 11.7 (continued)

To provide the detailed performance of the proposed QLSTM and the three competing methods for the 100 residential consumers, a scatter plot of the average pinball loss of the proposed QLSTM versus the three competing methods is provided in Fig. 11.7. It can be seen that most of the points fall under the line $y = x$, which

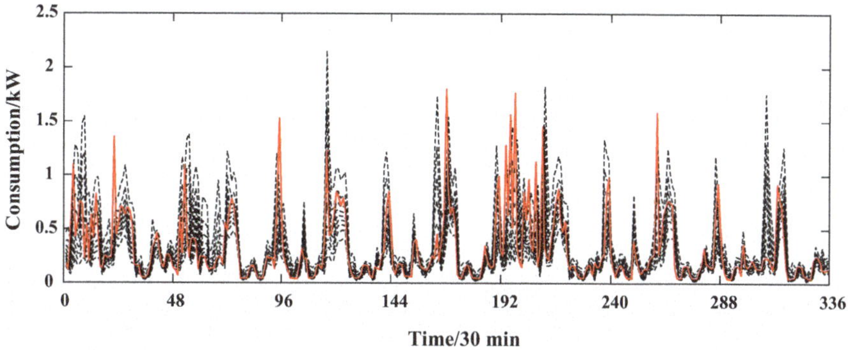

Fig. 11.8 Thirty minute ahead forecasts for one sample residential consumer over one week

means that QLSTM outperforms the other three methods for almost all residential consumers.

Figure 11.8 shows the 30 min ahead forecasts of one sample residential consumer over one week, from 17 October 2010 to 23 October 2010 (a total of 336 time periods), where the dotted lines denote a series of forecasted quantiles and the red line denotes the actual values. The quantiles can effectively capture the basic trends in the load profile, except for several sudden peaks.

We calculate the pinball loss for all 2000 time periods. Then, we draw the box-plots for the distribution of the pinball loss for the 48 time periods in a day. The distribution of the pinball losses is shown in Fig. 11.9. The pinball loss is higher and more dispersed from 7:00 to 8:00 and from 17:30 to 22:00. The time period from 7:00 to 8:00 corresponds to the time that people get up for work, and the time period from 17:30 to 22:00 corresponds to the after-work time. Consumers have higher load demands with larger uncertainties in these two time periods, and thus, the loads are more difficult to forecast. The distributions of the pinball losses for different forecasting lead times have similar trends. These results can be a good reference for demand response targeting, baseline estimation, and reliability assessments. In addition, the pinball load distributions for 1, 2, and 4 h ahead forecasting are similar; while the distribution for 30 min ahead forecasting is different from the rest three distributions and has smaller averaged pinball loss.

11.5.3 SME Load Forecasting Results

Similar to the residential forecasts, we summarize the average pinball loss for all 100 of the SME consumers in Table 11.3. For these data, the proposed QLSTM also gives the best performance. Similarly, QRNN has the best performance in addition to QLSTM. However, in contrast to the residential consumer test, it is interesting that the performance of LSTM+E is better than that of QGBRT. This may due to two

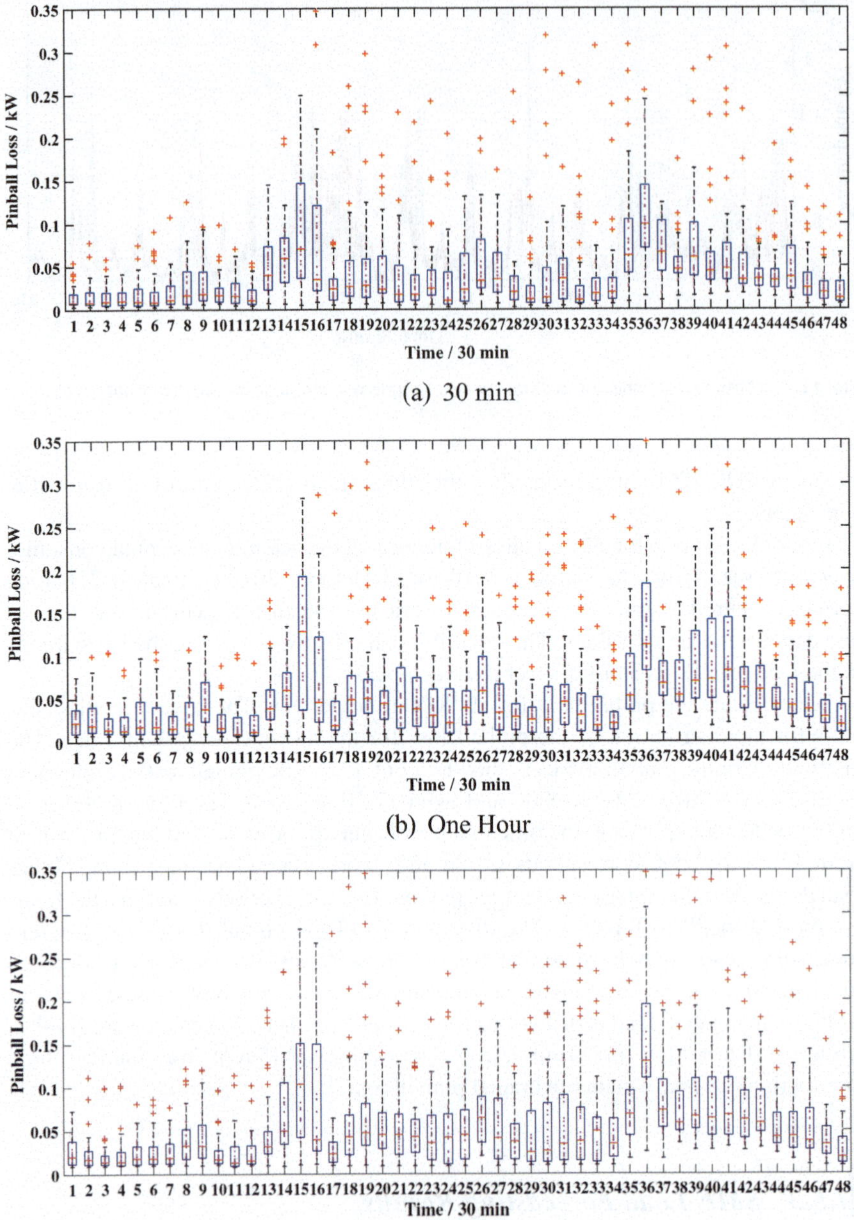

(a) 30 min

(b) One Hour

(c) Two Hours

Fig. 11.9 Distribution of pinball loss at different time periods for one residential consumer

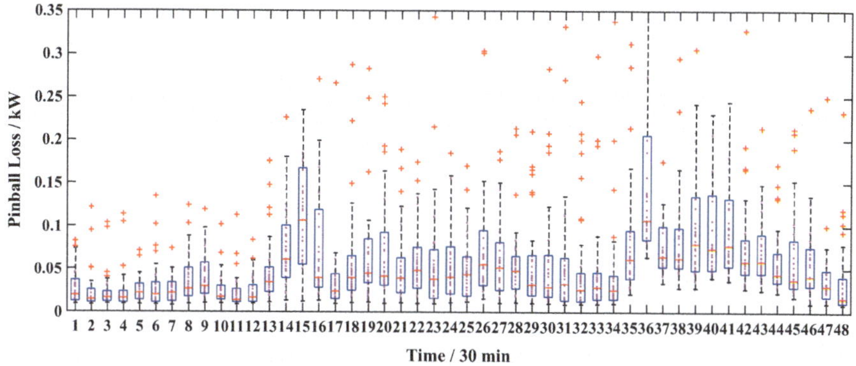

(d) Four Hours

Fig. 11.9 (continued)

Table 11.3 Overall performance of different methods for SME consumers

	Pinball loss (kW)				Relative improvement (%)		
	QLSTM	QRNN	QGBRT	LSTM+E	I_QRNN	I_QGBRT	I_LSTM+E
30 min	**0.1213**	0.1275	0.1461	0.1391	4.89	16.98	12.81
1 h	**0.1552**	0.1613	0.1975	0.1775	3.79	21.43	12.56
2 h	**0.1805**	0.1883	0.2381	0.2081	4.16	24.21	13.29
4 h	**0.1982**	0.2114	0.2671	0.2252	6.27	25.80	12.01

reasons: (1) the assumption that the forecasting errors follow a Gaussian distribution may be more reasonable for SME consumers than for residential consumers; (2) LSTM is able to provide more accurate point forecasts compared with GBRT. The improvements with respect to QLSTM are also greater than those for residential consumers for different forecasting lead times.

Figure 11.10 presents the scatter plot of the average pinball loss for the proposed QLSTM versus the three competing methods. We obtain similar results in that most of the points fall under the line $y = x$, which means that QLSTM outperforms the other three methods for almost all of the SME consumers.

Figure 11.11 shows the 30 min ahead forecasts for one sample SME consumer over one week, from 17 October 2010 to 23 October 2010, where the dotted lines denote a series of forecasted quantiles and the red line denotes the actual values. In contrast with the residential consumer, the SME consumer has more stable electricity consumption behavior and clearer patterns (there are no sudden peaks).

Accordingly, the box-plot of the pinball loss at different time periods is shown in Fig. 11.12. The pinball loss is higher and more dispersed between 9:30 and 20:00. This time period corresponds to working hours, which means that the consumer has large but highly uncertain, electricity consumption in this time period.

11.6 Conclusions

In this chapter, we proposed a pinball loss guided LSTM for probabilistic residential and SME consumer load forecasting. Comprehensive case studies are conducted on

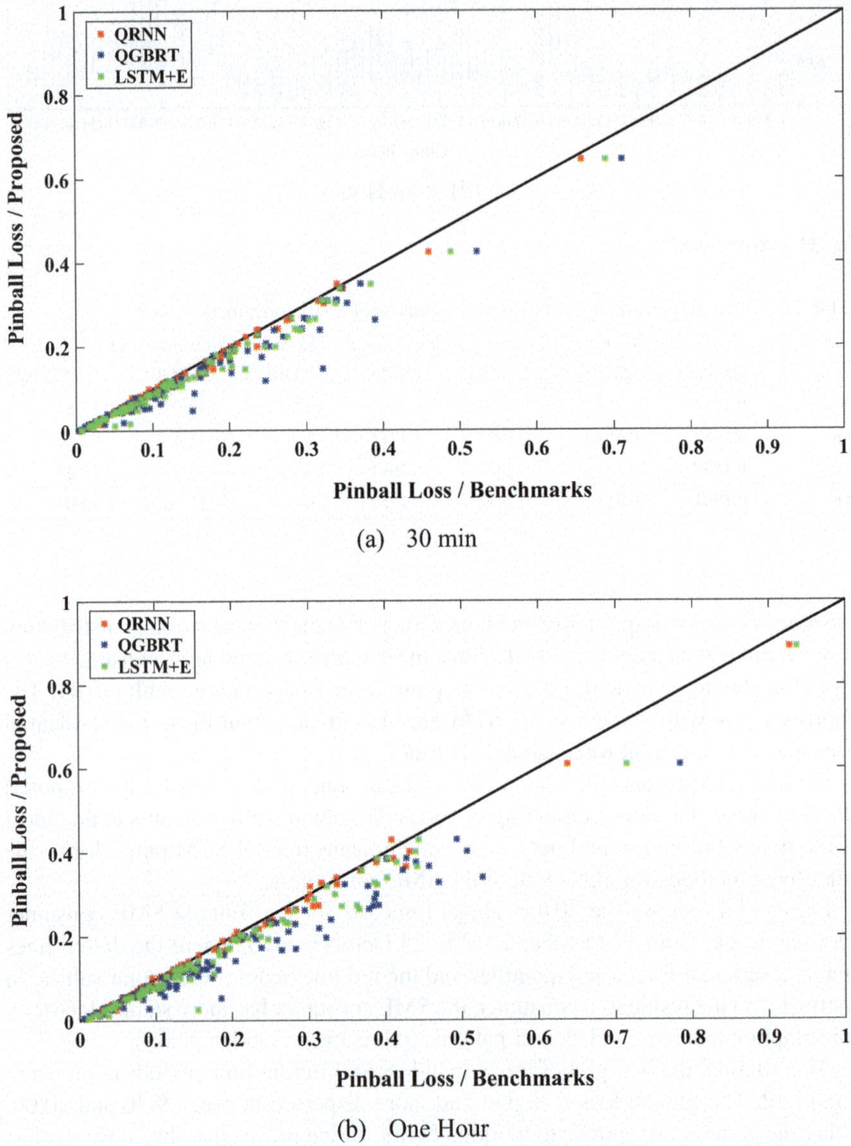

Fig. 11.10 Performance comparison between QLSTM and three benchmarks for all SME consumers

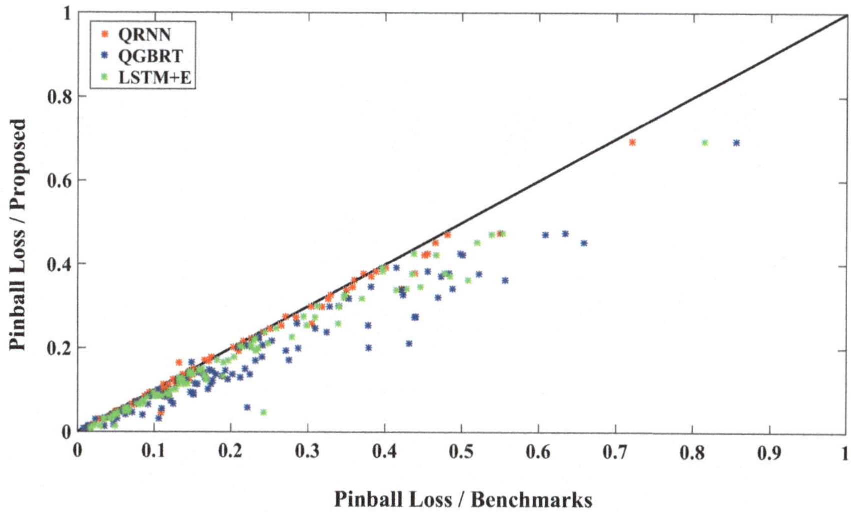

(c) Two Hours

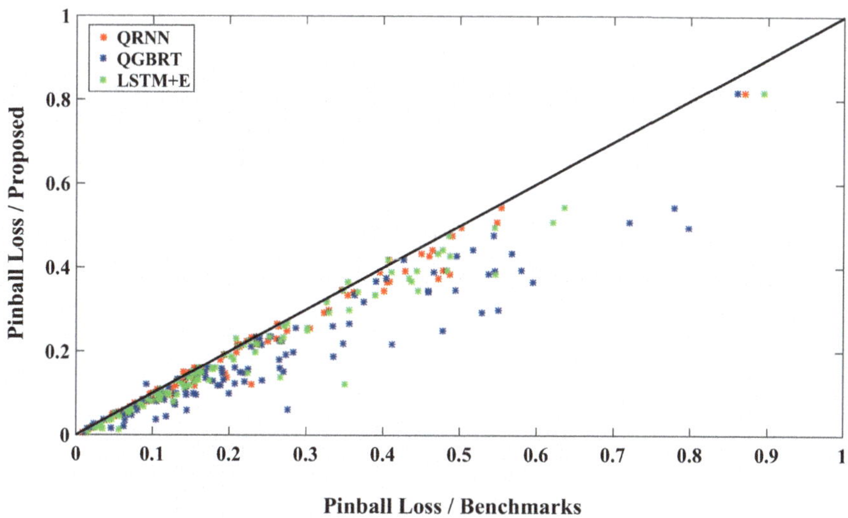

(d) Four Hours

Fig. 11.10 (continued)

different consumers with different forecasting lead times and with state-of-the-art competing methods. We can draw the following conclusions:

1. The proposed pinball loss guided LSTM has better performance than QRNN, QGBRT, and LSTM+E for almost all 100 residential loads and 100 SME loads.

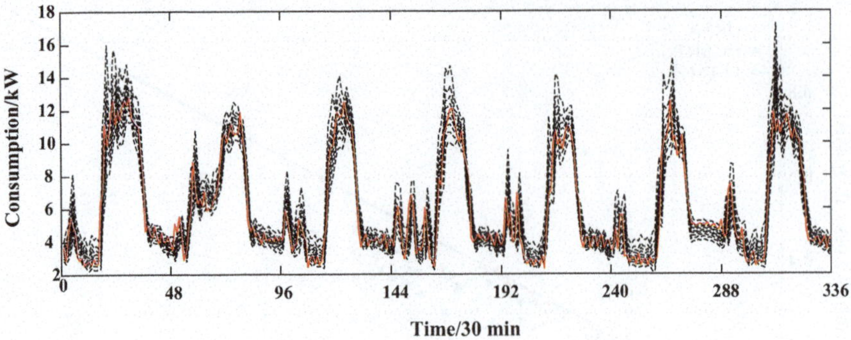

Fig. 11.11 Thirty minute ahead forecasts for one sample SME consumer over one week

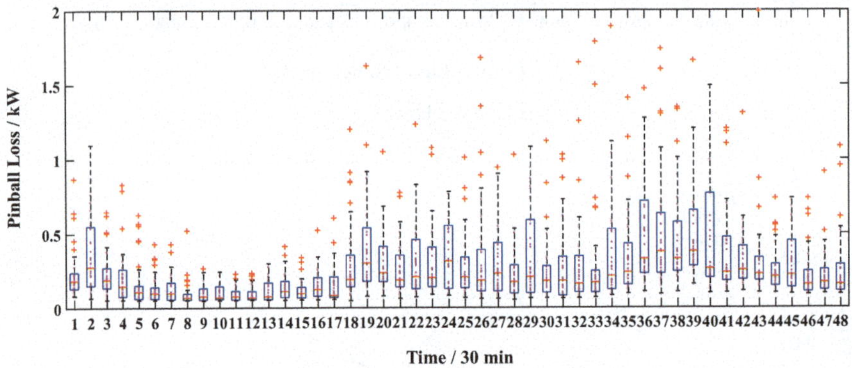

(a) 30 min

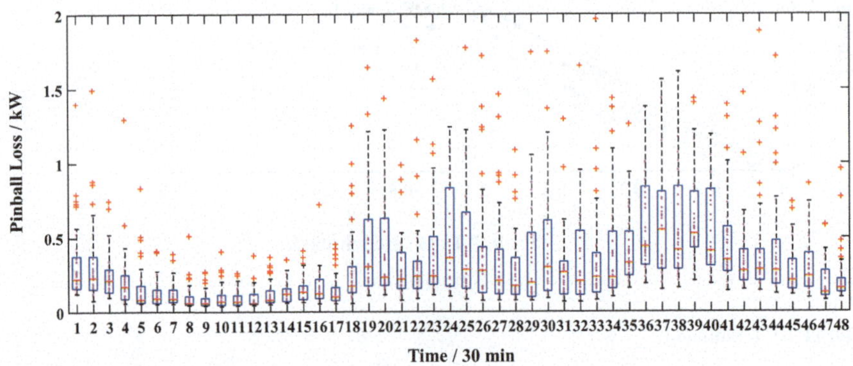

(b) One Hour

Fig. 11.12 Distribution of pinball loss at different time periods for one SME consumer

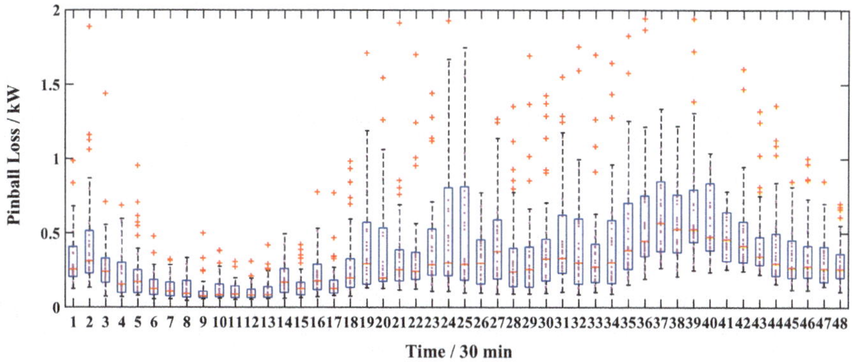

(c) Two Hours

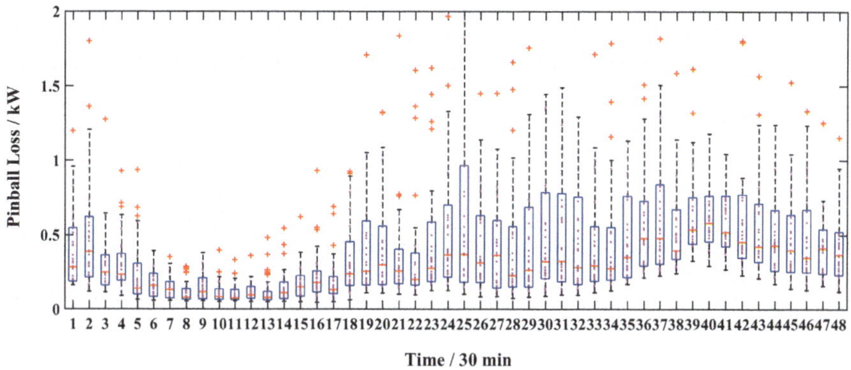

(d) Four Hours

Fig. 11.12 (continued)

2. Compared with the three competing methods, the proposed method has improvements ranging from 2.19 to 7.52% for residential consumers; while, the improvements range from 3.79 to 25.80% for SME consumers. The improvements for residential consumers are greater than the improvements seen for SME consumers, which means that QLSTM can more effectively capture the change patterns of SME loads.
3. The distributions of the pinball loss in different time periods are different. For residential consumers, the time periods that have the largest and most dispersed pinball losses are 7:00-8:00 and 17:30–22:00; while, for SME consumers, the time periods that have the largest and most dispersed pinball losses are 9:00–20:00; These time periods for residential consumers are complementary to those of SME consumers.

References

1. Hong, T., Pinson, P., Fan, S., Zareipour, H., Troccoli, A., & Hyndman, R. J. (2016). Probabilistic energy forecasting: Global energy forecasting competition 2014 and beyond. *International Journal of Forecasting*, *32*(3), 896–913.
2. Wang, Y., Chen, Q., Kang, C., & Xia, Q. (2016). Clustering of electricity consumption behavior dynamics toward big data applications. *IEEE Transactions on Smart Grid*, *7*(5), 2437–2447.
3. Keerthisinghe, C., Verbič, G., & Chapman, A. C. (2016). A fast technique for smart home management: Adp with temporal difference learning. *IEEE Transactions on Smart Grid*, *9*(4), 3291–3303.
4. Morstyn, T., Farrell, N., Darby, S. J., & McCulloch, M. D. (2018). Using peer-to-peer energy-trading platforms to incentivize prosumers to form federated power plants. *Nature Energy*, *3*(2), 94.
5. Li, P., Zhang, B., Weng, Y., & Rajagopal, R. (2017). A sparse linear model and significance test for individual consumption prediction. *IEEE Transactions on Power Systems*, *32*(6), 4489–4500.
6. Chaouch, M. (2014). Clustering-based improvement of nonparametric functional time series forecasting: Application to intra-day household-level load curves. *IEEE Transactions on Smart Grid*, *5*(1), 411–419.
7. Teeraratkul, T., O'Neill, D., & Lall, S. (2017). Shape-based approach to household electric load curve clustering and prediction. *IEEE Transactions on Smart Grid*, *9*(5), 5196–5206.
8. Shi, H., Minghao, X., & Li, R. (2017). Deep learning for household load forecasting–a novel pooling deep rnn. *IEEE Transactions on Smart Grid*, *9*(5), 5271–5280.
9. Kong, W., Dong, Z. Y., Jia, Y., Hill, D. J., Xu, Y., & Zhang, Y. (2017). Short-term residential load forecasting based on lstm recurrent neural network. *IEEE Transactions on Smart Grid*, *10*(1), 841–851.
10. Kong, W., Dong, Z. Y., Hill, D. J., Luo, F., & Xu, Y. (2017) Short-term residential load forecasting based on resident behaviour learning. *IEEE Transactions on Power Systems*, *33*(1), 1087–1088.
11. Yu, C-N., Mirowski, P., & Ho, T. K. (2017). A sparse coding approach to household electricity demand forecasting in smart grids. *IEEE Transactions on Smart Grid*, *8*(2), 738–748.
12. Tascikaraoglu, A., & Sanandaji, B. M. (2016). Short-term residential electric load forecasting: A compressive spatio-temporal approach. *Energy and Buildings*, *111*, 380–392.
13. Impact of calendar effects and forecast granularity. (2017). Peter Lusis, Kaveh Rajab Khalilpour, Lachlan Andrew, and Ariel Liebman. Short-term residential load forecasting. *Applied Energy*, *205*, 654–669.
14. Haben, S., Ward, J., Greetham, D. V., Singleton, C., & Grindrod, P. (2014). A new error measure for forecasts of household-level, high resolution electrical energy consumption. *International Journal of Forecasting*, *30*(2), 246–256.
15. Kim, S., & Kim, H. (2016). A new metric of absolute percentage error for intermittent demand forecasts. *International Journal of Forecasting*, *32*(3), 669–679.
16. Yang, Y., Li, S., Li, W., & Meijun, Q. (2018). Power load probability density forecasting using gaussian process quantile regression. *Applied Energy*, *213*, 499–509.
17. Wang, Y., Zhang, N., Chen, Q., Kirschen, D. S., Li, P., & Xia, Q. (2017). Data-driven probabilistic net load forecasting with high penetration of behind-the-meter pv. *IEEE Transactions on Power Systems*, *33*(3), 3255–3264.
18. Dahua, G., Yi, W., Shuo, Y., & Chongqing, K. (2018). Embedding based quantile regression neural network for probabilistic load forecasting. *Journal of Modern Power Systems and Clean Energy*, *6*(2), 244–254.
19. Liu, B., Nowotarski, J., Hong, T., & Weron, R. (2017). Probabilistic load forecasting via quantile regression averaging on sister forecasts. *IEEE Transactions on Smart Grid*, *8*(2), 730–737.
20. Hong, T., & Fan, S. (2016). Probabilistic electric load forecasting: A tutorial review. *International Journal of Forecasting*, *32*(3), 914–938.

21. Arora, S., Taylor, J. W. (2016). Forecasting electricity smart meter data using conditional kernel density estimation. *Omega*, *59*, 47–59.
22. Taieb, S. B., Huser, R., Hyndman, R. J., & Genton, M. G. (2016). Forecasting uncertainty in electricity smart meter data by boosting additive quantile regression. *IEEE Transactions on Smart Grid*, *7*(5), 2448–2455.
23. Shepero, M., van der Meer, D., Munkhammar, J., & Widén, J. (2018). Residential probabilistic load forecasting: A method using gaussian process designed for electric load data. *Applied Energy*, *218*, 159 - 172.
24. Amini, M. H., Karabasoglu, O., Ilic, M. D., & Boroojeni, K. G. (2015). Arima-based demand forecasting method considering probabilistic model of electric vehicles' parking lots. In *Power & Energy Society General Meeting* (pp. 1–5).
25. Abadi, M., Barham, P., Chen, J., Chen, Z., Davis, A., Dean, J., et al. (2016). Tensorflow: A system for large-scale machine learning. *OSDI*, *16*, 265–283.
26. Kingma, D. P., Ba, J. (2014). Adam: A method for stochastic optimization. arXiv:1412.6980.
27. Huber, P. J., & Ronchetti, E. M. (1981). *Robust statistics*. Series in probability and mathematical statistics. New York: Wiley.
28. Pedregosa, F., Varoquaux, G., Gramfort, A., Michel, V., Thirion, B., Grisel, O. et al. (2011). Scikit-learn: Machine learning in python. *Journal of Machine Learning Research*, *12*(Oct), 2825–2830.

Chapter 12
Aggregated Load Forecasting with Sub-profiles

Abstract With the prevalence of smart meters, fine-grained sub-profiles reveal more information about the aggregated load and further help improve forecasting accuracy. This chapter proposes a novel ensemble approach for aggregated load forecasting. An ensemble is an effective approach for load forecasting. It either generates multiple training datasets or applies multiple forecasting models to produce multiple forecasts. In this chapter, the proposed ensemble forecast method for the aggregated load with sub-profiles is conducted based on the multiple forecasts produced by different groupings of sub-profiles. Specifically, the sub-profiles are first clustered into different groups, and forecasting is conducted on the grouped load profiles individually. Thus, these forecasts can be summed to form the aggregated load forecast. In this way, different aggregated load forecasts can be obtained by varying the number of clusters. Finally, an optimal weighted ensemble approach is employed to combine these forecasts and provide the final forecasting result. Case studies are conducted on two open datasets and verify the effectiveness and superiority of the proposed method.

12.1 Introduction

Recent advances in load forecasting include probabilistic forecasting, hierarchical forecasting, ensemble forecasting, and etc. [1]. With the widespread popularity of smart meters, more and more fine-grained sub-profiles can be measured and collected. Consequently, research on individual load forecasting has also been investigated, such as the works in Chap. 11 and [2]. For aggregated load forecasting, a bottom-up approach, implemented based on the smart meter data, is proposed in [3]: forecast them individually and then aggregate the results. The individual load forecasting is implemented by modeling the conditional distribution of the profile labels and the transitions probabilities between profile labels. It has low computation burden and historical data requirements. To improve the efficiency of the forecasting procedure, a clustering-based aggregated load forecasting is proposed in [4]: different groups of consumers are first constructed based on their load patterns; afterward, forecast the load of each group separately; finally, sum the forecasts of different groups to obtain

Y. Wang et al., *Smart Meter Data Analytics*,
https://doi.org/10.1007/978-981-15-2624-4_12

the aggregated load forecast. The optimal number of clusters is determined by cross-validation. The results demonstrate that the clustering-based method outperforms the direct forecasting method.

Beyond the aforementioned single-output forecasting methods (i.e. only provide one final forecast value), a series of works have been done on ensemble forecasting methods, which can produce multiple forecasts from different models [5]. In general, ensemble forecasting can be classified as homogeneous and heterogeneous methods such as bootstrap aggregating methods and the combination of SVM and ANN [6]. This chapter tries to answer the following question: *Is it possible to utilize both ensemble techniques and fine-grained sub profiles to further improve the forecasting accuracy?*

This chapter first provides some primary experiment on the sub-profiles, including studying how the aggregation level affects the forecasting performance and a clustering-based forecasting approach to make full use of the fine-grained smart meter data. On this basis, this chapter proposes a novel ensemble forecasting method for the aggregated load with sub-profiles to answer this question. A brief summary of the ensemble method is as follows: First, the sub-profiles are grouped using hierarchical clustering method and forecasting is conducted on the grouped load profiles individually. Then, these forecasts are summed to form the aggregated load forecast. Thus, we can vary the number of clusters to obtain multiple aggregated load forecasts instead of a single forecast. Subsequently, an optimally weighted ensemble approached is used to combine these forecasts and provide the final result. Finally, case studies are conducted on two open datasets (residential and substation loads) to verify the effectiveness and superiority of the proposed method.

12.2 Load Forecasting with Different Aggregation Levels

12.2.1 Variance of Aggregated Load Profiles

The consumption behavior of individual consumer shows great uncertainties because individual consumer can be influenced by various internal and external factors. However, if more and more consumers are aggregated, their aggregated consumption behavior may show more clear patterns. This is the reason why the regional level load forecasting is much easier than individual level load forecasting. Figure 12.1 shows five weekly load profiles that are randomly selected with different aggregation levels, where the number of the aggregated consumer is 1, 20, 50, 100, 200, and 800, respectively. There is a clear trend that the pattern and periodicity and the load profiles are easier to be observed when the number of aggregated consumer increases.

In this section, we conduct short term load forecasting on load profiles with different aggregated levels. Since the specific forecasting model is not the main concern of this chapter, one of the most widely used forecasting models, Artificial

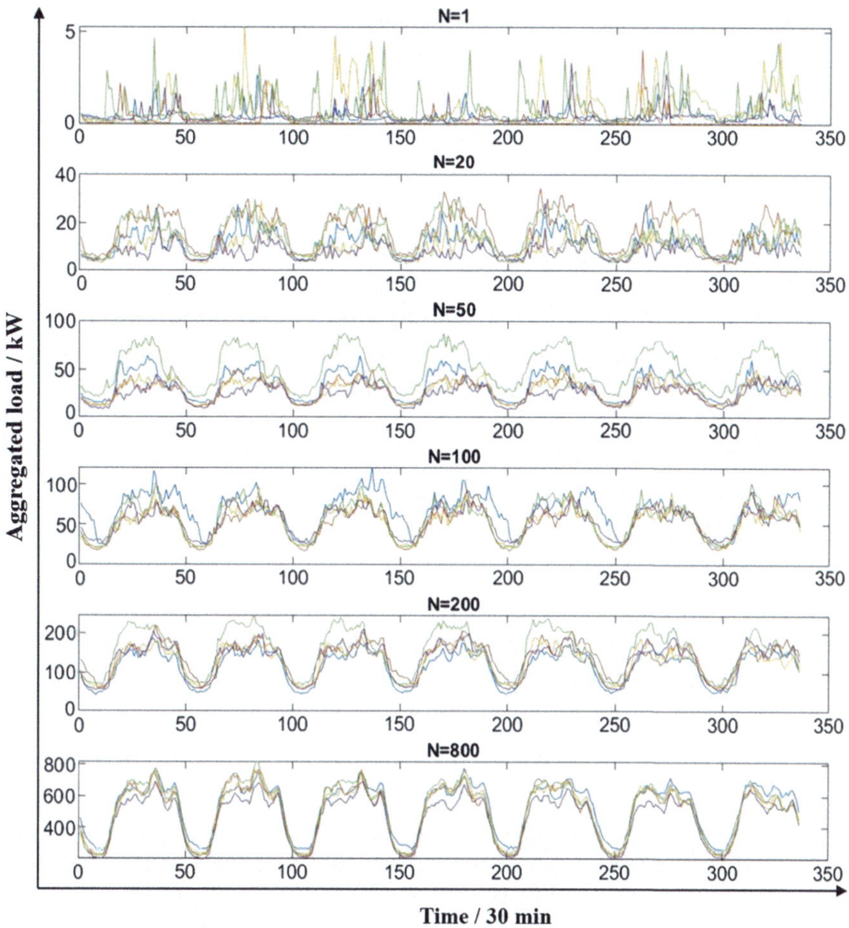

Fig. 12.1 Weekly load profiles at different aggregated levels

Neural Network (ANN), is applied to forecast different groups of load profiles. It consists of three layers: the input layer, the hidden layer, and the output layer where the hidden layer has four hidden neurons. It is implemented by "feedforward net" function in Matlab. We conduct 24 h ahead load forecasting here. Thus, the input of ANN is the lagged load values L_t (h denotes the number of time periods each day, $h = 48$ when the time resolution is 30 min) and calendar variables:

$$\mathbf{X}_t = [Week, Hour, L_{t-h}, L_{t-h-1}, L_{t-2h+1}, L_{t-2h}, L_{t-3h}] \quad (12.1)$$

Root-mean-square deviation (RMSE) and mean absolute error (MAE) are two frequently applied forecasting performance evaluation criteria. To make the load forecasting performances of different aggregation levels comparable, the relative root-mean-square deviation (R-RMSE) and relative mean absolute error (R-MAE) are used here:

$$\text{R-RMSE} = \frac{\sqrt{(\sum_{t=1}^{T}(L_t - \hat{L}_t)^2}}{\frac{1}{T}\sum_{t=1}^{T} L_t} \tag{12.2}$$

$$\text{R-MAE} = \frac{\frac{1}{T}\sum_{t=1}^{T}|L_t - \hat{L}_t|}{\frac{1}{T}\sum_{t=1}^{T} L_t} \tag{12.3}$$

where T denotes the total number of forecasting time periods; $\hat{L}_t$ denotes the forecasted load value. R-RMSE (or R-MAE) is the ratio between RMSE (or MARE) and the average load.

12.2.2 Scaling Law

Reference [7] provides a scaling law for short term load forecasting on varying aggregation levels. It is proven that the average R-RMSE can be approximated as a function of the aggregation level W.

$$\text{R-RMSE}(W) = \sqrt{\frac{\alpha_0}{W} + \alpha_1} \tag{12.4}$$

where W is the average load indicating the aggregation level; α_0 and α_1 are constants. The two constants can be estimated using the regression method based on experimental results. This scaling law can also be extended to R-MAE.

It clearly shows that R-RMSE will decrease with the aggregation level W increasing. When W is very small, $\frac{\alpha_0}{W} \gg \alpha_1$, the domain part is $\frac{\alpha_0}{W}$. Thus, the average R-RMSE can be estimated as:

$$\text{R-RMSE}(W) \simeq \sqrt{\frac{\alpha_0}{W}} \tag{12.5}$$

That is to say, when the aggregation level is very small, R-RMSE is approximately and linearly determined by $\frac{1}{\sqrt{W}}$.

When W is very large, $\frac{\alpha_0}{W} \ll \alpha_1$, the domain part is α_1. Thus, the average R-RMSE can be estimated as:

$$\text{R-RMSE}(W) \simeq \sqrt{\alpha_1} \tag{12.6}$$

It is interesting that when the aggregation level is very large, R-RMSE changes very slightly and approximately equals to $\sqrt{\alpha_1}$.

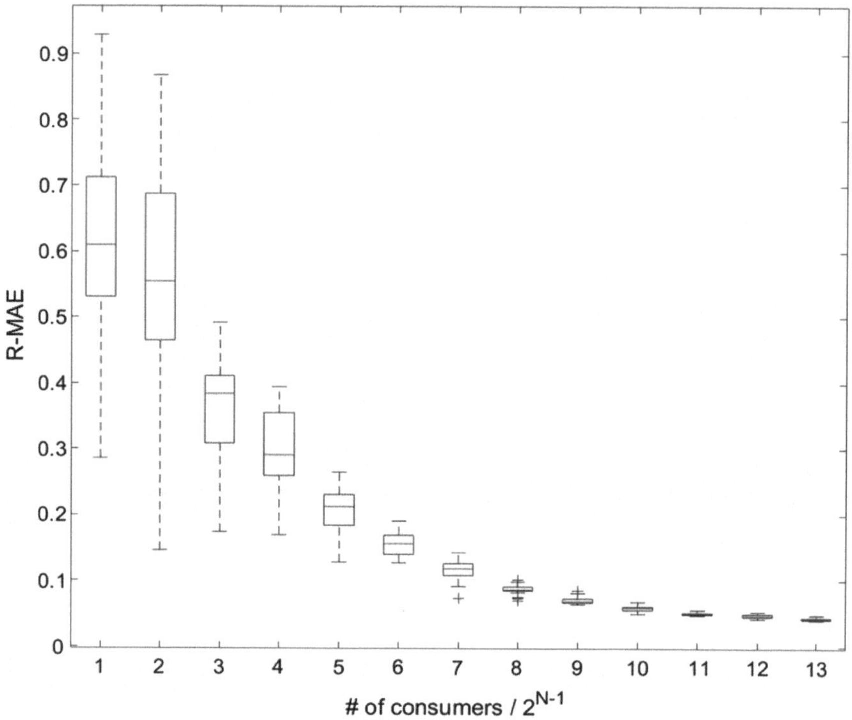

Fig. 12.2 R-MAE for different aggregated levels

To verify the scaling law, we conduct massive case studies by randomly selecting individual consumers. We define 13 aggregation level where the number of consumers at the nth aggregation level is 2^{n-1}. For example, there are $4096 = 2^{12}$ consumers randomly selected and aggregated at the 13th aggregation level. For each aggregation level, 20 experiments are conducted by repeatably selected the individual consumers. Figures 12.2 and 12.3 provide the boxplot of R-MAE and R-RMSE. We can find a clear trend that both average R-MAE and average R-RMSE decrease when the number of aggregated consumer increase. When the number of aggregated consumers is greater than $286 = 2^8$, both average R-MAE and average R-RMSE change very slightly. These trends are consistent with the scaling law provided in Eq. (12.4). Another observation is that the variances of R-MAE and R-RMSE also decrease when the number of aggregated consumer increase. For lower aggregation level, the aggregated load profile shows higher volatility, and thus the forecasting performance is unstable; for higher aggregation level, the aggregated load profile shows lower volatility, and thus the forecasting performance is much more stable.

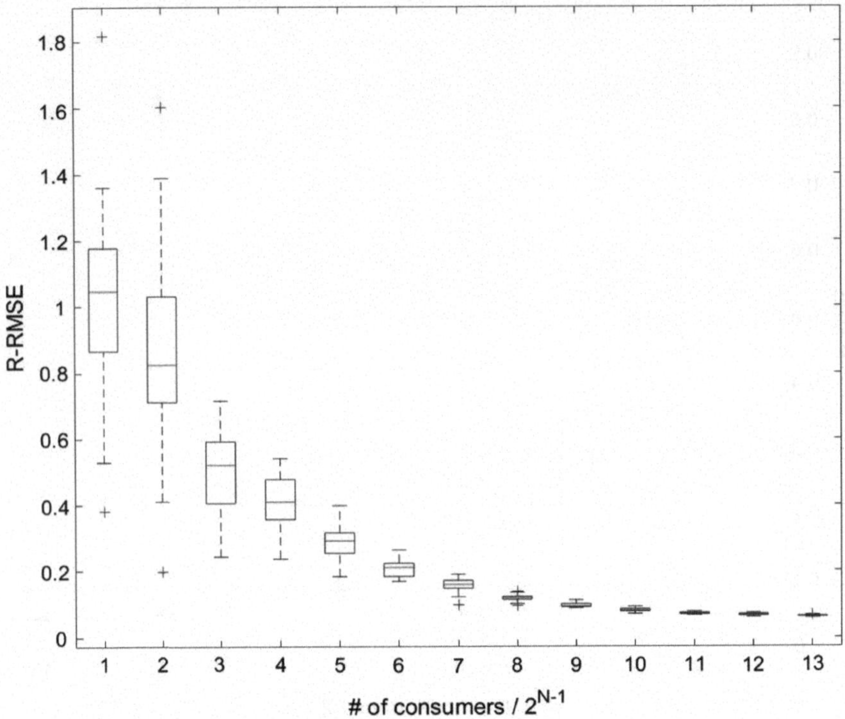

Fig. 12.3 R-RMSE for different aggregated levels

12.3 Clustering-Based Aggregated Load Forecasting

12.3.1 Framework

With fine-grained sub-profile, we have two intuitive ideas for aggregated load forecasting: (1) Directly train the forecasting model based on the final aggregated load. (2) Train the forecasting model for each individual consumer first and then obtain the summation of all the individual forecasts to form the final forecast. The first strategy is the traditional load forecasting approach but does not make full use of the fine-grained sub-profiles. The second approach can train specific forecasting model for each consumer. However, it suffers from two drawbacks: (1) training forecasting model for each consumer is time-consuming and needs more computing sources; (2) since individual load profile has great volatility, the trained model may over-fit, and their summation may even have worse performance.

Clustering is an effective approach to aggregated consumers with similar consumption behavior into the same group. Is it possible to first partition the consumers into different group first, then train the forecasting model for each group, and finally

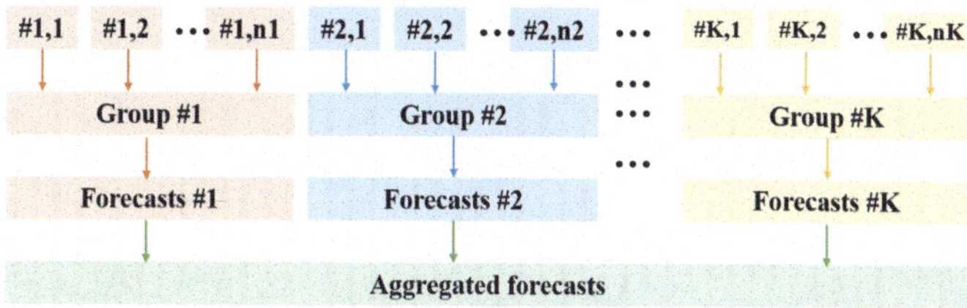

Fig. 12.4 Clustering-based aggregated load forecasting strategy

sum all the forecasts? This section studies the performance of the proposed forecasting strategy. The Cluster-based aggregate forecasting strategy is shown in Fig. 12.4.

For a region with M sub-profiles, let L_t and $L_{m,t}$ denote the total load and the mth sub load at time t, the matrix form of the sub-load profiles can be represented as $\mathbf{L}_{M\times T}$. Each column of $\mathbf{L}_{M\times T}$, $\mathbf{L}_{\cdot,t}$ denotes the load consumption of all N consumer at time period t; each row of $\mathbf{L}_{M\times T}$, $\mathbf{L}_{m,\cdot}$ denotes the load consumption of the consumer m at all time periods T.

First, we partition all the M consumers into K different groups $C{=}C_1, C_2, \ldots, C_K$; then the K aggregated load profile is:

$$l_k = \sum_{m\in C_k} \mathbf{L}_{m,\cdot} \tag{12.7}$$

On this basis, each forecasting model f_k is trained for each aggregated load profile l_k: $\hat{L}_{t,k} = f_k(\mathbf{X}_{t,k})$. Thus, the final forecast is:

$$\hat{L}_t = \sum_{k=1}^{K} \hat{L}_{t,k}. \tag{12.8}$$

12.3.2 Numerical Experiments

We apply two traditional clustering methods, k-means and k-medoids, to group the consumers according to their average weekly load profiles. The number of clusters varies from 1 to 50. All the forecasting models are ANNs. Figures 12.5 and 12.6 give the aggregated load forecasting performance with different numbers of clusters in terms of MAPE, MAE, and RMSE, based on k-means and k-medoids, respectively.

The optimal number of clusters is 2 for Irish dataset for both clustering methods. It seems that there is a clear trend or correlation between forecasting performance and the number of clusters. This observation is different from the results in [4] where we can easily find the optimal number of cluster according to the clear trend of the performance.

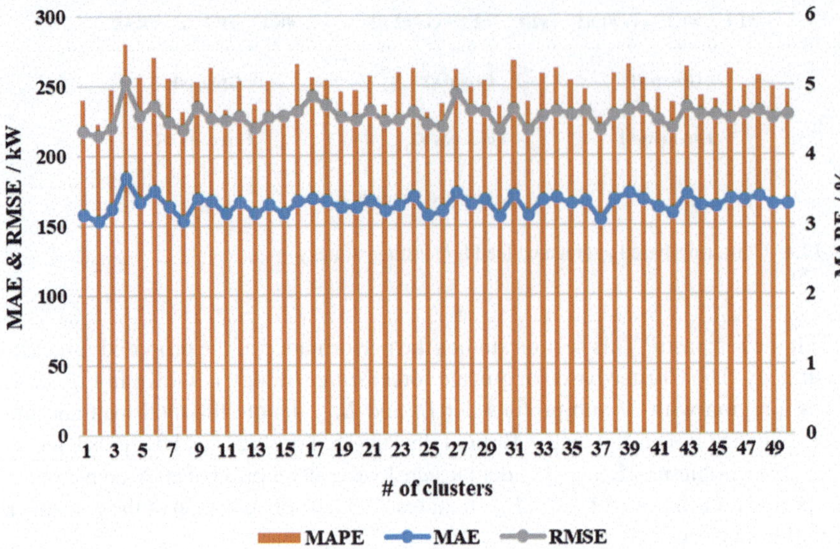

Fig. 12.5 Load forecasting performance with different numbers of clusters (*k*-means)

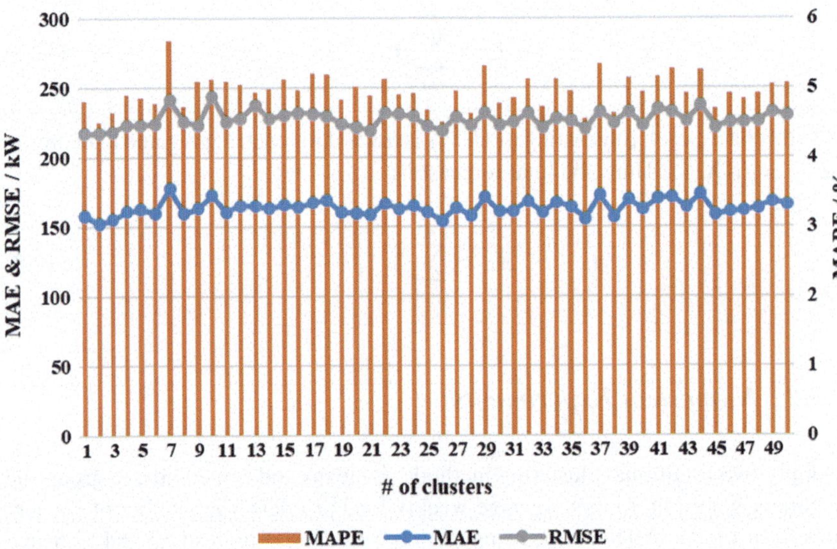

Fig. 12.6 Load forecasting performance with different numbers of clusters (*k*-medoids)

Since the forecasts are different with a different number of clusters, we can apply ensemble learning method on these forecasts. This idea inspires the work in the next section.

12.4 Ensemble Forecasting for the Aggregated Load

12.4.1 Proposed Methodology

To highlight the idea of the proposed method, only historical load data is employed as input features for constructing the forecasting model. Note that other relevant factors (e.g. temperature) can also be considered in the proposed framework. First, the sub profiles $\mathbf{L}_{M\times T}$ are segmented into three parts: the first part $\mathbf{L}_{tr}$ is used to train the forecasting model for each group load profile; the second part $\mathbf{L}_{en}$ is used to calculate the weights $\boldsymbol{\omega}$ for ensemble; the third part $\mathbf{L}_{te}$ is used to test the performance of the aggregated load ensemble forecasting model. The proposed method includes four main stages: the clustering stage, training stage, ensemble stage, and test stage.

12.4.1.1 Clustering Stage

This stage is to establish the hierarchical structure of consumers according to the similarities of their consumption behaviors. This stage is performed on $\mathbf{L}_{tr}$. First, the representative load profile $L^r_{m,t}$ for each consumer is obtained by normalizing the calculated average weekly load profile to [0, 1] domain. The subscript r means representative load here. Thus, the distance matrix $\mathbf{D}_{M\times M}$ among these consumers can be calculated based on Euclidean distance:

$$D_{m,n} = \left(\sum_{t=1}^{T_W}(L^r_{m,t} - L^r_{n,t})^2\right)^{\frac{1}{2}} \tag{12.9}$$

where T_W denotes the number of time periods over one week. It is important to notice that, in this stage, a large number of clustering procedures are required to be performed on different numbers of groups. Therefore, in this research, the agglomerative hierarchical clustering method with single linkage is selected to cluster the customers because of its capability to establish the hierarchical structure and the fact that it does not need to be performed repeatedly [8].

12.4.1.2 Training Stage

The purpose of this stage is to produce multiple forecasts by varying the number of clusters. This stage is also conducted on $\mathbf{L}_{tr}$. When the number of clusters is M, the forecasting is essentially the bottom-up approach; when the number of clusters is 1, the forecasting is performed directly based on historical aggregated load data. In order to diversify the forecasting results, we vary the number of clusters exponentially. Thus, a total of N forecasts will be obtained:

$$N = \left[log_2 M\right] + 1 \tag{12.10}$$

where $[\cdot]$ denotes the round-down function. For example, $N = 7$ when $M = 100$. The nth forecast is obtained by summing the forecasts of k_n grouped load profiles, where k_n is expressed as follows:

$$k_n = \min\left\{2^{n-1}, M\right\} \tag{12.11}$$

For example, the set of cluster number is $\mathbf{K} = [1, 2, 4, 8, 16, 32, 64, 100]$ when $M = 100$.

12.4.1.3 Ensemble Stage

As one of the main contributions in this work, ensemble stage is proposed to calculate the weights $\boldsymbol{\omega}$ for the N forecasts and combine them into final forecast. This stage is conducted on $\mathbf{L}_{en}$ instead of $\mathbf{L}_{tr}$ to reduce overfitting risk. The ensemble of N forecasts is formulated as an optimization problem where the objective function is to minimize the mean absolute percent error (MAPE) and the constraints include the equations of the combined forecasts, the summation of all the weights, and non-negativity of the weights.

$$\begin{aligned} \hat{\boldsymbol{\omega}} &= \arg\min_{\boldsymbol{\omega}} \sum_{t=1}^{T} \frac{1}{T} \frac{|L_{en,t} - \hat{L}_{en,t}|}{L_{en,t}} \\ s.t.\ \hat{L}_{en,t} &= \sum_{n=1}^{N} \omega_n \hat{L}_{en,n,t}, \ \sum_{n=1}^{N} \omega_n = 1,\ \ \omega_n \geq 0. \end{aligned} \tag{12.12}$$

The absolute percent error in the objective function can be easily transformed into linear programming (LP) problem by introducing auxiliary decision variables $v_{en,t}$, as follows:

$$\begin{aligned} \hat{\boldsymbol{\omega}} &= \arg\min_{\boldsymbol{\omega}} \sum_{t=1}^{T} \frac{1}{T} \frac{v_{en,t}}{L_{en,t}} \\ s.t.\ \hat{L}_{en,t} &= \sum_{n=1}^{N} \omega_n \hat{L}_{en,n,t}, \ \sum_{n=1}^{N} \omega_n = 1,\ \ \omega_n \geq 0. \\ v_{en,t} &\geq L_{en,t} - \hat{L}_{en,t},\ \ v_{en,t} \geq \hat{L}_{en,t} - L_{en,t} \end{aligned} \tag{12.13}$$

12.4.1.4 Whole Algorithm

The whole procedures of the proposed method are presented in Algorithm 2.

Algorithm 2 Aggregated Load Ensemble Forecasting

Require: Segmented sub profiles $\mathbf{L}_{tr}$, $\mathbf{L}_{en}$, and $\mathbf{L}_{te}$ for training, ensemble, and test forecasting models; set of cluster numbers $\mathbf{K} = [k_1, k_2, \ldots k_n, \ldots k_N]$.
Clustering Stage (based on $\mathbf{L}_{tr}$):
Obtain normalized representative weekly load profile for each consumer $L_{m,r,t}$;
Calculate the distance matrix $\mathbf{D}$ among the consumers;
Implement agglomerative hierarchical clustering.
Forecasting Stage (based on $\mathbf{L}_{tr}$ and $\mathbf{L}_{en}$):
for $n = 1 : N$ **do**
  Cluster the sub-load profiles into k_n groups;
  for $j = 1 : k_n$ **do**
    Train the forecasting model f_j for the jth group based on $\mathbf{L}_{tr}$;
    Forecast the jth grouped load profiles $\hat{L}_j$ for $\mathbf{L}_{en}$;
  end for
  Calculate the sum of the forecasts of grouped load $\hat{L}_n = \sum_{j=1}^{k_n} \hat{L}_j$.
end for
Ensemble Stage (based on $\mathbf{L}_{en}$):
Solve the optimization problem shown in Eq. (12.12).
Test Stage (based on $\mathbf{L}_{te}$):
Forecast the load profile in $\mathbf{L}_{te}$ and calculate the MAPE and RMSE;

12.4.2 Case Study

In this section, case studies are conducted on two open datasets. In particular, 50, 25 and 25% of the whole dataset are partitioned into a training dataset, test dataset, and ensemble dataset, respectively.

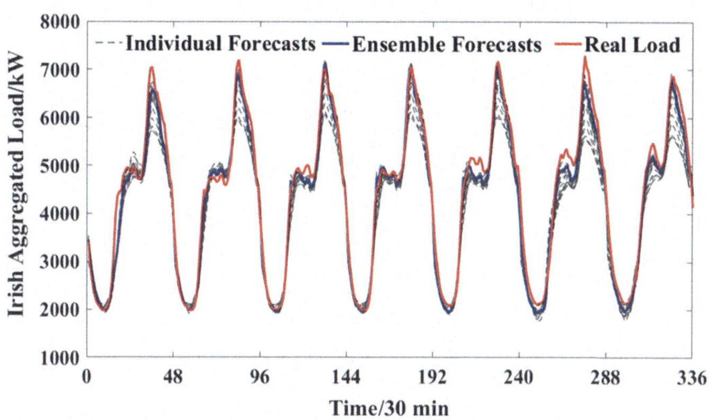

Fig. 12.7 Predicted and real aggregated individual load profiles

Table 12.1 Performance of individual and ensemble forecasts for Irish dataset

N	1	2	4	8	16	32	64	128	256	…	5237	Ensemble
ω	0.634	0	0	0.271	0	0	0.095	0	0	…	0	/
MAPE	***4.25%***	5.05%	5.29%	4.74%	5.55%	4.66%	4.79%	5.09%	5.59%	…	10.31%	**4.05%**
RMSE	***210.95***	229.73	228.01	217.68	244.9	217.64	227.36	232.61	250.27	…	441.33	**202.88**

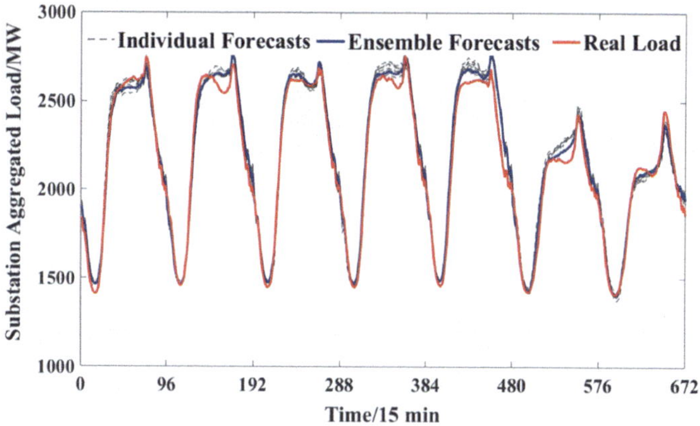

Fig. 12.8 Predicted and real aggregated substation load profiles

12.4.2.1 Irish Residential Load Data

The residential load data used in this section are obtained from the Smart Metering Electricity Customer Behaviour Trials (CBTs) initiated by the Commission for Energy Regulation (CER) in Ireland. It contains half-hour electricity consumption data of over 5000 Irish residential consumers and small and medium enterprises (SMEs) [9]. After excluding the consumers with a large number of zero values, the data of a total of 5237 consumers from July 20, 2009, to December 26, 2010 (75 weeks) are used for forecasting and testing. Figure 12.7 shows the weekly predicted and real load profiles from December 13, 2010 to December 19, 2010. As shown in the figure, the dotted lines are individual forecasts; the blue and red lines are the ensemble forecast and actual value, respectively. Table 12.1 provides the weights, MAPE, and RMSE of individual forecasts. Regarding the individual forecasts, it can be seen that instead of using the clustering strategy (i.e. $N > 1$), direct load forecasting based on the aggregated data (i.e. $N = 1$) exhibits the best performance. Nevertheless, the superior performance of the proposed ensemble method can be indicated by the 4.71% and 3.83% lower MAPE and RMSE values, respectively, than those of the best individual forecast method. Results also show that the performance of the bottom-up approach is much worse than the clustering-based method due to the large variety of individual load profiles.

12.4.2.2 Ausgrid Substation Load Data

We use the Ausgrid substation load data from May 5, 2014, to April 24, 2016 (103 weeks). After deleting the substations with a large number of non-value, a total of 155 substations data are retained [10]. Thus, nine individual forecasts are obtained by varying the number of clusters. The predicted load profiles from April 11, 2016 to

Table 12.2 Performance of individual and ensemble forecasts for Ausgrid dataset

N	1	2	4	8	16	32	64	128	155	Ensemble
ω	0	0	0	0	0.113	0	0	0	0.887	/
MAPE	5.68%	5.59%	5.47%	5.27%	5.15%	5.19%	5.13%	5.12%	***5.09%***	**5.08%**
RMSE	223.23	217.4	215.47	208.21	203.91	206.3	204.66	202.73	***202.65***	**202.55**

April 17, 2016 and performances are shown in Fig. 12.8 and Table 12.2, respectively. After the optimization procedure, the weights for forecasts #5 and #9 are 0.113 and 0.887 respectively, whereas the weights for other forecasts are zeros. When comparing the calculated MAPE and RMSE values, it is very interesting to find that, in contrast to Irish dataset, the bottom-up approach (i.e. $N = 155$) have the lowest forecasting errors. The reason for this phenomenon might be that the substation load profiles are more regular than residential load profiles.

12.5 Conclusions

This chapter proposes an ensemble forecasting method for aggregated load profile using hierarchical clustering and based on fine-grained sub-load profiles. It is a new way to make full advantages of fine-grained data to further improve the forecasting accuracy of the aggregated load. Case studies on both residential load data and substation load data demonstrate the superior performance of the proposed ensemble method when comparing with the traditional direct or bottom-up forecasting strategies.

References

1. Hong, T., & Fan, S. (2016). Probabilistic electric load forecasting: A tutorial review. *International Journal of Forecasting*, *32*(3), 914–938.
2. Yu, C.-N., Mirowski, P., & Ho, T. K. (2017). A sparse coding approach to household electricity demand forecasting in smart grids. *IEEE Transactions on Smart Grid*, *8*(2), 738–748.
3. Stephen, B., Tang, X., Harvey, P. R., Galloway, S., & Jennett, K. I. (2017). Incorporating practice theory in sub-profile models for short term aggregated residential load forecasting. *IEEE Transactions on Smart Grid*, *8*(4), 1591–1598.
4. Quilumba, F. L., Lee, W.-J., Huang, H., Wang, D. Y., & Szabados, R. L. (2015). Using smart meter data to improve the accuracy of intraday load forecasting considering customer behavior similarities. *IEEE Transactions on Smart Grid*, *6*(2), 911–918.
5. Li, S., Wang, P., & Goel, L. (2016). A novel wavelet-based ensemble method for short-term load forecasting with hybrid neural networks and feature selection. *IEEE Transactions on Power Systems*, *31*(3), 1788–1798.
6. Mendes-Moreira, J., Soares, C., Jorge, A. M., & De Sousa, J. F. (2012). Ensemble approaches for regression: A survey. *ACM Computing Surveys (CSUR)*, *45*(1), 1–10.
7. Sevlian, R., & Rajagopal, R. (2018). A scaling law for short term load forecasting on varying levels of aggregation. *International Journal of Electrical Power & Energy Systems*, *98*, 350–361.
8. Steinbach, M., Karypis, G., Kumar, V., & et al. (2000) A comparison of document clustering techniques. *KDD Workshop on Text Mining* (Vol. 400, pp. 525–526). Boston.
9. Irish Social Science Data Archive. (2012). Commission for Energy Regulation (CER) Smart Metering Project. http://www.ucd.ie/issda/data/commissionforenergyregulationcer/.
10. Ausgird. Distribution zone substation information data to share. http://www.ausgrid.com.au/Common/About-us/Corporate-information/Data-to-share/DistZone-subs.aspx#.WYD6KenauUl. Retrieved July 31, 2017.

Chapter 13
Prospects of Future Research Issues

Abstract Although smart meter data analytics has received extensive attention and rich literature studies related to this area have been published, developments in computer science and the energy system itself will certainly lead to new problems or opportunities. In this chapter, we discuss some research trends for smart meter data analytics, such as big data issues, novel machine learning technologies, new business models, the transition of energy systems, and data privacy and security. By the end of this book, we hope this chapter can help readers identify new issues and works on smart meter data analytics in the future smart grid.

13.1 Big Data Issues

Substantial works in the literature have conducted smart meter data analytics. Two special sections about big data analytics for smart grid modernization were hosted in IEEE Transactions on Smart Grid in 2016 [1] and IEEE Power and Energy Magazine in 2018, respectively [2]. However, the size of the dataset analyzed can hardly be called big data. How to efficiently integrate more multivariate data with a larger size to discover more knowledge is an emerging issue. As shown in Fig. 13.1, big data issues with smart meter data analytics include at least two aspects: the first is multivariate data fusion, such as economic information, meteorological data, and EV charging data apart from energy consumption data; the second is high-performance computing, such as distributed computing, GPU computing, cloud computing, and fog computing. It should also be noted that more data collection and analysis may bring more value as well as a larger cost. Collecting smart meter data without consideration of cost is unreasonable. How to make a balance between the value and the cost of data collection and analysis is also an interesting problem.

(1) Multivariate Data Fusion

The fusion of various data is one of the basic characteristics of big data [3]. Current studies mainly focus on the smart meter data itself or even electricity consumption

Y. Wang et al., *Smart Meter Data Analytics*,
https://doi.org/10.1007/978-981-15-2624-4_13

System/Data	Data Type
Economic Information	GDP、CPI、PMI（Purchasing Managers Index）、Sales Value、Prosperity Index
Energy Consumption Data	Electrical Load、Output、Power Quality、Temperature
Meteorological Data	Temperature、Humidity、Rainfall
EV Charging Data	Current、Voltage、Charging Rate、State of Charge

Multivariate Data Fusion

Fig. 13.1 Big data issues with smart meter data analytics

data. Very few papers consider weather data, survey data from consumers, or some other external data. Integrating more external data, such as crowd-sourcing data from the Internet, weather data, voltage, and current data, and even voice data from service systems may reveal more information. The multivariate data fusion needs to deal with structured data with different granularities and unstructured data. We would like to emphasis that big data is a change of concept. More data-driven methods will be proposed to solve practical problems that may traditionally be solved by model-based methods. For example, with redundant smart meter data, the power flow of the distribution system can be approximated through hyperplane fitting methods such as ANN and SVM. In addition, how to visualize high dimensional and multivariate data to highlight the crucial components and discover the hidden patterns or correlations among these data is a seldom touched area [4].

(2) High-Performance Computing

In addition, a majority of smart meter data analytics methods that are applicable to small datasets may not be appropriate for large datasets. Highly efficient algorithms and tools such as distributed and parallel computing and the Hadoop platform should be further investigated. Cloud computing, an efficient computation architecture that shares computing resources on the Internet, can provide different types of big data analytics services, including Platform as a Service (PaaS), Software as a Service (SaaS), and Infrastructure as a Service (IaaS) [5]. How to make full use of cloud computing resources for smart meter data analytics is an important issue. However, the security problem introduced by cloud computing should be addressed [6]. Another high-performance computation approach is GPU computing. It can realize highly efficient parallel computing [7]. Specific algorithms should be designed for the implementation of different GPU computing tasks.

13.2 New Machine Learning Technologies

Smart meter data analytics is an interdisciplinary field that involves electrical engineering and computer science, particularly machine learning. The development of machine learning has great impacts on smart meter data analytics. The application of new machine learning technologies is an important aspect of smart meter analytics. The recently proposed clustering method in [8] has been used in [9]; the progress in deep learning in [10] has been used in [11]. When applying a machine learning technology to smart meter data analytics, the limitations of the method and the physical meaning revealed by the method should be carefully considered. For example, the size of data or samples should be considered in deep learning to avoid overfitting.

(1) Deep Learning and Transfer Learning

Deep learning has been applied in different industries, including smart grids. As summarized above, different deep learning techniques have been used for smart meter data analytics, which is just a start. Designing different deep learning structures for different applications is still an active research area. The lack of label data is one of the main challenges for smart meter data analytics. How to apply the knowledge learned for other objects to the research objects using transfer learning can help us fully utilize various data [12]. Many transfer learning tasks are implemented by deep learning [13]. The combination of these two emerging machine learning techniques may have widespread applications.

(2) Online Learning and Incremental Learning

Note that smart meter data are essentially real-time stream data. Online learning and incremental learning are varied suitably for handling these real-time stream data [14]. Many online learning techniques, such as online dictionary learning [15] and incremental learning techniques such as incremental clustering [16], have been proposed in other areas. However, existing works on smart meter data analytics rarely use online learning or incremental learning, expect for several online anomaly detection methods.

13.3 New Business Models in Retail Market

Further deregulation of retail markets, integration of distributed renewable energy, and progress in information technologies will hasten various business models on the demand side.

(1) Transactive Energy

In a transactive energy system [17, 18], the consumer-to-consumer (C2C) business model or micro electricity market can be realized, i.e., the consumer with rooftop PV becomes a prosumer and can trade electricity with other prosumers. The existing applications of smart meter data analytics are mainly studied from the perspectives of

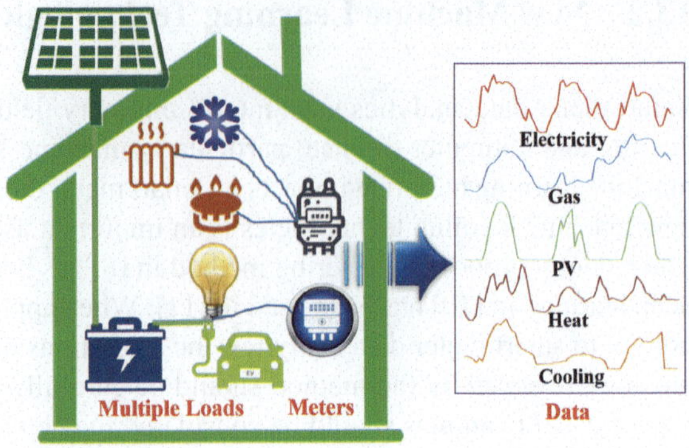

Fig. 13.2 Transition of energy systems on the demand side

retailers, aggregators, and individual consumers. How to analyze the smart data and how much data should be analyzed in the micro electricity market to promote friendly electricity consumption and renewable energy accommodation is a new perspective in future distribution systems.

(2) Sharing Economy

For the distribution system with distributed renewable energy and energy storage integration, a new business model sharing economy can be introduced. The consumers can share their rooftop PV [19] and storage [20] with their neighborhoods. In this situation, the roles of consumers, retailers, and DSO will change when playing the game in the energy market [21]. Other potential applications of smart meter data analytics may exist, such as changes in electricity purchasing and consumption behavior and optimal grouping strategies for sharing energy.

13.4 Transition of Energy Systems

As shown in Fig. 13.2, the integration of distributed renewable energy and multiple energy systems is an inevitable trend in the development of smart grids. A typical smart home has multiple loads, including cooling, heat, gas, and electricity. These newcomers such as rooftop PV, energy storage, and EV also change the structure of future distribution systems.

(1) High Penetration of Renewable Energy

High penetration of renewable energy such as behind-the-meter PV [22, 23] will greatly change the electricity consumption behavior and will significantly influence the net load profiles. Traditional load profiling methods should be improved to consider the high penetration of renewable energy. In addition, by combining weather data, electricity price data, and net load data, the capacity and output of renewable energy can be estimated. In this way, the original load profile can be recovered. Energy storage is widely used to suppress renewable energy fluctuations. However, the charging or discharging behavior of storage, particularly the behind-the-meter storage [24], is difficult to model and meter. Advanced data analytical methods need to be adopted for anomaly detection, forecasting, outage management, decision making, and so forth in high renewable energy penetration environments.

(2) Multiple Energy Systems

Multiple energy systems integrate gas, heat, and electricity systems together to boost the efficiency of the entire energy system [25]. The consumptions for electricity, heat, cooling, and gas are coupled in the future retailer market. One smart meter can record the consumptions of these types of energy simultaneously. Smart meter data analytics is no longer limited to electricity consumption data. For example, joint load forecasting for electricity, heating, and cooling can be conducted for multiple energy systems.

13.5 Data Privacy and Security

As stated above, the concern regarding smart meter privacy and security is one of the main barriers to the privilege of smart meters. Many existing works on the data privacy and security issue mainly focus on the data communication architecture and physical circuits [26]. How to study the data privacy and security from the perspective of data analytics is still limited.

(1) Data Privacy

Analytics method for data privacy is a new perspective except for communication architecture, such as the design of privacy-preserving clustering algorithm [27] and PCA algorithm [28]. Strategic battery storage charging and discharging schedule was proposed in [29] to mask the actual electricity consumption behavior and alleviate the privacy concerns. However, several basic issues about smart meter data should be but have not been addressed: Who owns the smart meter data? How much can private information be mined from these data? Is it possible to disguise data to protect privacy and not to influence the decision making of retailers?

(2) Data Security

For data security, the works on cyber-physical security (CPS) in the smart grid such as phasor measurement units (PMU) and supervisory control and data acquisition (SCADA) data attacks have been widely studied [30]. However, different types of cyberattacks for electricity consumption data such as nontechnical loss should be further studied.

References

1. Hong, T., Chen, C., Huang, J., Ning, L., Xie, L., & Zareipour, H. (2016). Guest editorial big data analytics for grid modernization. *IEEE Transactions on Smart Grid*, *7*(5), 2395–2396.
2. Hong, T. (2018). Big data analytics: Making the smart grid smarter [guest editorial]. *IEEE Power and Energy Magazine*, *16*(3), 12–16.
3. Lv, Z., Song, H., Basanta-Val, P., Steed, A., & Jo, M. (2017). Next-generation big data analytics: State of the art, challenges, and future research topics. *IEEE Transactions on Industrial Informatics*, *13*(4), 1891–1899.
4. Hyndman, R. J., Liu, X. A., & Pinson, P. (2018). Visualizing big energy data: Solutions for this crucial component of data analysis. *IEEE Power and Energy Magazine*, *16*(3), 18–25.
5. Baek, J., Vu, Q. H., Liu, J. K., Huang, X., & Xiang, Y. (2015). A secure cloud computing based framework for big data information management of smart grid. *IEEE Transactions on Cloud Computing*, *3*(2), 233–244.
6. Bera, S., Misra, S., & Rodrigues, J. J. P. C. (2015). Cloud computing applications for smart grid: A survey. *IEEE Transactions on Parallel and Distributed Systems*, *26*(5), 1477–1494.
7. Mittal, S. (2017). A survey of techniques for architecting and managing gpu register file. *IEEE Transactions on Parallel and Distributed Systems*, *28*(1), 16–28.
8. Rodriguez, A., & Laio, A. (2014). Clustering by fast search and find of density peaks. *Science*, *344*(6191), 1492–1496.
9. Wang, Y., Chen, Q., Kang, C., & Xia, Q. (2016). Clustering of electricity consumption behavior dynamics toward big data applications. *IEEE Transactions on Smart Grid*, *7*(5), 2437–2447.
10. Gers, F. A., Schmidhuber, J., & Cummins, F. (2000). Learning to forget: Continual prediction with lstm. *Neural Computation*, *12*(10), 2451–2471.
11. Marino, D. L., Amarasinghe, K., & Manic, M. (2016). Building energy load forecasting using deep neural networks. *IECON 2016-42nd Annual Conference of the IEEE Industrial Electronics Society* (pp. 7046–7051). IEEE.
12. Pan, S. J., & Yang, Q. (2010). A survey on transfer learning. *IEEE Transactions on Knowledge and Data Engineering*, *22*(10), 1345–1359.
13. Bengio, Y. (2012). Deep learning of representations for unsupervised and transfer learning. *Proceedings of ICML Workshop on Unsupervised and Transfer Learning* (pp. 17–36).
14. Diethe, T., & Girolami, M. (2013). Online learning with (multiple) kernels: A review. *Neural Computation*, *25*(3), 567–625.
15. Xie, Y., Zhang, W., Li, C., Lin, S., Yanyun, Q., & Zhang, Y. (2014). Discriminative object tracking via sparse representation and online dictionary learning. *IEEE Transactions on Cybernetics*, *44*(4), 539–553.
16. Zhang, Q., Zhu, C., Yang, L. T., Chen, Z., Zhao, L., & Li, P. (2017). An incremental CFS algorithm for clustering large data in industrial internet of things. *IEEE Transactions on Industrial Informatics*, *13*(3), 1193–1201.
17. Rahimi, F. A., & Ipakchi, A. (2012). Transactive energy techniques: Closing the gap between wholesale and retail markets. *The Electricity Journal*, *25*(8), 29–35.
18. Kok, K., & Widergren, S. (2016). A society of devices: Integrating intelligent distributed resources with transactive energy. *IEEE Power and Energy Magazine*, *14*(3), 34–45.
19. Celik, B., Roche, R., Bouquain, D., & Miraoui, A. (2017). Decentralized neighborhood energy management with coordinated smart home energy sharing. *IEEE Transactions on Smart Grid*, *9*(6), 6387–6397.
20. Liu, N., Xinghuo, Y., Wang, C., & Wang, J. (2017). Energy sharing management for microgrids with PV prosumers: A stackelberg game approach. *IEEE Transactions on Industrial Informatics*, *13*(3), 1088–1098.
21. Ye, G., Li, G., Di, W., Chen, X., & Zhou, Y. (2017). Towards cost minimization with renewable energy sharing in cooperative residential communities. *IEEE Access*, *5*, 11688–11699.
22. Shaker, H., Zareipour, H., & Wood, D. (2016). Estimating power generation of invisible solar sites using publicly available data. *IEEE Transactions on Smart Grid*, *7*(5), 2456–2465.

23. Wang, Y., Zhang, N., Chen, Q., Kirschen, D. S., Li, P., & Xia, Q. (2017). Data-driven probabilistic net load forecasting with high penetration of behind-the-meter PV. *IEEE Transactions on Power Systems*, *33*(3), 3255–3264.
24. Chitsaz, H., Zamani-Dehkordi, P., Zareipour, H., & Parikh, P. P. (2017). Electricity price forecasting for operational scheduling of behind-the-meter storage systems. *IEEE Transactions on Smart Grid*, *9*(6), 6612–6622.
25. Krause, T., Andersson, G., Frohlich, K., & Vaccaro, A. (2011). Multiple-energy carriers: modeling of production, delivery, and consumption. *Proceedings of the IEEE*, *99*(1), 15–27.
26. Yongdong, W., Chen, B., Weng, J., Wei, Z., Li, X., Qiu, B., & et al. (2018). False load attack to smart meters by synchronously switching power circuits. *IEEE Transactions on Smart Grid*, *10*(3), 2641–2649.
27. Xing, K., Chunqiang, H., Jiguo, Y., Cheng, X., & Zhang, F. (2017). Mutual privacy preserving *k*-means clustering in social participatory sensing. *IEEE Transactions on Industrial Informatics*, *13*(4), 2066–2076.
28. Wei, L., Sarwate, A. D., Corander, J., Hero, A., & Tarokh, V. (2016). Analysis of a privacy-preserving pca algorithm using random matrix theory. *IEEE Global Conference on Signal and Information Processing (GlobalSIP)* (pp. 1335–1339).
29. Salehkalaibar, S., Aminifar, F., & Shahidehpour, M. (2017). Hypothesis testing for privacy of smart meters with side information. *IEEE Transactions on Smart Grid*, *10*(2), 2059–2067.
30. Yan, Y., Qian, Y., Sharif, H., & Tipper, D. (2012). A survey on cyber security for smart grid communications. *IEEE Communications Surveys & Tutorials*, *14*(4), 998–1010.

The manufacturer's authorised representative in the EU is Springer Nature Customer Service Centre GmbH, Europaplatz 3, 69115 Heidelberg, Germany. If you have any concerns regarding our products, please contact ProductSafety@springernature.com

Printed and bound by CPI Group (UK) Ltd, Croydon, CR0 4YY
25/09/2026
02222901-0006